Project Management Tools and Techniques

The latest edition of this textbook fills a significant void in the realm of project management tools and techniques, where many existing resources fall short. By seamlessly integrating Microsoft Project into the content, students are equipped with the essential skills needed to excel in project management. The addition of screenshots featuring the most up-to-date Microsoft Project software enables readers to effectively visualize and implement the concepts discussed in the textbook.

The third edition of *Project Management Tools and Techniques: A Practical Guide* continues to prepare students for certification exams like the Project Management Professional Exam, ensuring they are well prepared for success in their future careers. This updated edition provides a contemporary perspective on this ever-evolving subject, with five new chapters covering project management and its tools, project variability, critical chain theory, and hybrid models.

For both students and instructors, this textbook offers practical examples and videos that bridge the gap between theory and practice in project management. It also includes case studies that provide hands-on experience in applying project management techniques and tools. Lab assignments are included to further enhance learning.

For qualified textbook adoptions, a test bank, PowerPoint slides, and sample data are available to support instructors in their teaching endeavors. To enrich the learning experience, this book is accompanied by a dedicated website designed for both students and instructors.

Project Management Tools and Techniques

A Practical Guide

Third Edition

Gary L. Richardson and
Deborah Sater Carstens

CRC Press
Taylor & Francis Group
Boca Raton London New York

CRC Press is an imprint of the
Taylor & Francis Group, an **informa** business

Third edition published 2026
by CRC Press
2385 NW Executive Center Drive, Suite 320, Boca Raton FL 33431

and by CRC Press
4 Park Square, Milton Park, Abingdon, Oxon, OX14 4RN

CRC Press is an imprint of Taylor & Francis Group, LLC

First edition published by CRC Press 2013
Second edition published by CRC Press 2020

Library of Congress Cataloging-in-Publication Data
Names: Richardson, Gary L. author | Carstens, Deborah Sater author
Title: Project management tools and techniques : a practical guide / Gary L. Richardson and Deborah Sater Carstens.
Description: Third edition. | Boca Raton, FL : CRC Press, 2026. | First edition published by CRC Press 2013. Second edition published by CRC Press 2020. | Includes bibliographical references and index.
Identifiers: LCCN 2025027972 (print) | LCCN 2025027973 (ebook) | ISBN 9781032874791 hardback | ISBN 9781032875095 paperback | ISBN 9781003532989 ebook
Subjects: LCSH: Project management--Methodology | Project management--Data processing
Classification: LCC HD69.P75 C374 2026 (print) | LCC HD69.P75 (ebook) | DDC 658.4/04--dc23/eng/20250905
LC record available at https://lccn.loc.gov/2025027972
LC ebook record available at https://lccn.loc.gov/2025027973

ISBN: 978-1-032-87479-1 (hbk)
ISBN: 978-1-032-87509-5 (pbk)
ISBN: 978-1-003-53298-9 (ebk)

DOI: 10.1201/9781003532989

Typeset in Minion
by SPi Technologies India Pvt Ltd (Straive)

Access the Instructor and Student Resources: http://www.routledge.com/cw/Richardson

To my wife, Shawn, who tolerated excessive nights in the Man Cave.
Gary L. Richardson

To my children, Ryan and Lindsay, I am incredibly proud of both of you and am your biggest fan. To my husband, Darrell, who dedicates so much love, time, and energy to support me. To my parents, Stan and Norma, who are no longer with me but provided me with a loving environment while teaching me the value of hard work and education. Lastly, I appreciate my coauthor, Gary, for allowing me to collaborate with such an expert in this field.
Deborah Sater Carstens

Contents

Preface

More than ever, automated tools and emerging technologies are increasingly supporting the process of managing projects. This trend is forecast to continue with a new generation of even more advanced offerings. There is a saying in the industry that you can't control what has not been planned, and the planning and control aspects are increasingly populated by tools that serve data capture, quantification, and communication goals.

The original goal for writing the first edition of the text was to provide more understandable descriptions of various life cycle management process mechanics and highlight some of the evolving techniques that seemed "too complex" for regular usage. At this point, Microsoft Project software had become the predominant tool for project planning. Our observations indicated that the average user wanted a greater understanding of the design assumptions related to the software model or the operational realities that challenged its proper usage. That observation remains with this third edition. Beyond dealing with the operational mechanisms of the various planning and control processes, the industry squabbles over which of two popular methods, waterfall and agile iteration, is best, when both have full life cycle shortcomings. Significant management improvement is needed to produce successful outcomes, which involves more than trying to select from the current offerings. In the meantime, there is a need to enhance motivation to improve existing processes. Beyond seeking better use of new technologies, there is value in understanding why techniques related to concepts such as the Critical Chain model theory, Earned Value mechanics, and simulation modeling are underutilized.

One of the tool areas that can be misused is Microsoft Project, which is the most popular automation tool in the industry. This tool makes it easy to enter task data and present beautiful, color-coded computer-generated plans but often doesn't present a valid view of the work requirement and misrepresents the total effort. This is a two-fold issue. First, the software code embedded in its design automates a 70-year-old model that has known shortcomings with reality. Second, even for the project types that closely match the model design, the typical user does not know how to utilize the software to accurately and properly represent the project. These strong statements will take many pages of the text to describe. Yes, it is time for a new generation of software, but the tactical issue is improving techniques to utilize what is available and recognize the impact of design gaps. There is a need in the industry for creative leadership to improve the modeling tools. It will require a strategic rather than tactical period, to improve the software available for the industry.

Over the past several years, there has been an increasing controversy in the project industry regarding delivery methodology. The agile/SCRUM iterative model school believes they have found the correct answer, while the traditional waterfall group resists their claims. This is reminiscent of a political campaign where neither side looks beyond their personal views. The text will claim that both are correct, and both are also wrong. The belief behind this obtuse point is that projects are not singular or binary. This means the proper management approach should be flexible and based on the underlying requirements and delivery environment. This statement suggests that the correct process involves balancing output objectives against scope, time, budget, resources, risk, and quality factors. Achieving this balance requires skill, experience, and improved support tools and techniques. To that end, this text explores techniques related to these areas. Microsoft Project is the primary delivery tool to illustrate multiple segments of the planning and control process examples, which are designed to show how the standard model can be tweaked to reflect reality better. As a supplement to this, tutorial videos and downloadable sample programs are available for the reader to further explore the text content. This text is meant to be more than theory based.

Since the 1950s, managing projects has progressed from a relatively simplistic theoretical view to the current state, where an increasing array of new theories and mechanics are emerging. Along with this evolution of knowledge, there has been an associated growth in the complexity of the supporting tools, particularly in computerized form. Machine learning technology seems to have matured to the point where it now stands as a vehicle to enhance this process further, provided that the industry view changes from arguing over which is best and focuses more on what is needed. The material described in this text is aligned with traditional model theory and modified as necessary to fit real-world gaps in the models. All topic sections will contain a theoretical core view and a practical mechanics perspective, hence the text title "Tools and Techniques."

The term *project* has become a universal statement for getting new things done. There is no longer doubt that this organizational format is a needed technical and intellectual subject, and one might also observe it as a process that is not well mastered in commercial or personal lives. The topic has been so broadly described through various models that it is now a well-recognized birth-to-death concept. These development stage views have also been compared to one's individual goal-centric stages, which can improve outcomes. That essentially means that the appropriate process for both should be to define and execute planned stage goals.

Even though one can be critical of the traditional model, it still represents this process' most mature life cycle overview. *Properly executed*, its defined view represents the life cycle's best management planning and control description. Based on this perspective, the first 17 chapters of the textbook describe contemporary planning and control process groups related to this model. Section V, "Advanced Techniques," opens the future door with a discourse in Chapter 18 outlining some of the major gap areas in the traditional model. Chapters 19 through 22 describe a sample of non-traditional modeling views that potentially improve delivery outcomes but do not fit the standard approach. Chapter 24 describes a hypothetical hybrid model that has built-in flexibility to handle multiple projects and delivery work types. If one follows the broad rationale and logic of this final section, major re-engineering of the project life cycle delivery model is needed, but there is no clear leader currently visible to guide that effort.

Given current project success rates, the serious project manager must be receptive to new ideas regarding techniques to improve the delivery process. The path to this future state starts with understanding current issues and then developing visions for a better approach. Throughout the text, examples of needed short-term patches are described, but much work is required to construct a more robust overall approach that deals with a model reality match in areas such as scope control, task estimating, status measurement, risk, and others.

It would not be fair to the reader to suggest that everything you will ever want to know about this topic is covered in this text, but the material that is covered will provide a good core understanding of the primary management process and related operational activity that every project team should be using or thinking about using. The serious reader will come away with a better understanding of this topic, and we humbly state that it is more reliable than one might find scattered throughout the Internet-based world of prescriptive advice.

Gary L. Richardson
Deborah Sater Carstens

Authors

Gary L. Richardson, PhD, retired from the University of Houston, College of Technology's graduate project management program as the Project Management Institute (PMI) Houston Endowed Professor. He chaired the program focused on Project Management Professional (PMP) certification, the university's project-related academic programs, and external organizations during his tenure. He also taught various project management programs to audiences in Finland, Russia, Iraq, China, and South America. He held professional certifications as a Professional Engineer (PE), PMP, and Certification in Earned Value Management (EVM). During his academic stints, he produced 13 professional texts and numerous articles on general project management and information technology topics.

The mid-level of his career started in 1979 when he moved from Texas A&M to Texaco's Central IT Department, where he held multiple senior management roles. In 1988, he became Managing Director of IT at a Texaco/Aramco joint venture called Star Enterprise. In 1991, he was a finalist for Outstanding IT Executive in the South Texas Region. He retired from the joint venture in 2002 to become the Managing Director of IT at Service Corporation International for three years. He then returned to academia at the University of Houston to institute a project management program.

During the early post-college phase of his career, Gary served as an officer in the U.S. Air Force, leaving as a captain. Following the military, he worked as a manufacturing engineer at Texas Instruments, consulting assignments in Washington, DC, at the Defense Communications Agency, Department of Labor, and the U.S. Air Force (Pentagon). Interspersed through these positions, he was a tenured professor at the University of South Florida and Texas A&M and an adjunct professor at the University of Houston and Sam Houston State University. He moved from Washington, DC, to Texas A&M in 1979.

Gary earned a BS in mechanical engineering at Louisiana Tech, an AFIT post-graduate program in meteorology at the University of Texas, an MS in engineering management at the University of Alaska, and a PhD in business administration at the University of North Texas.

Gary was a varsity basketball player at Louisiana Tech, and his love of the game never left him. He played competitively throughout his adult life, and his Senior Olympics team won numerous basketball awards over 25 years, including four national and two international awards. He played in the Senior Olympics until age 81, when the warranty on his knees expired. He is proud to earn the first team All–Air Force designation in 1963 at the worldwide Air Force Command tournament. He now lives in Kingwood, Texas, and can be reached at richagl0001@gmail.com.

Deborah Sater Carstens, PhD, PMP, is a Professor of Aviation Human Factors and Graduate Program Chair at the Florida Institute of Technology (Florida Tech), College of Aeronautics. She is the Director of the Human Factors: Aeronautics, Safety, Sociability, Interfaces, Stress, and Training (HF ASSIST) Laboratory. Further, she is the Site Director for the Federal Aviation Administration (FAA) Partnership to Enhance General Aviation Safety, Accessibility, and Sustainability (PEGASAS) Center of Excellence. She has worked for Florida Tech since 2003, where she instructs both graduate and undergraduate students, teaching on-campus and online programs. Formerly, she was a faculty member in information systems and the Academic Chair for the Project Management Track in the online MBA program.

Before Florida Tech, she worked as a civil servant for the National Aeronautics and Space Administration (NASA) at the Kennedy Space Center (KSC) from 1992 to 2003. During this time, she worked as a researcher and project manager on many studies related to the Space Shuttle program. She also worked at NASA KSC during her Fall 2024 sabbatical, working on MARS-related human factors research. In her career both at Florida Tech and at NASA KSC, she has been the principal investigator or team member of funded research for NASA, Small Business Administration (SBA), Florida Department of Education (DOE), Department of Health and Human Services (DHHS), Defense Advanced Research Projects Agency (DARPA), Oak Ridge Associated Universities (ORAU), Virginia Space Consortium, and other local organizations. Her research has been in project management and human factors.

She earned a PhD in industrial engineering and a BS in business administration at the University of Central Florida. She also holds an MBA from Florida Tech. She has over 90 peer-reviewed publications; this is her fourth project management book. She has had the honor of being the faculty advisor on several first-place teams and honorable-mention teams for the Airport Cooperative Research Program (ACRP) Transportation Research Board (TRB) National Competition since 2018. She lives in Melbourne, Florida, and can be reached at dscarstens@yahoo.com.

I

Background

CHAPTER 1

Introduction to Project Management and Its Tools

Project N Motion Videos for This Chapter

1INTRO	Library Introduction	Library Overview
1W	Waterfall Model	Review Waterfall Model

1.1 READER SUPPORT RESOURCES

External supplementary items related to the text material are offered in the form of both video and problem-related data. This section will explain some of the key mechanics related to the access process for this library of supporting material.

The video library is called *Project N Motion* and contains a broad set of subject tutorial topics related to theory and mechanics. The goal of providing these is to improve the reader's understanding of the theory and mechanics related to core project management processes. Each of the library topics is linked to a companion chapter subject area. These additional resources enhance the reader experience beyond the standard text

DOI: 10.1201/9781003532989-2

material. Tutorial-type support video and sample problem data can be accessed on demand, as described below.

1.1.1 Video Files

The *Project N Motion* videos were created initially to answer student questions regarding various project management theories and mechanics. They are not fancy commercial videos but are technically accurate regarding content. In that regard, these are more like "home movies." The goal for these is to provide an easy-to-follow audio/visual dialogue for various topics that are somewhat hard to explain in text format. Subjects selected for this library support various chapter materials. In each case, these are meant to add additional perspective to the standard text. These can be selected as the reader desires.

The *Project N Motion* video files are in MP4 format. These items can be accessed using the standard Internet protocol outlined below. This requires a standard browser and an Internet connection. Individual lessons can be selected using the target file names outlined in Table 1.1.

Selecting the file code name shown invokes the video to play. The files have been tested using Microsoft Media Player, but other standard video utilities should work as well.

1.1.2 Data Files

The data library contains various problem setup items that are related to various chapters. These consist primarily of Excel spreadsheets, Microsoft Project MPP files, and sample problems.

1.1.3 Access Mechanics

The method of accessing each of the library groups require a browser and a specific access code, called a URL or Uniform Resource Locator. Think of the URL as an international mailbox address. The following library references can be inserted into a browser to access:

- Videos Library: https://routledgetextbooks.com/textbooks/9781032875095/videos.php
- Data Library: https://routledgetextbooks.com/textbooks/9781032875095/data-library.php

TABLE 1.1 *Project N Motion* Library

Chapter	File Code	Topic	Content
1	*1INTRO*	Library Introduction	Explain the Library
1	*1W*	Waterfall Model	Review Waterfall Model
3	*3S1*	Stakeholder Management	Stakeholder Overview
5	*5W1*	WBS to MSP	Translate WBS to MSP
5	*5W3*	WBS	WBS Overview
5	*5S1*	Scope Management	Scope Overview
5	*5S2*	Scope to WBS	Scope Migration
6	*7MSP1*	MSP Keystrokes	Setting up MSP
6	*6MSP3*	Pool Project	Pool Project Schedule
7	*7T0*	Time Management Overview	Time Overview
7	*7T2*	Plan Checklist	Plan Creation Steps
7	*7T4*	MSP Plan Parameters	MSP Initiation Parameters
7	*7T6*	MSP Predecessors	MSP Predecessors
8	*8C*	Cost Management	Cost Overview
8	*8C1*	Cost Management	MSP Mechanics
9	*9PLAN1*	Finalize Plan	Viable Plan Concepts
9	*9PLAN2*	Reschedule the Plan	MSP Rescheduling
13	*13R0*	Report Planning	Overview of Status Planning
13	*13R1*	Reporting Status	Project Metrics
13	*13R2*	Reporting Overview	Status Concepts
13	*13S2*	Status Tracking	Tracking Overview Theory
13	*13S3*	Status Tracking	Tracking Techniques
13	*13S1*	Status Tracking	MSP Reporting Overview
13	*13S3*	Status Tracking	MSP Mechanics
14	*14EV1*	Earned Value A	Intro—Part A Mechanics
14	*14EV2*	Earned Value B	Intro—Part B Mechanics
14	*14EV3*	Earned Value	EV Conceptual Overview
14	*14EV4*	Earned Value	EV in MSP Mechanics
15	*15RL*	Resource Leveling	MSP Leveling Mechanics
15	*15R2*	Procurement	Procurement Overview
16	*16R1*	Risk Management	Risk Management Overview
17	*17C2*	Project Closing	Project Closing Process
18	*18VAR*	Variability Concepts	Reviewing Variability
19	*19AG*	Agile Model	Agile Overview
20	*20S1*	Simulation Modeling	Simulation Mechanics
21	*21CC2*	Critical Chain Model	CC Overview
21	*21CC2*	Critical Chain Model	PowerPoint Slides
22	*22VAR*	Project Variability	Variability Overview

- Main Publisher Website: www.routledge.com/cw/richardson

1.1.4 Data Access Mechanics

- The *video library* can be accessed by pasting the URL shown. This will produce a listing of files in the library. Clicking on the desired option will invoke execution. Select the forward arrow to continue. Selecting the ESC (Escape) key will terminate execution.
- The *problem data library* access is identical to the video library, except the retrieval will produce a chapter collecton of sample items.
- The Publisher's website also has a similar access. Major topics will be shown for selection.

1.2 INTRODUCTION TO PROJECT MANAGEMENT HISTORY

The evolution of project management practices since World War II (WWII) can be described as somewhat haphazard. Kabeyi (2018) discusses the origin and evolution of project management, starting with the term coined in the 1950s and how software for projects came about in the 1980s. The project environment has many factors that are difficult to control, such as scope definition, task estimating, risk, resource availability, and team productivity. There are four evolutionary stages of project management (Max, 2024). In the first stage, before 1958, the key management advancements consisted of the Gantt chart as a mechanism to visualize data, the Precedence Diagramming Method (PDM) and Project Evaluation Review Technique (PERT) to map out predecessor-successor relationships of project tasks, and the Critical Path Method (CPM) for planning the longest-duration path to destablish the planned project schedule. The second stage, 1958 to 1979, is considered the modern project management era. In fact, the Project Management Institute (PMI) and International Project Management Association (IPMA) were founded during this stage (IPMA, n.d.; PMI, n.d.).

During stage two, the key project management advancements identified consist of the Work Breakdown Structure (WBS), discussed in Chapter 5, which lays out the steps to achieve project objectives (Max, 2024). Additionally, topics such as conflict management to optimize team members' interaction and iterative project planning to refine and improve a project continuously became important within the project management field. The third stage, 1980 to 1994, evolve a model called the Waterfall method, which surfaced to plan and control project initiatives. Key project management advancements during this stage, such as risk management, discussed in Chapter 16, were introduced by PMI, along with the first certification test in project management. Project management software utilities were launched around the mid-1980s as technology advancements continued to soar. Stage four ushered in a new breed of methods. Sor example, iterative models called agile, Scrum, Rapid Application Development, and Crystal Methods were touted as productivity enhancers. From 1995 to the present, the field of project management continued to evolve through various development methods through the Internet, transforming project work methods and even increasing remote work. Iterative methodologies, discussed further in Chapter 19, were launched to help with team collaboration and the work process. In the current stage, formal project management degrees from undergraduate to graduate programs have become available.

The Mosaic website also presents a fascinating historical chronology of project management (Weaver, n.d.; Mosaic Project Service PTY LTD, n.d.).

It is important to note that ffour classic and strategic core management support tools emerged in the early period: WBS, CPM, PERT, and Earned Value (EV). These four tools were collectively designed to define the project scope, outline the work units required, estimate the work duration, and evaluate the ongoing status. In a perfect world, these represented all that was needed; however, it was not long until the arithmetic mechanics required for these items caused a hiccup in the evolution. For moderate and large projects, these related computational mechanics were not feasible and awaited better support tools to technically evolve. Unfortunately, the historical result was the omission of PERT and EV concepts from the ongoing evolutionary process. Because of this computational gap, the early 1900s Gantt bar chart remained the standard status display format. By the 1960s, the project model had been recognized as an effective organizational model for developing complex products. Major management processes began to be formalized after 1960. Many of these were sponsored by the Department of Defense for use in their procurement practices. The typical target project for this period was to develop large, tangible, and high-technology products (i.e., airplanes, ships. Significant time and cost overruns became quantified and visible during this era. By the 1980s, broadly available computing power triggered the emergence of a more rigorous analytical approach to these management issues. During this period, formal management models were in place, and various technical industry sources were beginning to become involved in documenting the management process. The waterfall model was the predominant view but after 2000, there was increased attention focused on various iterative approaches, resulting in a growing school of iterative methodologies.

1.3 CURRENT PROJECT MANAGEMENT PERSPECTIVES

In reviewing project management evolution from the 1940s to the present, one is struck by the ad hoc nature of current practices and approaches. Some notable characteristics of this evolution are as follows:

1. Creating projects seems to be more departmental in origin than strategic and competitively oriented.
2. Until recently, the majority of management approaches essentially followed a single life cycle view with minimal integration or crossover to new paradigms.
3. Project managers are often not trained in formal project management models or other academic aspects. In many cases, they are selected more on experience related to the technical side of the equation and not so much for their knowledge of formal management model theory.
4. The waterfall model and its various defined work processes have been the most common project management approach. Project deliverables are assumed to fit this model.
5. Most organizations would not be judged as having a mature project support structure.

As a result of these trends, the typical project management approach is characterized as immature regarding deliverable results and poor success rates. That said, even organizations that are mature in this arena also have problems with repetitive success. This characteristic speaks to the subtle complexity of projects. One interesting phenomenon regarding project culture is observing a similar process in two projects and finding two utterly different success outcomes. There is more to this topic than following a list of required practices. Described another way, there is an implicit assumption that using some fixed prescription of processes and tasks will achieve successful outcomes.

There are many ways in which the current project environment can be described, but the following list represents frequent descriptions of environmental characteristics:

1. Before projects are approved, senior decision-makers require schedule and cost estimates using high-level views of deliverable requirements.
2. There is often a management conflict dilemma related to project deliverable goals regarding time, cost, functionality, resources, and customer satisfaction, which is assumed if the deliverables are produced.
3. During execution, shortcomings in pursuing the initially defined and approved project scope create change requests that increase schedule and cost. However, the original plan may remain fixed (i.e., more scope), but the schedule and cost plan remain fixed.
4. Task estimates for defined work are challenging to produce accurately, so estimates are padded to protect the plan, yet overruns remain for those tasks.
5. Projects contain varying degrees and sources of risk events that are hard to anticipate, yet these unknown events can significantly affect the project outcome. Techniques to manage this aspect of the project remain varied and are considered immature in accuracy.
6. Planned resources are often unavailable as defined or assumed by the formal plan, which often creates schedule and cost overruns (i.e., even a valid plan fails here).
7. Many projects use transient resources from multiple sources and skill groups, creating a complex personnel management issue (i.e., training, morale, communication, and coordination are related issues here).

More examples of these difficult-to-handle environmental situations exist, but this sample list provides a reasonable starting point to deal with management issues.

1.4 HISTORICAL BACKGROUND

To set the stage for describing improved management processes, it will be useful to take a quick historical look at the path of key milestones that have led to the current view. Each of these significant classic tools, techniques, or processes was created to deal with some perceived management aspects of the problem related to some deliverable need. In other words, the content of current popular models contains both assumed solutions and processes to deal with a particular management area (i.e., schedule, budget, risk, resources, and scope change). As a result, managing projects has evolved through a fragmented trial-and-error set of management models, tools, and processes. The section below briefly describes some memorable project-related historical management model milestones.

1.4.1 Evolution of Management Toolsnagement View

During the early 1900s, the management school led by Frederick Taylor developed various tools to improve outputs (Witzel & Warner, 2015). This culture of "scientific management" was focused on work process improvement. As a part of this movement, Gilbreth's formal time-estimating technique and Henry Gantt's famous graphical bar chart schedule work plan have remained visible in the management culture for over one hundred years. Large product-oriented project experience from the WWII era created a formal awareness of the project's role in producing complex deliverables. Following this trend into the 1950s, the U.S. Department of Defense (DoD) began documenting and prescribing various project-oriented tools and techniques often viewed as the first "silver bullet" solutions for project management (e.g., the answer to how to manage a project). One of the most notable examples of process definition from this stage came from the very complex and successful U.S. Navy Polaris missile development project (U.S. Naval Institute, 2024). Two strategic management tools were introduced in this project. First, a variable-time task estimating tool called the PERT was introduced to define task estimates. Second, a task network model for planning and defining the overall schedule and the critical path was introduced. Both new approaches were credited with making this very complex project a success (that conclusion was later refuted, by the way). Over time, the underlying arithmetic complexity related to the PERT algorithm calculations caused its usage to decline in favor of simple single-parameter task time estimates, even though this is now recognized as a backward trend in modeling project schedules. However, the critical path network concept for schedule planning remains an accepted planning and control tool, typically called CPM for Critical Path Management (or Model). Two other strategic tool techniques emerged in this time frame—WBS and EV. Mil Std 881 continues today as a definitional source standard for DoD's WBS family of products (i.e., airplanes, ships, and tanks) (Department of Defense, 2022). The WBS schematic format is now a ubiquitous tool across all industries, although there is no commercial movement to standardize this format. Meanwhile, EV usage as a status parameter is fragmented, and organizations are often not operationally mature enough to produce these parameters despite evidence that this concept represents the best status-tracking tool available for the traditional management model. In recent years, EV has been accepted as an international standard, but it still has weak industry sponsorship. Nevertheless, these four classic tools have been widely recognized as management thought leaders, even though only two are actively used (WBS and task networks).

By around 1980, d, the concept of a project life cycle of tasks became the accepted view of projects. Another evolutionary thread over this time was the publication of formal project specification documentation by various organizations. Most of these "how-to" specifications were based on a predictive cascading task waterfall design model structure characterized by specifying the project scope and work definition before execution. From this specification, a formal schedule and budget can be produced before the project is started, even though experience shows this to be poorly correlated to the actual result. The concept of predefining project work and, from that, producing schedules and budgets remained a key concept in the evolution.

1.4.2 Modern Management Trends

Starting after WWII, the emergence of the Cold War brought an almost panicky period of increased military spending for a wide array of new weapons. As an example of this period, three high-technology airplanes, X-24A, X-24B, and X-70, were initiatives undertaken during the 1948 to 1966 period, none of which went into production. The X-70 project was eventually canceled with the claim that the cost was excessive for its planned function. Numerous other less exotic projects were also being pursued during this period. The government's goal was to find the proper method to control the schedule and cost of such projects. The management process emerging here was to produce a completion prediction for the schedule and budget from deliverable goals and then track those variables through the development process. This cultural approach to project management is visible today, and the design roots of the classic waterfall model are found in these historic projects. If one wanted to be critical of this early era, it would be in the strategic visioning side, as well as the immature approaches to planning and control.

By the 1960s, publicly available data from large DoD projects for the first time revealed tremendous variances in planned time, budget, and functionality. More serious efforts were undertaken from this initial recognition to define and formalize improved management and control models and processes. Five notable initiatives are selected here to highlight key "silver bullet" solutions for solving the project management problem. Since their inception, none of these have solved the problem of avoiding failed outcomes, but many are part of the relics found in current methods. The five areas outlined below provide insight into a formal management starting point. It is essential to recognize that this evolutionary process has spawned new insights into this complex process. The following five epoch management specifications collectively represent the basic thought ideas driving management since the 1960s. The date-grouped, summarized list follows:

1. 1960s to present—Various *DoD initiatives* to define and document project management.
2. 1970s and 1980s—*Process definition tools.* Various manufacturing and software industry initiatives focused on speed and structure. Just-in-Time (JIT) delivery process techniques and structured programming represent this domain group.
3. The 1970s to present—*Professional methodologies.* This segment relates to the emergence of professional organizations that published prescriptive models to improve the management of projects.
4. 1950s to present—*Quality management programs.* In response to successful Japanese experiences. Various programs were introduced to improve the customer satisfaction aspects of project initiatives. This improvement activity took on a general character of project management.
5. 1960s to present—Organizations began to recognize that projects should be viewed similarly to stock investment by evaluating alternatives and making a more strategic selection. This school of thought became known as *project portfolio management* (PPM).

Each of these evolutionary threads left its mark on the current environment, and for that reason, it is essential to describe them in more detail.

1.4.2.1 DoD Initiatives

There are many examples of DoD documentation related to project management, and much of this source material has migrated into the commercial environment through their contractor providers. One significant example of a strategic document first introduced in the 1960s is the recognition that a valid project architectural view for organizations is part of the required view of the overall management requirement. The initial DoD specification document, DoD 5000.1, was titled the Cost Schedule Control Systems Criteria (CSCSC), and it defined 32 management "criteria" (processes), which by 1981 became 35 criteria, that government contractors had to meet as evidence of their project management control maturity (Hallum, 1981). Later versions of this document focused more on EV project status certification, and the underlying specifications related to the original 32 key processes remained in view (DCMA, 2012). This document represents a classic definition of an organizational, architectural view of project support control processes and can still be used to define that portion of organizational project support maturity.

1.4.2.2 Process Reengineering

The linkage between reengineering a process and project management is not immediately obvious. However, if one views a process as a collection of linked tasks, that overall view begins to resemble a project life cycle process. From this perspective, it is possible to see how this movement began to also influence project management practices.

1.4.2.3 Professional Management Models

One of the most popular international management models visible today is sponsored by the PMI. In 1986, this organization introduced the initial version of a formal project description document titled A Guide to the Project Management Body of Knowledge, known in the industry as the PMBOK. This view of project management is generally considered the best-known and respected overview approach to this activity. However, the document does not describe how to do the processes indicated. A similar effort from England created a comparable methodology known as PRINCE2, Projects in a Controlled Environment (PRINCE2, 2022). These two internationally recognized efforts and other similar ones represent a broad attempt to standardize project management, and many of these include formal proprietary training and certification programs. Until recently, one could characterize the basis for these documents as modeling the project life cycle by assuming it to have a predictive scope and a cascading stream of defined work tasks; however, recent revisions are softening that highly predefined model structure view, and that is a key activity to understand in the scope of this text.

Industry experience indicates that increased usage of such models has had a positive effect on project success, even though failure rates remain high in many areas. There is growing recognition that these documented project descriptions are bloated and cumbersome to execute, which results in extra-long delivery cycle times. In the author's opinion, there is also a feeling that the formal models are too static in their view and often require excessive documentation for no additional value, particularly in the planning phase. In examining the defined models, one can conclude that project overruns are not caused by the model being wrong but more by the project not fitting the tenets of the model. In other words, if the scope cannot be clearly defined before execution, the rest of the model logic does not fit the problem very well.

Recognize that this statement now enters the realm of a religious-type argument regarding how to best manage the project life cycle process. The conclusion from this historical review is the need to recognize the variability of projects and that none of the previous modeling efforts have satisfactorily embedded that concept into their design. The model needs to fit the problem and have flexibility with other processes. The thesis of this text is that the most popular management models do not fit the broad view of project types and the related management process; however, there are positive attributes that need to be understood. The traditional view of project management that emerged from this historical evolution remains the prevalent view regarding managing a project.

1.5 QUALITY MANAGEMENT PROGRAMS

After WWII, the United States witnessed operational success in creating improved products using quality management processes initially developed in Japan and copied in the United States to capture these ideas. Some example names for these programs are Zero Defects, TQM, ISO 9000, and, most recently, Six Sigma. These efforts focused organizations more on customers and efficient processes to deliver desired results. The quality movement's additional tool, which focuses on organizational process improvement steps, is important for this discussion, like a product-oriented project. The quality project management models are not as broad or complex as the traditional models, but they highlighted and sensitized an increased focus on lean principles (no waste) and customer. Also, there is an implicit focus on organizational learning and continuous improvement. Because of this trend, the quality management school is now having both a tool and a cultural impact on how projects should be executed.

1.6 STRATEGIC PROJECT VIEW

Projects have been historically viewed as discrete, standalone initiatives. As organizations began to formalize and proliferate this mechanism for "getting things done," it became apparent that organizational projects were competing for the same pool of scarce resources. To deal with this situation, many organizations created a formal oversight group to examine the collection of proposals and attempt to maximize their collective value. This activity is called PPM (i.e., processes to evaluate the prospective value of competing proposals). A companion organizational decision process to this activity is often titled PMO or Project Management Office. These two linked functions combine to assess, prioritize, and approve projects from the portfolio of options. The need for formal processes to select project targets is now well recognized as part of the strategic management view and must be recognized in any modern formal model. Each of the key evolutionary items described here has significantly impacted how projects are viewed and managed. This high-level project view represents an essential thread of the management tapestry.

1.7 UNIVERSAL MODELING LANGUAGE

One of the more rigorous and interesting methodology attempts in the early 1990s was the development of the Unified Modeling Language (UML). This attempt to design a technically correct approach to modeling software development was headed by three well-respected gurus in the software industry—James Rumbaugh, Grady Booch, and Ivar Jacobson. They aimed to design a "single-unified specification method" (Visual Paradigm, 2022). By the latter 1990s, this effort was felt to be the beginning of automated software delivery based on sophisticated UML design models. Even though this model is still considered technically accurate, its use is declining. There have been attempts to link it to iterative development, but there is still no visible broad acceptance. This appears to be the case of a technically superior initiative sponsored by respected technicians that was still not accepted. In this same period, the industry accepted the iterative model associated with, a much less sophisticated process. The iterative development school has been less aggressive in the use of rigorous modeling from a purely technical viewpoint, but focuses more on the project team and its less-defined sprint-oriented approach to producing outcomes. Whereas UML models are focused on software and detailed technical characteristics, so there is little crossover seen in using these models together for traditional product development, other than recognizing that schematic models play a part in visualizing technical aspects. The history and evolution of UML are noted here primarily because of its characteristics regarding sponsors and technical sophistication, which logically are prerequisites for successful acceptance. It would be worthwhile for the reader to browse the comparative histories of both Agile and UML, recognizing that both were focused on the same development target. Other than this history lesson, UML falls outside the scope of the integrated management model target.

1.8 VARIABLE WORK STRUCTURES

There is a tendency to pursue project management with a singular, biased view of the work. We will call this the current organizational belief culture. However, there are now five alternative disconnected views that help describe how the industry now views project work. These are as follows:

Predictive—Cascading work groups and linked tasks that fit the waterfall design.

Iterative—Utilizing the Agile school of thought with iterative sprint cycles of development, searching for customer satisfaction.

Work chains—The view of work described by Goldratt's Theory of Constraints views tasks as a coupled chain to be executed as quickly as possible (Goldratt, 1997). The analogy is that the project should be managed without a fixed schedule and viewed as a track relay race with no schedule inhibitors.

Process-oriented—This project class is focused on delivery systems with heavy customer interactions. These are often pursued using the iteration model or Six Sigma quality school tools.

> *Portfolio management*—This evolving high-level project view involves the strategic aspect of the management process with a tight linkage to the ongoing set of projects. Strategic portfolio management—"connects the projects, programs, and initiatives that exist across an organization to its overall corporate strategy with the goal to optimize business outcomes while adapting to disruptions".
>
> (Itonics, 2024, para. 2)

Each of these work-type goal views contains valid management strategies that merit being included in a re-engineered and integrated management model based on a project's profile and goal characteristics. Recognize that these strategies have a place in the future of project management theory.

The process of homogenizing these views into a single management structure will require that the reader understand what each of these basic models entails concerning work management. Understanding how various management gaps in current practices impact successful delivery is essential. In this case, the model is not in error but rather the proper work management selection for the related process. In other words, given the character of a specific task work unit, what is the best way to execute it to produce its output goals (i.e., scope definition, delivery speed, control, resource management, customer satisfaction, or risk)? More elaboration on this variable work selection idea will be offered in upcoming sections, but this represents one of the core components of the integrated approach. An improved model view must have the capability to be flexible regarding the way various work units are executed based on the project profile's priority goals. The expectation is that there will be multiple options within the project and not a single approach.

Here is a thought question. What if the project schedule and budget are not the main determinants of success? For example, the competitive goal is to achieve a strategic market position, and the output product must be widely accepted or may need to be produced in the shortest time possible. This project has a vague future requirement and thus requires a lot of user involvement and prototypes to review. Control is a lesser concern than customer satisfaction. Because of these variable success views, it is no longer feasible to pour the project tasks into a previously defined model structure and follow that to success. Recognizing this variable success criteria view will significantly impact the proper management approach.

1.8.1 Delivery Forecast Projections

Accurately forecasting the future for any subject is challenging, but ne should have sense of why changes are occurring. A data-oriented person, is amused when weather forecasters predict that will start at 2:15 p.m. (six hours from now). Given the complexity of weather forecasting, does this level of quantification make sense, and is it accurate? Quite often, not only does it not rain that day, but also not at the time predicted. Project managers often fall into this same quantification data minefield trap. Management requires that completion dates be defined in the project case, so the plan says that the project will be finished on June 4, and the budget will be $1,110,000. Neither occurs; therefore, the project manager must not know what they are doing, just like the weather forecaster. Both know what they are doing but are reporting their data the wrong way based on the complexity of the topic. Given this, it would be better to get the receiving audience to understand the uncertainty and report accordingly. In the case of projects, this implies using probability distributions or range values rather than discrete values. We will leave the weather reporting issue for another group to solve. More of this thought process will be reviewed later in this textbook.

Interest in changing how projects are managed seems destined, as other methods are now being documented as showing improved success, even though they do not appear to contain a full breadth of life cycle view. In response to all we are describing here, project managers today are band-aiding their models in reaction to the user and senior management prodding for quicker completions or answers that do not make sense given the situation. In response to recognized shortcomings in current delivery methods, band-aid patches are used in a notable way to help traditional methods recognize assumption gaps better.

One of the common complaints about projects is the "excessive" time they take. Indeed, this variable is an essential and critical success item. There are many examples where a project team took a shortcut in the interest of time, which resulted in project failure. All project managers would choose to go slower, while all users would press to go faster. It is important to remember the *story of the Tortoise and the Hare* (remember that the tortoise won the race). Is it better to be fast and later crash the ship, or plan carefully and arrive safely later? The *need for speed* in the project world can be disastrous if a poor pilot guides the process. There are many industry examples of project managers accepting quick silver bullets that did finish quicker but then failed because the new technology had some unknown flaws. There is no answer to this question, but it simply represents one more stressor for the project manager.

An attempt to define a more flexible management model requires that current shortcomings be recognized and processes be derived that better match reality. An assessment of the current practice shows an ineffective understanding of why the model is not working and a lack of understanding of the proper project priorities. A well-skilled project team may use a patched traditional model to achieve better success rates, but few project managers have this depth of understanding. Hopefully, the pieces outlined here will sensitize you to the various negative issues resulting from this. Much of what is described here will be where to look and how to effectively integrate the pieces into a viable solution package. We have begun recognizing the various causes associated with delivery success, but there is much more to understand regarding why projects fail and some management drivers that can help mitigate these factors. One currently accepted management specification model defines ten important knowledge areas (i.e., PMI's PMBOK). This treatise is a helpful background theory, but the question remains as to what degree the components are essential in a specific case. The question of how thorough we should be in these areas is a crucial success issue. It may be reasonable to skip formal risk assessment, but is that wise? Many projects implicitly do just that and hope all will turn out right. How does one decide on factors such as this? Which of the management factors is most important? These unresolvable tidbits are dropped here to answer the question of why the answer cannot be a fixed cookbook or step. The future approach requires mature technical and management professionals to understand techniques to match different work profiles to the defined delivery goal. These priorities must be ranked for each project and not assumed to be static variables. Concern for risk may be the number one priority, while the schedule may fall to number four. Admitting that all factors are different is just one of the cultural changes implicit in a new view.

The standard waterfall model offers a significant development approach for a project if the underlying assumptions are valid. There is still a question about how thorough one must be with the defined processes. Some might believe that the proper way to build a house is to collect a pile of lumber and other resources in the yard and work with the owner to decide what the house will look like; however, most would agree that some level of predefinition is best, especially if schedule and budget are also factors of interest. In essence, the philosophical question related to project management is based on one's belief regarding the planning rigor required in known target areas of interest (i.e., level of definition in scope planning, task estimating, resource analysis, risk analysis, project approval process, status reporting requirements, and stakeholder concerns). Higher levels of plan definition for each of these add additional time and cost to the project. As an example, the following decisions made regarding the rigor required from proper project management will significantly influence the appropriate approach:

- The level of scope predefinition before execution—the house-building example
- Deciding the level of work management delegation to the project team—this can impact scope definition and other management roles
- Specifying project deliverable priority goals and constraints (i.e., speed, cost, risk, customer satisfaction)
- Quantifying resource capacity needs (quantity and quality)
- Team management techniques to apply (morale, productivity, professional growth)
- Defined level of stakeholder support and communication

Projects are now viewed as any structured work activity that changes the existing state of a product or process. The output goal must no longer be viewed as a tangible product. Focus on this broader view by envisioning a significant reorganization project involving the movement of people, facilities, physical infrastructure, technical infrastructure, customer interfaces, and other factors. This is taking a current state and moving it to a future state, thus a project. The new management model needs to fit this sort of activity as well.

This historical overview-related technique and philosophical section is the first step in focusing attention on the essence of global project management and offering clues as to where one might look for better methods to execute those ventures.

This overview has described how the evolution of tools, processes, and documentation has been ad hoc, at best. Many of the items developed were niche targets unsuitable for every situation. These historical examples would not represent the universal best guidance regarding approaching a specific project. The critical point has been made multiple times here in that a fixed cookbook-type guidance model is not the answer to this complex environment. The future will require a more customized model based on the particulars of a specific project. This view is not currently the norm, but the increased recognition of customer satisfaction as a project goal is now highlighted. That has been a somewhat secondary goal on the traditional delivery goal list in the past, but it is now recognized as requiring much more active concern.

The wide variety of real-world projects and the various factors outlined here are intended to support the justification for proposing a more flexible model for the management process. This will have to include more than a listing of the essential things. The future model must contain a broader view of the management process, including birth analysis, strategic and tactical planning, execution techniques, and customer links. Based on the project profile, This design approach must also contain a decision architecture and critical success factor techniques.

A visible laboratory of project management practice lies all around us, and these case scenarios are the best mental models to observe the impact of both good and bad techniques. Hopefully, examining this real-world view will make you more aware of poor management and improve your understanding of the process. That is the learning lesson that may well be the significant reader benefit. For instance, does it bother you to wait on the phone for 30 minutes for *customer service*? What about sitting in a line of cars watching road construction with traffic locked up for miles and seeing only a few workers? Is it a good service goal to allow customers to wait an hour to spend ten minutes with the doctor, or stand in line when other workers could leave their non-time-critical tasks to help? Once you get the spirit of project management, being critical of ineffective designs will bother you more. Also, you will instantly see what you would do to improve the situation. It is time to start looking around at real-world projects and build a management model based on the ideas described here.

1.9 SUMMARY

The early evolutionary period (1945–1960) legitimized the role of formal projects as the best organizational strategy to "get something done." Today, the term *project* is ubiquitous in our vocabulary. The management vision for this activity became defined as a sequential-step life cycle view, often grouped into macro-stages related to defining, executing, and implementing the desired goal. This view remains the most utilized high-level definition today, although storm clouds are on this horizon seeking new methods. Associated with the traditional view, projects are recognized to have a birth-to-death life cycle of decision and task-level execution work steps. The authors recommended reading the Kabeyi (2018) article because it also provides an overview of project management, as referenced at the end of this chapter. Some require these to be clearly defined, while others are less formally defined. The thesis of this text is that project profiles need to be used to customize the management process using a broader set of flexible options to achieve the prioritized deliverables.

REFERENCES

Defense Contract Management Agency (DCMA). (2012, October). Earned Value Management System: Program Analysis Pamphlet, DCMA-EAPAM200.1. https://www.dcma.mil/Portals/31/Documents/Policy/DCMA-PAM-200-1.pdf?ver=2016-12-28-125801-627

Department of Defense. (2022, May 13). Department of Defense Standard Practice Work Breakdown Structures for Defense Materiel Items. https://cade.osd.mil/Content/cade/files/coplan/MIL-STD-881F_Final.pdf

Goldratt, E.M. (1997). *Critical Chain.* Great Barrington: The North River Press. CIO. The Scaled Agile Framework (SAFE). https://www.cio.com/article/220569/what-is-safe-the-scaled-Agile-framework-explained.html

Hallum, M.M. (1981, December). *A Technical View of Cost/schedule Control System Criteria (C/S CSC).* US Army Missile Laboratory. Retrieved August 20, 2024, from https://apps.dtic.mil/sti/tr/pdf/ADA120005.pdf

History, The First Years, Internet. (n.d.). International Project Management Association (IPMA). Retrieved August 17, 2024, from https://ipma.world/ipma-governance/history-of-ipma/

History of PMI. (n.d.). Project Management Institute (PMI). Retrieved August 18, 2024, from https://www.pmi.org/about/learn-about-pmi/history-of-pmi

Itonics (2024). Definition of Strategic Portfolio Management. Retrieved August 21, 2024, from https://www.itonics-innovation.com/strategic-portfolio-management-guide

Kabeyi, M.J.B. (2018). Evolution of Project Management, Monitoring and Evaluation, with Historical Events and Projects that Have Shaped the Development of Project Management as a Profession. *International Journal of Science and Research (IJSR)*, 8(12), 63–79. DOI: 10.21275/ART20202078

Max (2024, June 19). *The History of Project Management: Planning the 20th Century.* Management Library. https://management.org/history-of-project-management

Polaris Program. (2024). U.S. Naval Institute. Retrieved August 17, 2024, from https://www.usni.org/press/oral-histories/polaris-program

Prince2. (2022). What is PRINCE2? Retrieved August 13, 2024, from https://www.prince2.com/usa/what-is-prince2?gclid=Cj0KCQjw8uOWBhDXA

RIsAOxKJ2ENUip9QohrZbSI-x7fYkrflxRnEDSJ_qmgmYiiK9q-YH7QiaWM6akaAqLPEALw_wcBPRIINCEREF

The evolution of project management. (n.d.). Mosaic Project Service PTY LTD. Retrieved August 18, 2024, from https://mosaicprojects.com.au/Mag_Articles/AA021_The_evolution_of_project_management.pdf

Visual Paradigm. (2022). The Unified Modeling Language. Retrieved August 13, 2024, from https://www.visual-paradigm.com/guide/uml-unified-modeling-language/what-is-uml/

Weaver, P. (n.d.). *The Real Gantt Charts Were Smarter Than You Think*. Mosaic Project Service PTY LTD. https://mosaicprojects.com.au/PDF_Papers/P158_Henry_Gantt_PPT.pdf

Witzel, M. & Warner, M. (2015). *Taylorism Revisited: Culture, Management Theory & Paradigm-shift*. University of Cambridge Judge Business School. Working paper. 1–21. Retrieved August 13, 2024, from https://www.jbs.cam.ac.uk/wp-content/uploads/2020/08/wp1501.pdf

CHAPTER 2

Role of the Project Manager

2.1 PROJECT MANAGER

The PM is the person who is formally charged with leading the initiative forward through its life cycle. There are three main routes to developing a PM skill, as identified by Kilkelly (2009). One of the least desired of these is to allow a new PM to learn by making trial-and-error mistakes. However, most organizations are not very forgiving of errors made. So, a better strategy for new PMs is to be mentored through formal or informal methods using formal programs. Another method to avoid career-limiting errors is to begin as an assistant to an experienced PM. A third method is to learn the basic mechanisms of the management process through some form of training before taking on the formal role. This can be a formal academic degree in project management, focused training courses, or self-study. It is also realistic to suggest that this be a combination of more than one of these forms of training.

Research suggests that the skills necessary to guide a project successfully are not necessarily bred and can be improved through formal methods (Edmonds, 2010). PMs are often selected based on their recognized reliability and technical expertise within their specific field. As a result of this action, a new PM may find themselves responsible for a project team of individuals yet lacking an appropriate background to deal with the complexity of this new role. A lack of appropriate training could result in a gap in the skills required to manage the project effectively. For example,

DOI: 10.1201/9781003532989-3

a new PM may not understand specific essential details and focus too much time on unimportant details. There is a saying about not being able to see the forest for the trees, and this is another way of describing a PM focusing on the wrong details. Blending mentorship and training offers the best chance of producing a successful PM. Many organizations focus only on the view of producing project deliverables, but they also need to focus on the growth of successful PMs (Nixon et al., 2012). Project deliverables may be deemed successful if they meet customer expectations; however, an organization must also grow stable and high-quality PMs, which is essential in strategic goal alignment.

2.2 POWER

Projects are natural breeding grounds for conflict, often leading to ineffective human behavior. These situations must be handled effectively for a project's success and health (Richardson & Jackson, 2022). Every organization, whether small or large, will have unique power struggles due to changing parameters of the project resources, environment, goals, and deliverables (Vidal & Marle, 2008). Power can be in terms of which projects senior management supports and is committed to bringing them to fruition (Maddalena, 2012). Trust between the stakeholders and the project team is a powerful asset as it can foster loyalty from critical stakeholders and result in a positive project environment (Brewer & Strahorn, 2012). Identifying and managing stakeholders is necessary to help sell and continue to sell the project throughout its life (Little, 2011). Another critical aspect of the project is assembling the right team of individuals (Clark & Colling, 2005; Maddalena, 2012). These struggles are rooted in resources being in limited supply.

Key questions related to resource limitations become paramount. For many organizations, these issues come from labor, monetary, facilities (space), and equipment resources. If you are a manager and your boss asks you to spare two of your current labor resources for a year to work on a particular project, which resources will you choose? Will the two-star performers or the two lowest performers be chosen? If you choose your two-star performers, your boss may be pleased because these star performers will make that project more likely to succeed. However, without the star performers, will your office dynamics change so much that you are left with a dysfunctional group of employees that will leave you having to pick up the slack by working extra hours weekly? If you choose your two lowest performers, will your boss be back in your office a month from now, saying that the workers you offered were unsuitable for the project, and he now requires you to let go of your star performers?

These are the decision-making issues going on constantly inside the head of every lead, manager, and executive in the work environment. So, did you decide which employees to send to the project yet? Maybe you decide to compromise and send just one star and one average performer so your boss will succeed and your office is not interrupted as well. As a manager, it is also your responsibility to help your employees' career development. So, if this project has high visibility for the employees sent, then maybe both star performers deserve that opportunity to make the decision much more complex. We call this a conundrum!

Now, let us divert our focus from labor resources for a moment and discuss monetary resources. There is not a single organization that exists that has unlimited money to spend. So, executives spend time planning which efforts or projects their company should invest in that will result in the maximum growth of a company's profits. This is a political game as there may be limited resources yet unlimited power struggles within an organization at any given time. Different communication channels go up and down the organization as employees ask their bosses for monetary resources, such as equipment upgrades. In return, a specific organizational level then asks their managers for funding for the employees' needs, resulting in managers asking their bosses and their bosses asking the company's executives for funding. This chain of demands permeates the organizational structure. Different employee leads, managers, and executives must continuously choose the best way to use their limited monetary resources to address the endless funding requests. Sometimes, the choices made regarding how to spend the funds have to do with what is best for a company, but often, choices have less to do with a company's needs and more to do with funding the manager with the most significant degree of power. There can also be hidden agendas, such as a business unit creating a need for a specific project for its value without aligning it with organizational-level goals. These hidden agendas have more to do with returning favors to individuals to keep conflicts to a minimum. This is never the recommended approach to leadership and management. Managers might feel pressured to provide funding for one project over another based on a favored personal relationship or promised favors in the future. Ideally, decisions on how to use the limited amount of funding should be based on which projects or needs are best in global alignment with organization goals. However, this is a true challenge because of the abovementioned interpersonal factors. Another problematic decision topic is outsourcing.

The decision to outsource is often made purely based on current cost, and middle segments of the organization are typically most impacted by this decision. The third area of resource conflict comes from equipment and space considerations. Depending on the organization, specific rules or guidelines may be distributed regarding how much space one-level manager gets for their office versus another manager. An organization can even further define this by one-level manager getting a desk of a specific size and amenities, such as window views or even who gets the expensive ergonomic chairs. Even conflicts over which department gets the better conference room to conduct a workshop occur as, again, the number of needs in an organization is unlimited. In contrast, the resources to fulfill those needs are limited. The character of this decision environment may change across industries, but the effect is similar. This can be anything from professors at a university wanting specific classroom space because one classroom may have better computer stations for students, whereas other classrooms are absent from technology. Different product managers could also need the same factory equipment to produce their product line. As with people and monetary resources, equipment is also physically limited, so not everyone's needs are met within the time frame that best fits them.

With these examples as a conceptual review, let us return to the labor resource question discussed earlier in this chapter. Have you resolved yet which resources to give away to the other project? What type of reward will your project receive if you allow your star performers to be part of the other project for a year? Perhaps your boss will provide a reward if the star performers are sacrificed on your project for your boss' project. Will you then be given the new computer resources requested for all employees in the office? Will you then get to hire a new employee or recruit an employee from another department to fulfill a skill need currently lacking? After all, the most significant rewards tend to be resources in the form of labor, monetary resources, equipment, or facilities. Moreover, we begin to see the fuzzy and complex world of resource management.

2.3 CONCLUSION

The chapter discussed the role of the PM. The PM leads the project through its life cycle. There are three main routes to developing a PM skill. Power and political conflicts occur for many organizations due to labor, monetary, facility (i.e., space), and equipment resources. Every organizational culture is different, but regardless of the culture, every project needs the organization's support to succeed. There will be different power struggles and politics because no project is without these issues.

QUESTIONS

Student Review Questions

1. Describe ways that a PM can be developed.
2. Describe an example of each power and political conflict that falls under labor, monetary, facility, and equipment resources.

In-Class Question

1. Identify and summarize a news article for a power-and-political conflict for a company project.

EXERCISE

1. What type of power struggles may occur within an industry of your choice?

REFERENCES

Brewer, G., and Strahorn, S. (2012). Trust and the Project Management Body of Knowledge. *Engineering, Construction and Architectural Management*. 19(3), 286–305.

Clark, I., and Colling, T. (2005). The Management of Human Resources in Project Management-led Organizations. *Personnel Review*. 34(2), 178–191.

Edmonds, J. (2010). How training in project management can help businesses to get back on track. *Industrial and Commercial Training*, 42(6), 314–318. https://doi.org/10.1108/00197851011070695

Kilkelly, E. (2009). Blended learning: Pathways to effective project management. *Development and Learning in Organizations*, 23(1), 19–21. https://doi.org/10.1108/14777280910924081

Little, B. (2011). The Principles of Successful Project Management. *Human Resource Management International Digest*. 19(7), 36–39.

Maddalena, V. (2012). A Primer on Project Management. *Leadership in Health Services*. 25(2), 80–89.

Nixon, P., Harrington, M., and Parker, D. (2012). Leadership performance is significant to project success or failure: A critical analysis. *International Journal of Productivity and Performance Management*, 61(2), 204–216. https://doi.org/10.1108/17410401211194699

Richardson, G.L., & Jackson, B. (2022). *Project Management Theory and Practice*, 4th ed. CRC Press, Taylor & Francis Group: New York.

Vidal, L.-A., and Marle, F. (2008). Understanding Project Complexity: Implications on Project Management. *Kybernetes*. 37(8), 1094–1110.

CHAPTER 3

Project Management Environment

Project N Motion Videos for This Chapter		
3S	Stakeholder Management	Stakeholder Overview

3.1 INTRODUCTION

A project is

> A temporary endeavor undertaken to create a unique product, service, or result. The temporary nature of projects indicates a beginning and an end to the project work or a phase of the project work. Projects can stand alone or be part of a program or portfolio.
>
> (PMI, 2021, P. 4)

Organizations need change so they continue to grow and maintain a competitive advantage. The changes tend to respond to meeting new demands, such as new customer demands, market demands, and regulatory demands. These demands become the trigger mechanism resulting in organizational change. The way to implement change is to have projects because these projects are, in fact, the lifeblood of an organization. Let us look at the success of Amazon.com. We can see how the company

DOI: 10.1201/9781003532989-4

began in the owner's garage selling books and then sold products from A to Z through a series of complex projects in areas such as inventory, web design, customer relationship, and innovation, which led to a competitive advantage (DePillis & Sherman, 2019).

Over the past 70 years, the waterfall model has been the de facto project management model for delivering project results. Therefore, it is important to understand what it is and the underlying assumptions implicit in its internal processes. Let us take a moment to look at the Waterfall model assumptions. The following list contains 11 implicit project design assumptions worthy of note before we dive further into the project environment:

1. Project *delivery requirements and scope* can be reasonably defined before execution.
2. *Reasonable time estimates* can be produced for defined work units.
3. *Future risk events* can be adequately evaluated in advance, and proper measures can be taken to mitigate those items. A risk contingency reserve may be used.
4. *The task can be executed* according to the defined estimates and sequence.
5. *Project success* is measured by the planned time, cost, and deliverables.
6. *Timely quantity and quality resources* will be committed by the support organization according to the plan.
7. *Senior management and users are readily available* and standing by to help when problems emerge.
8. *Key stakeholders fully support* the project needs concerning technical and physical resources.
9. The project *team members understand basic project management processes.*
10. *Scope changes will be formally managed*, and the plan will be adjusted based on approved changes.
11. The defined set of *user and technical documentation* will be produced.

If we look at actual projects, we find that many of these assumptions are suspect. You must look at the project's specific characteristics to judge what changes are logical to fit that environment better. Examples in the textbook will stay consistent with this model but also show methods to improve its accuracy if an assumption is inappropriate. The role of project models is to help envision the delivery requirement and provide a roadmap to achieve the stated goals. Please recognize that this model has been the survivor of many who have attempted to improve it.

Projects have internal and external environments in which they exist and operate (PMI, 2021). These environments influence project planning and activities, with the impact being favorable, unfavorable, or neutral regarding project characteristics, stakeholders, or project teams. The internal environment consists of organizational artifacts, practices, or internal knowledge. Internal knowledge consists of process assets such as methods and templates, data assets such as databases and metrics, and knowledge assets such as subject matter assets and tacit knowledge among team members. It includes resource availability, people and material resources, and contracting and purchasing providers or constraints. Resource expertise also falls under the umbrella of the internal environment, which "include[s] general and specialized expertise, skills, competencies, techniques, and knowledge" (p. 17). It also includes government documentation, such as policies and processes, and information technology software ranging from scheduling software to collaboration tools to web interfaces. Infrastructure is part of the internal environment and includes equipment, facilities, information technology, hardware, and capability. This includes security and safety measures such as proprietary secrets, data protection, and even facility access practices. Furthermore, organizational culture, structure, and governance are the "aspects of an organization [that] include the vision, mission, values, beliefs, cultural norms, leadership style, hierarchy and authority relationship, organizational style, ethics, and code of conduct" (p. 17). Lastly, the geographic distribution of facilities and resources comprises the internal environment such as "work locations, virtual project teams, and shared systems" (p. 17).

The external environment consists of factors ranging from enhancing to constraining a project. This involves a regulatory environment that "include[s] national and regional laws and regulations related to security, data protection, business conduct, employment, licensing, and procurement" (PMI, 2021, p. 18). Also, the physical environment, which consists of working conditions and weather, can impact a project. Marketplace conditions "include competitors, market share, brand recognition, technology trends, and trademarks" that also affect a project (p. 18). Social and cultural influences and issues include "political climate, regional customs and traditions, public holidays and events, codes of conduct, employment, licensing, and procurement" (p. 18). Commercial databases contain industry risk, cost-estimating data, and industry standards "related to products, production, environment, quality, and workmanship" (p. 18). These databases, standards, and academic research, from industry studies to publications, are also part of the external environment. Lastly, financial considerations such as "currency exchange rates, interest rates, inflation, taxes, and tariffs" can also affect a project" (p. 18).

The project lifecycle is a series of phases that a project goes through from the beginning till the end. The life cycle is a basic framework for everything involved in project management, regardless of the type of business to which the project is connected. The PM or organization will adapt the phases of a project life cycle to fit each specific project. These generic life cycle phases consist of starting the project, organizing and preparing for the project's work, carrying out the actual project work, and finishing the project, such as project close-out documentation. This chapter addresses the project environment and success factors that lead a project to success, bringing about change in an organization. Hence, it continues to grow and respond to internal and external demands.

3.2 PROJECT ENVIRONMENT FACTORS

The environment for a project involves the internal and external environment, as discussed earlier in this chapter. Enterprise environmental factors (EEFs) and organizational process factors (OPAs) influence projects. EEFs are "conditions, not under the immediate control of the team, that influence, constrain, or direct the project, program, or portfolio" (PMI, 2021, p. 239). OPAs are "plans, processes, policies, procedures, and knowledge bases specific to and used by the performing organization" (p. 243). Internal or external conditions in an organization are outside the project's control. On the other hand, OPAs are inside the overall organization. These factors influence a project, resulting in a favorable or unfavorable impact.

TABLE 3.1 Examples of Internal EEFs (PMI, 2017)

Internal EEFs	Examples
Organizational Culture	Organizational Values Organizational Beliefs Cultural Norms Ethics
Organizational Structure	Organizational Style Hierarchy Authority Relationships
Organizational Governance	Code of Conduct Leadership Style Mission Vision
Geographic Distribution of Facilities	Factory Locations Headquarters Site Locations
Geographic Distribution of Resources	Team Member Locations Virtual Teams Shared Systems Cloud Computing
Infrastructure	Facilities Equipment Capacity
Information Technology Software	Scheduling Tools Configuration Management Systems Websites Work Authorization Systems
Resource Availability	Labor Pool Subcontractors Collaboration Agreements Contracting and Purchasing Constraints
Employee Capability	Knowledge Skills Expertise Competencies

EEFs can be internal or external to an organization. Examples of internal EEFs are listed in Table 3.1 and those of external EEFs in Table 3.2. These EEFs must be considered throughout the life cycle of a project, and although they are not found in the seventh edition of the PMBOK, they can be found in the sixth edition (PMI, 2017).

TABLE 3.2 Examples of External EEFs (PMI, 2017)

External EEFs	Examples
Marketplace Conditions	Competitors Market Share Trademarks
Social and Cultural Influences/ Issues	Political Climate Code of Conduct Ethics Perceptions
Legal Restrictions	Laws Regulations
Commercial Databases	Benchmarking Results Cost Estimating Data Industry Risk Study Information Risk Databases
Academic Research	Industry Studies Publications Benchmarking Results
Government or Industry Standards	Regulatory Agency Regulations Product Standards Production Standards Environment Standards Quality Standards Workmanship Standards
Financial Considerations	Currency Exchange Rates Interest Rates Inflation Rates Tariffs Geographic Locations
Physical Environmental Elements	Working Conditions Weather Constraints

OPAs are grouped into two categories of factors (PMI, 2017). The first category is processes, policies, and procedures that are not updated during the project work but are already established within an organization. Examples of these are displayed in Table 3.3. The second category is organizational knowledge bases, updated throughout the project as project information becomes available. Examples of organizational knowledge bases are displayed in Table 3.4.

TABLE 3.3 Examples of Processes, Policies, and Procedures OPAs (PMI, 2017)

OPAs: Plans, Processes, Policies	Examples
Guidelines	Work Instruction Guidelines Proposal Evaluation Criteria Performance Measurement Criteria
Standards	Human Resource Policies Health and Safety Policies Security and Confidentiality Policies Quality Policies Procurement Policies Environmental Policies
Methods and Procedures	Project Management Methods Estimation Metrics Process Audits Improvement Targets Checklists Standardized Process Definitions Change Control Procedures Financial Control Procedures Issue and Defect Management Procedures Work Authorization Procedures
Templates	Change Log Issue Log Project Management Plans Project Documents Project Registers Report Formats Contract Templates Risk Categories Risk Register Risk Statement Templates Probability and Impact Definitions Probability and Impact Matrices Stakeholder Register Templates
Preapproved Supplier Lists and Contractual Agreements	Preapproved Supplier Lists Fixed-Price Contract Cost-Reimbursable Contract Time and Materials Contract

EEFs and OPAs can influence a project either positively or negatively. Considering the examples provided in Tables 3.1 through 3.4 throughout the project life cycle is essential. In addition to these factors, numerous other factors can impact a project, as these serve as only examples. Other factors range from psychological to ecological factors.

TABLE 3.4 Examples of Organizational Knowledge Bases OPAs (PMI, 2017)

OPAs: Organizational Knowledge Bases	Examples
Configuration Management Knowledge Repositories	Versions of Software and Hardware Components Baselines of Organizational Standards Baselines of Organizational Policies Baselines of Organizational Procedures Baselines of Project Documents
Financial Data Repositories	Labor Hours Incurred Costs Budgets Project Cost Overruns
Historical Information and Lessons Learned Knowledge Repositories	Project Records and Documents Project Closure Information and Documentation Information on Previous Project Selection Decisions Information on Previous Project Performance Information Information on Risk Management Activities
Issue and Defect Management Data Repositories	Issue and Defect Status Control Information Issue and Defect Resolution Action Item Results
Data Repositories for Metrics	Process Measurement Data Product Measurement Data
Previous Project Repositories	Scope Cost Schedule Performance Measurement Baselines Project Calendars Project Schedule Network Diagrams Risk Registers Risk Reports Stakeholder Registers

3.3 PROJECT SUCCESS AND FAILURE FACTORS

There are many ways in which a project can succeed or fail. This section will focus on the success factors of a project. Table 3.5 provides examples of project success factors and failure factors. There are many more success and failure factors, but this table identifies some of these as a starting point for PMs. The success factors serve as a guideline of factors to

TABLE 3.5 Examples of Project Success and Failure Factors

Categories	Examples of Success Factors	Examples of Failure Factors
Measures	Select measurable project objectives (PMI, 2021) Select financials that are measurable are discussed by PMI (2021): • Payback period • Internal rate of return (IRR) • Return on investment (ROI) • Net present value (NPV) • Cost-benefit analysis Track customer/end-user adoption (PMI, 2021) Provide deliverables on-time Provide deliverables within the scope Accepted and on time deliverables (Roseke, 2016) Provide quality deliverables (PMI, 2021)	Lack of measurable objectives Lack of project metrics to measure Late deliverables Unaccepted deliverables Lack of end-user adoption
People	Recruit a project sponsor if one does not already exist Employ a skilled PM Recruit skilled team members Attract intelligent people (team members, suppliers, and stakeholders) (Palmer, 2018) Cultivate team member commitment (Palmer, 2018) Cultivate a shared vision among team members (Palmer, 2018) Cultivate teams that work well together (Roseke, 2016)	Lack of a project sponsor Employ an untrained PM Lack of skilled team members Experience team member turnover Experience inept leadership (Palmer, 2018) Foster an out-of-sync team (Palmer, 2018)

(Continued)

TABLE 3.5 (Continued) Examples of Project Success and Failure Factors

Categories	Examples of Success Factors	Examples of Failure Factors
Satisfaction	Create an environment where stakeholders are satisfied. A Net Promoter Score (NPS) is "the degree to which a stakeholder (usually the customer) is willing to recommend a product or service to others. It measures a range from -100 to +100" (PMI, 2021, p. 103). In addition to measuring satisfaction, it also indicates customer loyalty. A mood chart or board can also be used to gauge the project team's mood or reactions at the end of each day, using colors, numbers, or emojis to identify potential issues and areas of improvement.	Create an environment where there is a lack of satisfied stakeholders
Project Planning	Stakeholder involvement (Palmer, 2018) Create a schedule with realistic timeframes (Palmer, 2018) Create clear documentation of milestones and deliverables (Palmer, 2018)	Lack of stakeholder involvement Create a schedule with unrealistic timeframes Lack of documentation of milestones and deliverables
Communication	Identify project expectations from all key stakeholders Achieve sufficient and on-time communications with stakeholders (Roseke, 2016) Clear roles and responsibilities for team members Keep the team well-informed (Palmer, 2018) Communicate project status with all stakeholders at the level each stakeholder group desires	Lack of project expectations from all key stakeholders Lack of clear roles and responsibilities for team members Lack of communication among stakeholders
Risk Management	Create a risk log with an action plan if risks surface (Palmer, 2018) Mitigate risks successfully (Roseke, 2016) Manage risks that occur well (Roseke, 2016)	Lack of risk management throughout the life cycle of the project
Project Closure	Agree upon project closure (delivery, testing, and release) and have it signed off so the project does not continue to consume resources (Palmer, 2018)	Lack of documentation on what constitutes the end of a project

consider that contribute toward a successful project. Failure factors are a guideline for avoiding factors because they contribute to unsuccessful projects. We must also understand stakeholders' position on and perspective of the project, such as "power, impact, attitude, beliefs, expectations, degree of influence, proximity to the project, interest in the project, and other aspects surrounding stakeholder interaction with the project" (PMI, 2021, p. 11). The project vision should be a collaborative effort involving the project team members and key stakeholders to answer questions such as the following:

- "What is the project purpose" (PMI, 2021, p. 23)?
- "What defines successful project work" (PMI, 2021, p. 23)?
- "How will the future be better when the project outcomes are delivered" (PMI, 2021, p. 23)?
- "How will the project team know that it is drifting from the vision" (PMI, 2021, p. 23)?

The answers to these questions will help the PM ensure clear expectations are established of what is expected of the project team and deliverables and which factors need to be tracked and measured throughout the project.

3.4 CONCLUSION

Many factors influence a project's environment and ultimate project success. There are EEFs and OPAs that can positively or negatively influence projects. Examples of internal EEFs, such as organizational culture and infrastructure, and external EEFs, such as legal restrictions and financial considerations, were discussed. Examples of OPAs were also discussed. Examples of the first category of OPAs—processes, policies, and procedures—are guidelines and templates and that of the second category of OPAs—organizational knowledge bases—are financial data repositories and data repositories for metrics.

Success factors and failure factors for projects were also discussed, for example, measures, people, satisfaction, project planning, communication, risk management, and project closure. There are an endless number of factors to consider, but Table 3.5 is a starting point for PMs to include throughout the life cycle of their project.

Overall, it is essential to consider the different factors discussed in this chapter. Some factors fall inside or outside the organization's control, but all factors can influence a project. If there is no control over a factor, such as a company mission statement, the PM must accept this and align the project accordingly. Other factors may be some degree of control for the PM, but control is still not guaranteed. The authors hope that this chapter provides some guidance to PMs on the project environment and different factors to consider throughout the life cycle of a project.

QUESTIONS

Student Review Questions

1. Describe internal and external EEFs.
2. Describe the two types of OPAs.
3. Identify project success factors.
4. Identify project failure factors.

In-Class Questions

1. If you were a PM, how would you structure your project team around two of the internal and two of the external EEFs?
2. If you were a PM, how would you use two examples of the first category of OPAs, processes, policies, and procedures to help your project?
3. If you were a PM, how would you use two examples of the second category of OPAs, organizational knowledge bases, to help your project?
4. How would you incorporate two of the success factors if you were a PM over a new project?
5. How would you avoid two failure factors if you were a PM over a new project?
6. Provide one example of a situation where an OPA or EEF can enhance project success.

REFERENCES

DePillis, L., & Sherman, I. (2019, February 14). Amazon's Extraordinary Evolution: A Timeline. CNN.com. https://www.cnn.com/interactive/2018/10/business/amazon-history-timeline/index.html

Palmer, E. (2018, October 5). Five Factors That Lead to Successful Projects. Project-Management.com. https://project-management.com/five-factors-that-lead-to-successful-projects/

Project Management Institute (PMI). (2017). *A Guide to the Project Management Body of Knowledge*, 6th edition. Newtown Square, PA.

Project Management Institute (PMI). (2021). *A Guide to the Project Management Body of Knowledge*, 7th edition. Newtown Square, PA.

Roseke, B. (2016, April 4). 39 Project Success Factors. ProjectEngineer.net. https://www.projectengineer.net/39-project-success-factors/

CHAPTER 4

Gaining Support for a Project

4.1 INTRODUCTION

Chapter 3 discussed the factors influencing a project environment and project success factors. This chapter focuses on critical processes in moving the business case for a project to become a viable project structure. This customization process aims to select a project structure to produce a balance of target goals.

Without a good business case for approving a new project request, an organization opens its door to selecting projects that do not align with the organization's goals, resulting in the selection of low-value and weak projects. The first question to resolve once the project target has been selected is what should occur next. From that base view, this chapter will address how stakeholder support for the project is gained, including the decision process to officially launch a new project and the related purpose of a project charter. Processes to produce a successful project require a high degree of planning and skilled leadership; some would even argue that it also takes a small amount of luck (Little, 2011). The tools and techniques described in this text should help achieve the desired outcome, even if good luck is not on their side.

4.2 MANAGEMENT APPROACH

Every organization has a different way of approaching the official launch of a project. Still, the model approach requires that this be done using

 DOI: 10.1201/9781003532989-5

a formal charter signed by management and supported by a business case outlining the value of the project. The initial target vision is used to produce a business case that then leads to the charter issuance, which formally approves further evolution of the process. The primary goal of the business case is to validate that the initial project vision has value and is aligned with organizational objectives. At the end of this process, management will show approval and support for the effort by formally signing a charter document. Following this, the project team's first task usually revolves around expanding the definitional details of the project (Anonymous, 2002; Little, 2011). This should also involve the identification of the PM since the effort at this stage has minimal management guidance. Some organizations may wait to assign this person until further scope planning and work definition are complete; however, there is a risk that delaying this can cause the project direction to drift aimlessly.

4.3 PROJECT MANAGEMENT OFFICE

PMI (2021) defines a Project Management Office (PMO) as "a management structure that standardizes project-related governance processes and facilitates the sharing of resources, tools, methodologies, and techniques" (p. 211). Now, with every organization, the role of PMO takes on different roles. There are some commonalities, though, between PMOs, which are worthy of additional discussion. To begin with, PMOs have a shared goal of improving project management within their organizations (i.e., schedule, cost, quality, and risk). One of the most critical roles of PMOs is to have organizational projects that align with strategic goals. This helps ensure that organizations meet their companies' big-picture goals. Different PMOs provide varying levels of project management support. For instance, some provide templates and examples of good practices. Therefore, there may be standardized formats for different project management documentation. Coaching could even be offered to individuals in project management roles, especially for companies wanting to expand their capabilities. Some PMOs also offer support services ranging from assistance in risk management to project performance tracking. Many PMOs try to assist their organization in keeping a portfolio of projects by requiring a business case to initiate a project. This way, the clear benefits of the project are outlined in addition to the project being strategically aligned. They can also assist with allocating resources for the project (i.e., monetary, labor, and space). Overall, PMOs can assist by producing high-quality results through "fostering delivery and outcomes-oriented capabilities" (p. 213). PMOs also help with a project in different ways, such as providing "project teams, senior management, and business leaders with Information and guidance" (p. 213). Lastly, PMOs assist with "continuous Improvement, knowledge transfer, and change management" (p. 213).

4.4 STAKEHOLDERS

One of the early definitional tasks in the life cycle is formally identifying project stakeholders. A stakeholder is "an individual, group, or organization that may affect, be affected by, or perceive itself to be affected by a decision, activity, or outcome of a project, program, or portfolio" (PMI, 2021, p. 250). Stakeholders come in many different categories, as there will be project sponsors, functional managers, future users, external entities, and others who may not be identifiable until the charter's signing. This definition helps identify targets to aid in defining overall requirements. Additional stakeholders may surface as project requirements are further elaborated in the planning process. "Identifying, analyzing, and proactively engaging with stakeholders from the start to the end of the project helps to enable success" (PMI, 2021, p. 33). Some stakeholders will be very active with the project, providing financial and political support, whereas others will be less involved. Stakeholders are generally considered individuals who help provide positive outcomes for the project. However, stakeholders could harm the project and work against the desired goals, so the PM should continuously ensure that all stakeholders work toward the project's desired outcome.

Communication between all stakeholders and the project team is essential. The PM should ask each stakeholder to identify the preferred form and frequency of communication desired. For instance, some stakeholders may prefer monthly email updates, whereas others may prefer quarterly or bi-monthly formal status meetings. The only way to ensure the stakeholder gets the desired level of communication is to check in regularly to see if the current method and frequency of communication need to be adjusted. During status meetings or status emails, a simple statement such as "Please let me know if you would prefer more or fewer status updates or if a different form of communication is preferred." Therefore, the best way to evaluate how satisfied a stakeholder is with the current level of communication is to ask each stakeholder directly.

4.5 PROJECT SPONSOR

Someone within the organization should be the primary sponsor (Kloppenborg et al., 2011). A sponsor is typically a person in the organization who has a significant amount of influence and is most interested in achieving the deliverable benefits of the project. A formal definition for the sponsor is "a person or group who provides resources and support for the project, program, or portfolio and is accountable for enabling success" (PMI, 2021, p. 250). "Research shows that an active project sponsor is a critical success factor in achieving positive outcomes from projects" (p. 206). They support a project through support to the team in assisting with a project vision, business value, customer focus, decisions, motivation, and accountability. Furthermore, they "can help teams perform effectively and thus improve project outcomes" (p. 209). Sponsors can serve as liaisons to help project teams have access to physical resources and skill sets necessary to complete project deliverables successfully. They can help guide and motivate a team, as well as partner with the team through coaching or mentoring. They can also assist in removing impediments in resolving issues outside of the authority of the PM or team, which is called "unblocking." Sponsors can also perform a check function where they engage with team members by asking questions, fostering innovation, and challenging team assumptions. Sponsors assist in influencing stakeholders who can help the team with post-project operations. They also communicate information to and from the organization and team.

4.6 PROJECT CHARTER

The project charter has been previously discussed as initiating the formal existence of the project, and it also has specific implications for the project team. A project charter is "a document issued by the project initiator or sponsor that formally authorizes the existence of a project and provides the project manager with the authority to apply organizational resources to project activities" (PMI, 2021, p. 245). It formally *authorizes* a project and documents the stakeholder's needs and expectations. The basic role of a charter is to formally give delegation of authority to the PM or the person identified by the management to lead and to formally display management approval for the project. The term *PM* will be used here. However, it is essential to note that instead of identifying a PM, the charter may state that the functional departments A, B, and C will share in support of the approved core planning process and that a named individual is the team/

project leader. What is being said here is that the management believes this effort is worth more time, but they have not signed off on the final plan. Also, the execution team is not likely to be in place yet and, therefore, cannot be named. The items that should be covered in the charter document are a statement of project goals, general project time and cost goals, purpose, project management goal, and any other relevant guidance the management wishes to define. It is important to note that every organization has a different viewpoint regarding what a charter contains, which yields different definitions and the intermingling of these definitions for the project vision, business case, goal, and scope.

Chartering is an agreement between the PM, selected at this project stage, and the project sponsor (Miller, 2001). The project charter needs to be established, signed, and documented. The critical point is that the signing of a charter symbolizes approval of the project. It is easy for different individuals or departments to attempt to add additional needs to the project charter unrelated to the original charter (Peled, 2000). The PM, project sponsor, or both must manage the evolving scope from undue expansion to only including the appropriate target goals. If any part of the project is being outsourced, the role of the outsourced segments must be carefully defined. Unclear objectives, unrealistic deadlines, unclear management roles, and lines of authority represent weaknesses in project management (Anonymous, 2002). A clearly defined line of authority, including an accountability framework, must be established, beginning with the PM (Maddalena, 2012).

The sponsor establishes the vision for the output that the project will produce. From this, the PM turns the vision into a work plan to make the desired output. The original goal is translated through an iterative planning process into a project scope containing the work units necessary to produce the desired output.

A schematic tree diagram called a Work Breakdown Structure (WBS) is often used to identify the major scope elements of the project. The higher levels of the WBS represent the plan deliverables, while the lower levels represent more of the work required to produce those deliverables. The original charter typically does not define how the vision will be accomplished. Instead, it focuses more on the output side of the equation. It is up to the project team to translate this high-level view into a work plan. More details on this activity will be covered in Chapter 5.

The charter provides the guiding definition for further pursuit of the goal. The existence of a charter also assists in making individuals affiliated with the project more cohesive, with a shared sense of project ownership (Peled, 2000). Once the project charter is signed by management, it signifies that the organization formally believes this effort is worthy of more expenditure of time. Recognizing that the formal development plan has not been approved is essential. That event will occur at the end of the planning stage, and upcoming chapters will elaborate on more details regarding the future ongoing planning steps.

The charter information is an estimate to be validated by the sponsor and project team when developing further details regarding the project deliverables and associated work. Figure 4.1 displays the bare minimum details that a project charter should contain, and the remainder of the chapter will discuss different charter specification sections. Beyond this, a charter can even include specific statements related to PM actions regarding authority and responsibility, such as the following:

Communicate directly with the executive team regarding the project

Communicate directly with all subcontractors regarding the project

Authority to negotiate contracts up to some defined limit

Authority to expend up to a defined limit

***Date:**
***Project Title:** Facility 3B Landscaping Project
***Project Manager:** John Smith
***Project Sponsor:** Alana Bell
Tentative Budget: $35,000
***Project Objectives:** Design and install all landscaping for Facility #3B.
Tentative Team Resources:
***Project Approval Signatures:**

____

____

____

***Project Manager/Date** ***Project Sponsor/Date**
***Required Fields**

FIGURE 4.1 Project charter PM role specifics.

Authority to hire subcontractors per defined limits

Authority to manage approved resource levels

4.6.1 Background

The background section of the charter describes why the project has value and is at least perceived to fill an existing business need or resolve a problem within the organization. Also, the project is envisioned to fill a perceived gap in the organization's strategic mission. The project may emerge to address a new law, a new regulation, or a social need that has arisen within the local community. This project may create a demand for a new product. Alternatively, the goal could be a project to update a company's website, or new technology may need to be integrated into the existing business process. There may be any number of reasons why a project was selected, so the background section within the charter helps identify some of the key reasons why the project has been approved.

4.6.2 Purpose

The purpose section of the project charter in the template can be used to provide a high-level description or low-level details. For example, it can provide a high-level project goal or highlight its objectives. This type of information provides a much-expanded understanding of the underlying rationale for the project.

4.6.3 Deliverables

The deliverables section of the project charter will include information highlights relating to the project's scope. This will consist of project objectives and deliverables but also describe how the project is to be managed and the constraints that must be adhered to during the project. The charter may be a statement that identifies the main tasks to be performed by the team. For clarity, the deliverables section should identify the specific deliverables and significant milestones the team must produce.

4.6.4 Budget

The budget section of the project charter displays the overall project budget objective. The budget is just a rough order of magnitude (ROM) estimate because the required project work has not been fully identified at the charter stage. The ROM level of accuracy can be in error by as much as +/–100%, while the goal of the final planned project is often desired to be within 10% of the actual detailed budget that will be established later and described in Chapter 8. This section may contain budget-related details for different aspects of the project, such as equipment and labor. It may also address where the project funding comes from, such as which specific program, department, or combination of different programs or departments. In some projects, the charter may state how all or some of the project funding comes from the project itself. For example, sales may generate income from selling tickets for an event or a project deliverable. The event may be able to pay for entertainment, food, and so on based on the income that comes in from ticket sales. In other cases, the budget will define the funds required to execute the life cycle plan.

4.6.5 Team Membership Assumptions

The team membership section of the project charter may list the type of resources needed so that the project sponsor or PM, if already named, will not need to beg management for additional skilled resources once the project has begun. It may also give the PM the authority to communicate directly with company board members, specific customers, or subcontractors. The entire project team will likely not yet be established, but limited resources may be provided to help perform the upcoming detailed planning phase. This section should outline enough detail to define the skill composition needed to accomplish the project goals. Examples include naming job positions that must be filled, such as computer programmers, electrical engineers, and paralegals. Many organizations will not staff the entire team until more specific details are elaborated in the scope management plan, as discussed in Chapter 5 (Maddalena, 2012).

4.6.6 Perceived Risks

The perceived risks section of the project charter is an essential component of planning. The project charter may include a section on high-level risks that apply to the project. For instance, a project may have a weather-related risk that can be identified as part of the nature of the event. Another easily identifiable risk can be associated with key suppliers who can negatively impact the project schedule. Before the execution phase, project risks need to be identified and reviewed. More specific details regarding risk will be pursued in the formal planning stage, and these mechanics will again be discussed in Chapter 16.

4.6.7 Project Approval

The project approval section in the basic project charter template has signature blocks for the PM and the project sponsor. This section may also outline the criteria for which the project will be approved and categorized as completed. An example of acceptance criteria is submitting all project close-out documentation and all program code to the customer. The criteria that are identified should be written in very concise statements so that no two individuals would interpret the requirement differently. This clarity requirement is for the protection of all parties. Once the approval signatures are obtained, the project can progress through the life cycle.

4.7 CONCLUSION

This chapter addressed the general process, outlining how support for the project is gained, which includes the approach for officially launching a new project and the purpose of a project charter. When the project was selected to be performed, it was validated as a target vision that had value and aligned with organizational objectives. At the end of this process, management must show approval and support for the effort by formally signing a charter document. A determination is made to identify the PM or person selected by management to lead the project at this stage in the life cycle and identify key stakeholders such as the project sponsor(s). The basic template for the charter was discussed by outlining the purpose of the charter in providing delegation of authority to the PM and formal management approval for the project to exist. Additional charter items that may be part of an organization's charter were also discussed. Chartering assists in moving the project definition and process forward to the start of official scope planning and scope definition, which will be addressed in the next chapter. This will include discussing the scope management plan, scope statement, and the subsequent documents to be developed once the project charter is completed.

QUESTIONS

Student Review Questions

1. Describe the different steps involved in officially launching a project.
2. List and describe each section of the basic charter and expanded project charter.

In-Class Question

1. Find a charter template online and describe the major sections.

EXERCISE

1. Create a project charter for your class project using the charter sections described in this chapter and/or found online.

REFERENCES

Anonymous, (2002). Redesigning Project Management at Kimberly-Clark Europe. *Training Strategies for Tomorrow* 16(6), 8–11.

Kloppenborg, T., Tesch, D., & Manolis, C. (2011). Investigation of the Sponsor's Role in Project Planning. *Management Research Review* 34(4), 400–416.

Little, B. (2011). The Principles of Successful Project Management. *Human Resource Management International Digest* 19(7), 36–39.

Maddalena, V. (2012). A Primer On Project Management. *Leadership in Health Services* 25(2), 80–89.

Miller, C.W. (2001). Meeting Real Needs with Real Products. *Strategy & Leadership* 29(1), 15–20.

Peled, A. (2000). Creating Winning Information Technology Project Teams in the Public Sector. *Team Performance Management* 6(1), 6–15.

Project Management Institute (PMI). (2021). *A Guide to the Project Management Body of Knowledge*, 7th edition, Newtown Square, PA.

II

Planning Processes

CHAPTER 5

Scope Management

Project N Motion Videos for This Chapter

5W1	WBS to MSP	WMSP Setup
5W3	WBS	WBS Overview
5S1	Scope Management	Scope Overview
5S2	Scope to WBS	Scope Migration

Scope management can be considered the process of converting a fuzzy vision of a future product or process into a clear understanding of that vision's attributes and defining the technical work required to produce the item. This process has elements related to both an output goal and the project work related to that goal. Scope is defined by PMI (2021) as "[t]he sum of the products, services, and results to be provided as a project" (p. 249).

Project scope lists specific project goals, deliverables, tasks, costs, and constraints. Some say a picture is worth a thousand words. Figure 5.1 illustrates that idea as it outlines the role of the project.

One of the most critical management processes in this evolution is deciding the work requirements necessary to produce the "To Be" state and how to effectively define that to a broad array of stakeholders and technical experts who will collectively make the product or process. This definitional process is called scope management.

DOI: 10.1201/9781003532989-7

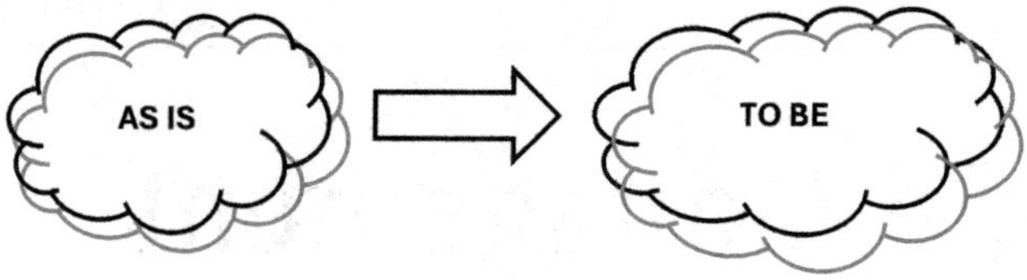

FIGURE 5.1 Evolution of a project [5S1].

5.1 INTRODUCTION

The following list summarizes three important scope management outputs (Wrike, 2024):

1. The project's functions and features are defined for the future process or product.
2. The work required to produce target item one is defined.
3. The deliverables and related project work specifications are communicated to appropriate stakeholders to achieve a common understanding of the project.

Knowledgeable individuals, who best define the proper deliverables, collect this information in various ways. On the surface, this activity appears to be straightforward; however, history suggests quite the opposite. Failure to define the project goals dooms the effort as the later technical decisions are made from flawed requirements. Some have labeled this activity as the Achilles' heel of project planning, and it must be effectively executed to achieve a successful outcome. Project scope definition represents the initial stage of project planning, and all subsequent processes flow from this description of the venture. The scope description contains the following details (PMI, 2021):

Project and product objectives

Product or service requirements and characteristics

Documentation of these parameters provides a common delivery goal for the project team members, stakeholders involved or interested in the outcome, and a general roadmap guide for developing further details regarding work, cost, schedules, and risk. Multiple supporting documents are mentioned in the literature as useful to support this activity. The one artifact that seems most worthy to include here is the Scope Management Plan (SMP) [5S2]. This document is designed to outline the resources required to produce the deliverables and describe the control process that will be used to keep the project within defined boundaries. PMI (2021) defines the scope management plan as "a component of the project or program management plan that describes how the scope will be defined, developed, monitored, controlled, and validated" (p. 249).

Many techniques are used to collect the data outlined above, but the primary method recommended here is a face-to-face collaboration among key stakeholders who are most knowledgeable about the end requirement. The theory here is to emphasize that the project's success is the responsibility of the project team as well as the organization. Also, the face-to-face process shortens the data collection cycle time. More examples of this will be described in the chapter.

A scope support tool that is most visible in the industry is the Work Breakdown Structure (WBS). Many PMs grade this tool as the most important tool in the planning and control role of the project. Based on this logic, this chapter will use the WBS as the core artifact to define the project deliverables and technical approach.

The trigger to initiate this activity is varied, but from a model theory viewpoint, the initial step in a project is triggered when a formal project Charter has been approved by senior management, which means no project will be permitted without a Charter. This restrictive statement requires some explanation. Senior management is responsible for the organization's direction and success. As part of this responsibility, they are awarded the authority to make decisions that dictate the organization's direction. Projects consume significant resources and can impact the direction of the organization. It would be naive to say that this rule is never broken, but there are legitimate reasons to follow that approach.

Two important vocabulary terms help set the stage for understanding this process. The first term is *alignment*. Projects are powerful activities to mold the future direction of an organization. To maximize their value, the goals of such efforts need to support organizational goals. That is the essence of alignment and is the acid test for project approval. The second term is *elaboration*. This is more difficult to explain clearly. A project does not leap out of some magic basket ready to go. Quite the contrary. The entire life cycle is one of layered decisions and actions. In the early stage, the goal is fuzzy and incomplete. As each stage is executed, the *what* aspect becomes more clearly defined. As the process continues, technical discussions outline *how* the work will be performed. The concept of elaboration essentially says that a project evolves iteratively and becomes more understandable through the scoping layered decision processes. Both terms are mental models relevant to the concept of project scope.

Chapter 3 reviewed success strategies and rated the inability to define project scope in the planning phase as one of the top reasons for project failure. It is important to note that the foundation of the traditional waterfall model is based on a clear scope definition. The linkage between scope and various downstream processes erodes from poor scope definition in the following ways:

- Work tasks cannot be properly defined or estimated.
- Resource requirements are linked to erroneous work task specifications.
- Budget estimates are linked to erroneous work specifications.
- The project schedule is constructed from erroneous work units.

This is an operative example of the adage "garbage in, garbage out." Incremental decisions made in the planning step *elaborate* on the work required. This is a somewhat fancy but important idea in project planning. One simple way of looking at this concept is to recognize that the fuzzy idea at the beginning is defined a step at a time until there is agreement for its full definition. That process is elaboration. Think of it as increasing levels of detail as decisions are made.

5.1.1 Initial Scope Planning

The following documentation steps summarize the early scope-related activities and key decisions that will need to be accomplished before serious technical consideration can begin:

1. Seek agreement on objectives
2. Develop high-level delivery milestones (phases, dates, functions, etc.)
3. Evaluate project risk elements
4. Define success characteristics and priorities

5. Produce a *preliminary* scope management document outlining the project's deliverables and guidance
6. Obtain approval from stakeholders and appropriate management

If one reads between the lines of the list above, it should be easy to understand the complexity of internal communications required to agree on any of the points. If only one person were involved, this would be a minor chore, but add hundreds to the list, and it becomes a formidable effort that is seldom entirely successful. Project success or failure takes its first step because future elaboration decisions will be based on the decisions represented (or not represented) in the scope documentation.

The first step in the process is triggered by an idea that someone believes is worthy of more analysis. Moving that idea to the next step requires a sponsor or someone who will wade through bureaucratic resistance to obtain the necessary resources. Invariably, any complex idea will have disagreements. One of the most difficult barriers to overcome is proving (or selling) that this idea is more valuable than others seeking resources. Organizations typically have many such ideas floating around at any time, and it is difficult to get one selected for a more formal review by senior management decision-makers. The idea must be embraced through various organizational approval processes at a level where resources are formally available. Human and monetary resources are necessary to move the fuzzy vision into a planning phase that will further define what is needed and the benefits that can be derived from the venture. The beginning point of such analysis starts with a vision-type statement, often beginning with the phrase "I believe..." The next step is to get others to understand why this might be a worthy idea. The initial project effort typically focuses on why the vision might be a worthwhile output. Too often, the early process jumps to how to produce such an output. Benefits, not necessarily new ideas, drive organizations. This means that the early work phase should be on what the merits of this vision can do for the organization and general impressions about the associated cost and risk feasibility. New ideas always seem to bring controversy, and this is a reaction one must be ready to face. Much of the early work involves selling the idea to various stakeholders, notably senior management. One of the early initiation success factors is having a senior sponsor at the highest possible level. The selling and analysis work continues as resources are approved for further examination. Estimates must be made outlining the future regarding organizational benefits, along with some quantification of schedule, cost, risk, resources, and the like. Large organizations often have a standard format for documenting these views, usually a business case. Many organizations have a Project Management Office (PMO) function that reviews project proposals and screens them for senior decision-makers. The role of the proposal document is to outline the vision in a somewhat standard format. Regardless of the initial screening process, senior management has the role of approving project proposals and sprinting the movement to a second-level planning process.

5.1.2 Phase Two Scope Planning

Details from the previous phase continue to be elaborated into more specific elements. This phase is the gap between *what* and *how*, as the various dictions have that flavor.

Once the project has passed through the initial high-level review, the definitional process can move forward more orderly, with less selling involved and more work focused on defining what the project will entail and the specific functions it will have at completion. Some would call this the *what* phase. Various stakeholders are involved in his activity and the related aggregation of decisions. As this effort moves forward through the elaboration process, high-level decisions are made regarding how to execute the effort that has been defined. To give an example of a high-level scope decision, suppose it was decided early in the deliberations to focus first on certain high-benefit capabilities and defer others for later (i.e., phase two). This decision breaks the project into multiple phases and has major implications for the next step. The tactical effort will now focus on the selected subset of capabilities and leave the rest for later consideration. Work now moves toward expanding details regarding the phase capabilities. As the initiative is reviewed, major work groupings begin to emerge, and this process continues to add more details to the effort. Many parallel efforts may be underway as technical experts begin to make decisions related to some subset of the overall effort. With these descriptions, you should see the elaboration term in action again.

As specific work elements are defined, the focus moves into the *what* portion of the scoping effort. Technical *Subject Matter Experts* (SMEs) begin to define additional specifics of the work further. Here is a memory example of this phase. If the requirement specifies the device is to "jump tall buildings," the technical specifiers have many related questions to resolve (i.e., how high, cost constraints, risk). This phase is characterized by moving the vague requirements into more specified work units and technical details regarding achieving the desired output. The *ibilities* factors described later in the chapter become part of the technical decisions related to this segment.

Once the technical design packaging for the work unit is completed, it is then necessary to define the internal resources required to execute the work, along with the estimated duration and cost of that work unit. To define work requirements, it is necessary to translate the output requirements into associated work units that will produce those deliverables.

To ensure that the functional requirements, technical details, and operational resource specifications are met, the users and the technical producers must have the same general understanding of the defined work required to produce a defined capability. This is a critical communication activity that will stress all projects to execute.

5.1.3 Work Packaging

The overall process of gathering these dispersed specifications into a single work envelope is called scope definition. The result is frequently translated into a WBS as a planning and communications tool. This hierarchical structure helps produce a clear understanding of the project work units, deliverables, assumptions, and constraints. A detailed work outline emerges through this layered decision process, representing the total project delivery specification. A WBS and its associated data repository represent the primary scope definition artifacts for the project. When completed, the scoping process must define the following items:

1. Stakeholder *needs, wants, and expectations* translated into prioritized requirements.
2. *Deliverables* related to the primary product or service, including supporting artifacts such as documentation.
3. Outlining the *project boundaries* (i.e., what is included in the project scope and what is not included).
4. Identifying various project *constraints* that the project must observe (i.e., cost, schedule).
5. *Assumptions* made during the process (i.e., resource availability, environmental approval).
6. *Acceptance process* for the various deliverables.

Italicized terms above highlight the key areas that are essential elements needed for a clear project scope definition. A mental image of the scoping activity described here is like *peeling layers of an onion*, maybe even realistically, along with the associated tears. The scope development steps fit the elaboration concept in that each step adds detail to the previous ones. The tearful metaphor relates more to achieving internal agreement, as those steps require selling and negotiating diverse opinions and views containing different ideas regarding how or even if the project should be pursued.

The output of the scope planning step is a documented and approved view of the desired project deliverables and the related work activities required to produce them. This process unfolds from the initial fuzzy vision into a series of increasingly detailed steps.

5.2 PARTITIONING THE PROJECT

One of the primary artifacts that aids in guiding the scoping process is a WBS. This classic tool for envisioning a project is often recognized as the most important tool in the project management kit. As seen here, it is used to help "envision work packages (WPs)" or collections of major project groups, such as phases. These initially vague box elements could represent tasks, subsystems, or any other work grouping associated with the project. One common metaphor for the initial view of a project is to see it as a collection of work or goal boxes. Each box has three key dimensions at the lower work unit level—function, time, and cost. The higher-level view is more function or capability-oriented, while the lower level is more operationally oriented. Collectively, this structure outlines both the capability and work aspects of the project. Figure 5.2 illustrates a high-level house construction project view, which requires further elaboration of the boxes into work details.

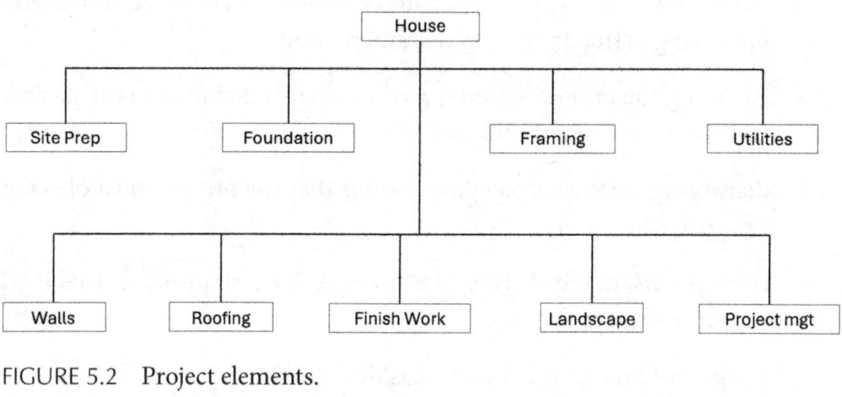

FIGURE 5.2 Project elements.

If one were to want to build a house, these boxes represent high-level scope segments of the related work requirements for the effort. Data attached to this structure would also reveal other functional and technical factors related to the effort (i.e., cost, resources, material). The example WBS-style format is called a product structure, as it represents components of the physical deliverable. Other box formats would be used for different project types, but the overall goal is similar. Notice that this view outlines the major work groups required. At an early point in the process, specific details are limited, but that changes as various decisions are made regarding each box. As the project view evolves, additional detail boxes will emerge to show more specifics. At the early stage, a WBS view outlines how the project is structured, while the lower layers of the structure represent increasing details of the work required to produce the higher levels. Combining those two ideas yields the name of this view as a WBS. More specific details regarding the use of his tool will be visible throughout the text. For now, think of this as a schematic view showing how the project is envisioned, decomposed, and communicated to various parties. Once again, get comfortable with the term elaboration process! It is a fundamental concept in project management and continues through the life cycle.

Using the WBS from Figure 5.2, we can illustrate the decomposition process. At this point, foundation experts would begin to specify details of that segment and record details in the WBS data store. These decisions are not limited to that one box, as its decisions may well affect others, so an overall view must be kept synchronized and communicated. For example, if the working definition of the house were to change, this could have major implications for each of the lower-level boxes. Think of the WBS as a communications mailbox with coded addresses.

At this point in the scope evolution description, the linkage between the WBS and the full project structure is not fully defined. Still, the key point is to recognize that it is one of the primary documentation artifacts used to represent the project scope and its associated work deliverable components. It should also be mentioned at this point that the industry usage of the WBS is quite varied, and some specifications differ from what is described here, but from a scope descriptive view, we will stay with the current approach.

The next view of the scoping process is to define when it is over, and that is a somewhat vague boundary. Note that Chapters 7 and 8 describe the development of schedule and budget plans for the project. These are also part of the scope definition but are decoupled here to focus on the main internal elements. For this discussion, the boundary for scope will be reached when the work units have been defined, and estimates are made for each. The definitional vagueness comes from stakeholder expectations. It is one thing to describe the project goals, but invariably, there is a requirement view and a constraint regarding the associated schedule and cost information. At this point, neither of these elements has yet been defined. The more appropriate process would be to obtain general approval for the project and then develop the schedule and cost for that scope. Even though this sounds logical, the problem emerges when the estimated schedule or cost exceeds their constraints. If these are violated, the original scope plan will have to be modified in some way to fit. The linkage between *what* and *how much* cannot be ignored. One option might be to initially give a ROM (rough order of magnitude) estimate for both schedule and cost before the initial scope process, then move on later to refine those estimates. But that still does not deal with the constraint issue, and organizations have a bad habit of forgetting that the earlier ROM value was not a detailed estimate. Confusion of this type can leave the project team with an incompatible scope-time-cost plan. Mature organizations work through this incrementally until the variables match, regardless of how the overall project planning process is executed, the goal. There is one additional planning variable that is often ignored until it surfaces in the execution stage. It is one thing to construct a plan that balances the goals and constraints of the user side of this question, but there is one other limiting factor that can derail such a plan. Resources to support the plan must exist. This includes material and human labor, plus other support elements such as facilities and infrastructure. When these were unavailable, the earlier plan was a wish list that could not be produced. So, the final view of scope planning also needs to recognize external constraints from the production side of the equation. When finally approved, both sides of this need to be recognized and agreed upon by all parties.

5.3 WORK PLANNING CONCEPTS

As the WBS is expanded downward, technical work units begin to be recognized. A common rule of thumb for these is that they represent approximately 80 hours of work and/or two weeks duration; however, that definition might not be appropriate for some project types. The starting point for this process is first to define high-level decisions and major capability grouping, then evolve that level of decision-making into associated work unit specifications. From a control perspective, short work units are better for estimating work and measuring status. Work unit definition specifications will be produced for execution time, cost, risk, human staffing, contractual, and material elements related to the work effort. Beyond the direct work estimates, additional data should be collected regarding risk events, support equipment, and other factors related to the work. This collection of specifications paints a technical and work-oriented picture of the project. Each of these individual planning parameters is an important foundation element of the project plan. The original vague vision is now described in such a fashion that the desired deliverables can be understood by all relevant groups. To obtain final approval, the sponsoring entity will use the final WBS view to approve the continuation of the project. Beyond evaluating the project's organizational goal alignment, the final decision will be based on the compiled plan parameters of deliverables, cost, time, human elements, risk, and other factors.

5.3.1 Requirements "ibilities"

On the surface, defining work requirements seems straightforward: identify the goal and then the technical work details to execute that goal. This is an area where the user's mind and the technical one diverge. Users tend to be general in their specifications, while the technician's needs are more specific regarding that view. There are ten often-ignored vagaries in the requirements specification. The following list summarizes these as the requirements "ibilities:"

1. Traceability
2. Affordability
3. Feasibility
4. Usability
5. Producibility
6. Maintainability
7. Simplicity
8. Operability
9. Sustainability
10. Predictability

Each "ibility" represents a work unit attribute to be considered in the requirements definition. It must be recognized that the project goal is not just trying to produce a stated deliverable. It must also consider a more definitive look at the attributes of the result. It is necessary to review the target item and adjust the scope statement according to each of the nine *ibility* attributes to ensure that the approach chosen appropriately matches the real requirement. In many cases, a particular solution will involve a trade-off of one or more of these attributes based on time, quality, functionality, or cost constraints. These decision alternatives will present themselves along the following general lines:

1. Present versus future time aspects
2. Ease of use versus cost or time
3. Quality versus time or cost
4. Risk of approach
5. Use of new strategic technology versus a more familiar tactical approach, etc.

As the project moves through its life cycle processes of scope definition, physical design, and execution, each of these considerations should be reviewed. All too often, one or more of the *ibilities* is ignored or overlooked, resulting in downstream frustration by someone in the chain of users or supporters of the item. The section below will offer a brief definition and consideration review for each of the specific *ibility* items:

Traceability relates to the ability to follow a requirement's life span in a forward and backward direction (i.e., from origin, development, and specification, its subsequent deployment and use, and periods of ongoing refinement and iteration in any of these phases). Envision traceability this way. If a design element or WP specification is changed, the configuration management process will document this and be able to ensure that the proper version is used.

Affordability relates to a match of the design approach to the budget. There is always pressure to cut costs through the design, but many of those decisions cause some adverse impact on other *ibilities*.

Feasibility can wear many hats in the project environment. The most obvious of these is the technical feasibility of the approach. Stretching to achieve some performance goal often exceeds the existing technical capabilities and creates additional risk. Similarly, the lack of available critical skills can adversely affect the outcome. One can view feasibility as anything that can get in the way of success, whether technical, organizational, political, resource, risk, or otherwise.

Usability is similar in concept to operability, except that it involves the resulting value generated by the output. It is what the process or product does in the hands of the future user. This can be reality or perception-based, but it is certainly a concern for the project team.

Producibility is an attribute associated with how the actual item will be created. In many cases, there is a gap between the designer and the builder, so the key at this stage is to be sure that the building entities are represented in the design and probably even in the initial requirements process. Think of this as a "chain" of events that need to be linked together, not just thrown over the wall to the next group. Each component in the life cycle needs to consider this attribute.

Maintainability deals with the item in production. The consideration here is how much effort is involved in keeping the device operational. In the case of high-performance devices there is often a significant downtime for maintenance. Having a device capable of "jumping tall buildings with a single bound" sounds good, but what if it can only do that about 10% of the time, with the remaining period being down for some required maintenance? There is often a trade-off consideration in design strategies. In this case, the design trade-off is to design a way to perform the maintenance quicker, cheaper, or with less downtime. The spectrum between operations and maintenance times is not a topic to ignore.

Simplicity is an overarching concept. A complex design is often required to produce the desired output, but that creates other profiles that are not designed. The scope goal must define the simplest approach to achieve the required output. This is somewhat of a motherhood statement, but a real requirement to keep in perspective.

Operability involves the future user's ability to quickly and safely use the product or device. Many years ago, aircraft designers found that the location of gauges, switches, and knobs had a lot to do with the safe operation of the airplane. Every device has characteristics like this. A real-world example is Boeing's 737-800 MAX navigation trim system issues that caused two crashes. A user-friendly design specification was lost in that case. Think of this attribute as not changing the requirement but making its functionality easier to use and safer. Automobile designers in recent years have found this to be an issue with some of the new dash functions being installed in the modern car (i.e., how do I turn on the radio or the windshield wipers?). As the device's functionality expands, so does the complexity of operation.

Sustainability is likely the least understood of the *ibilities*. This goes beyond all the other attributes in that it evaluates the ability of the process or product to exist in the long term. Will the underlying technology survive? Will the scope design be functional for as long as required?

Predictability is the final scope review *ibility*. In many projects, the volatility of organizational, customer, and related technology can involve some of the most difficult decisions. Errors in this category will lead to deliverables that fail because their design environment changes. Ford introduced a large car in the 1960s when the market was looking for a small car. Blockbuster only had physical videos when Netflix introduced online marketing. This type of market mismatch list goes on. If the project team has an accurate view of these environmental elements, the scope definition of the deliverable will be more successful. Too often, an underlying future assumption regarding the operational environment is made, only to find later that this was not valid, and the deliverable value suffers.

The final word on the *ibilities* set is that they represent factors that are important to both the project's short- and long-term success. One of the action items in both requirements and design reviews is to review the ten *ibilities* list and resolve the trade-offs outlined here. This process may be equally important to getting the user requirements correct because if the correct choices are not made at this stage, the future project success will be negatively impacted, regardless of whether the project is successfully developed according to the defined plan that did not recognize that element.

5.3.2 WBS Dictionary

In addition to the visible project activities related to product delivery, other supporting activities should be captured in the WBS data repository, now named the WBS Dictionary. This data store should capture project documentation in at least the following areas:

1. Approvals for various dictionaries are made through the planning process
2. Formal meetings and decision sessions
3. Customer and supplier communications
4. Quality inspections and defect repair processes
5. Training activities (team and users)
6. Communication plan for status reporting and presentations
7. Various project-related management processes (i.e., change control, quality assurance)
8. Issues Log (daily problem events)
9. Operational service level agreements with outside support groups
10. Project closeout log

Many of these items represent key life cycle management decisions more than work unit specification; however, if issues such as this are not accounted for, many supplemental data tidbits will be missing. This means the WBS dictionary must also include various environmental work activities that, on the surface, do not look like requirements but represent the management of the work. For example, project management and risk management data are considered part of the scope definition since both activities and events consume resources and, therefore, are part of the project scope.

As a control measure, it is common to have additional fields in a WP record to show approval by the various responsible organizational groups and the PM. With this set of information completed, the defined scope is assumed to be approved at the working level.

5.3.3 Work Sequencing

Defined work units often have a technical sequence constraint. The requirement here is to define the order of execution for this. The memory example for this is to envision two work units. One work unit is to pour concrete on a floor, and the second is to paint the floor. For planning purposes, a proper sequence to execute this is desirable. The waterfall model assumes a defined sequence between work units. Any changes made to this sequence would also change the related plan parameters. Sequencing information and other work unit data would be recorded to develop the schedule and budget. The mechanics to demonstrate these processes will be deferred until Chapters 7 and 8.

By defining the technical sequence of a task box to other boxes, a schedule for the overall boxes can be shown in model format. There are multiple vocabulary terms used to specify the scheduling relationships. The primary ones are as follows:

- Predecessor—the work cannot be started until the named unit is completed
- Successor—work for this unit must be done before the predecessor can commence
- Finish - Start (FS)—the predecessor starts immediately after its predecessor
- Start - Start (SS)—this work unit can commence at the same time as its predecessor
- Lag—the work unit has a time lag after the predecessor is completed (time for concrete to dry)
- Lead—the work unit can start before the predecessor is complete

As an example of box relationships, assume that two work units are related, meaning one box needs to be executed after the other in two sequential steps. The default view for this is called an FS relationship. Metaphorically, we are stacking our two boxes end to end, as shown in Figure 5.3.

For this example, assume that all workers involved have the same skill levels and can do all tasks required. They are also paid the same. These handy assumptions allow us to avoid many messy issues in the schedule calculation process, but they help simplify the focus on the basic raw mechanics of scheduling and budgeting. From a model perspective, stacking the two boxes head-to-head, or FS, will define not only the schedule but also the cost of those two boxes. Carrying through the full list of boxes yields the same result. There are still some complexities associated with

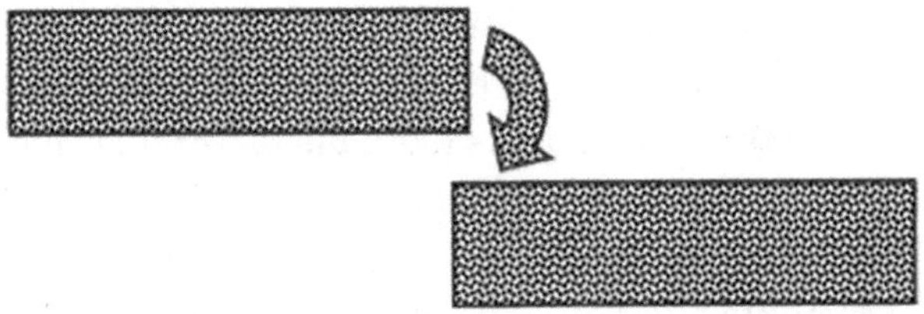

FIGURE 5.3 Multiple WPs in FS sequence.

scheduling and budgeting, but this view represents a good place to leave the topic for now.

5.4 DEVELOPING THE TOTAL PROJECT VIEW

To this point, the scoping process has been described as translating the original vague project vision into a format that can be produced. Many interrelated decisions are required to lay out a complete project plan. The primary WBS role in this process is to aid in defining and communicating the technical and managerial aspects of the effort. This tool is the best-known method to provide a visual view of a complex project that stakeholders and technical participants can relate to and understand. At the highest level, the WBS outlines broad structural groups of the effort. This could be chronological phases or major components of the overall product, as shown in Figure 5.2 for a house. Lower layers of the structure represent increasingly detailed views of the higher levels. At the bottom tier, specific work units represent the work required to execute the project. The proper perspective for this view is a top box representing the target result and a lower tier representing how to achieve that result.

Note that the top box in this structure represents the project. Note that high-level scoping decagons divided the project into four chronological phases. Phase 2 appears to be the active phase of the project since active work units appear in this segment. Phase 1 could be completed and shown here for reference purposes. The format here includes deliverables and WPs (work units) at the lower levels. This is an arbitrary format. Deliverables represent defined scope items, while the associated WPs represent the work required to produce the deliverable. There are multiple "standards" for the design of a WBS, and this example is purposefully different from all of them to make a point. The bias here is that this structure should define the problem from the user's perspective. The value of a WBS is not in satisfying some underlying standard; it is the communication value found in the format used. The standard used here represents that view. If it is a hierarchical structure that communicates a message, it is a WBS. As one might say, "If it quacks like a duck, we will call it a duck." In fairness to the WBS legacy, it should be recognized that multiple sources describe "standard" methods to construct this artifact. Project Management Institute's (PMI's) standard (PMI, 2021) and the Department of Energy Specification Execution Guide (DOE, 2014) are notable examples of standards. The grandparent of all WBS documentation is found in DoD's MilStd-881, which contains over 350 pages of product-related WBS specifications (DOD, 2024). Furthermore, it is a great source to cure insomnia if one gets the point.

One new standard term shown in Figure 5.4 is the WP. It is defined here as a box in the WBS representing a planned, manageable grouping of work designed to produce some identified component in the project. This means the unit should be of reasonable size. Some industry sources define this as an approximate duration of work requiring 80 hours (two normal work weeks).

The WP concept is fundamental to the scope definition. These represent the results of defining what needs to be done to deliver the defined project goal. In the traditional model, a WP has a physical connotation. It is a three-dimensional box, as shown in Figure 5.5.

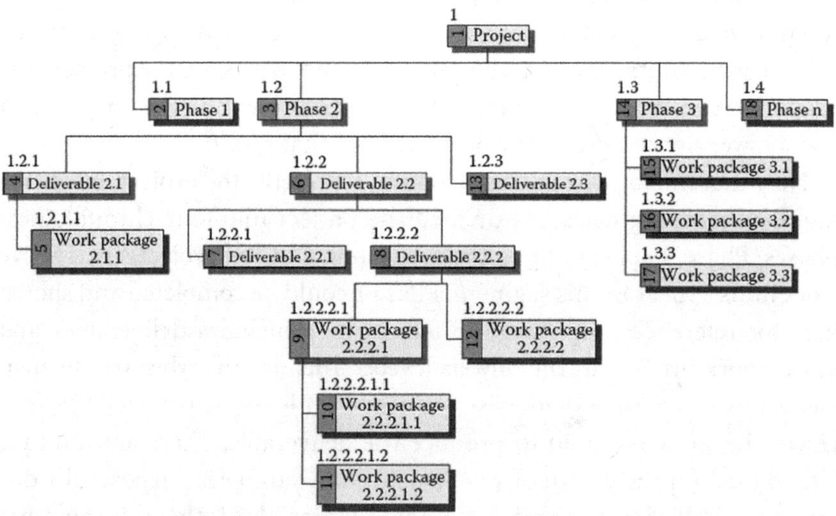

FIGURE 5.4 Abstract WBS.

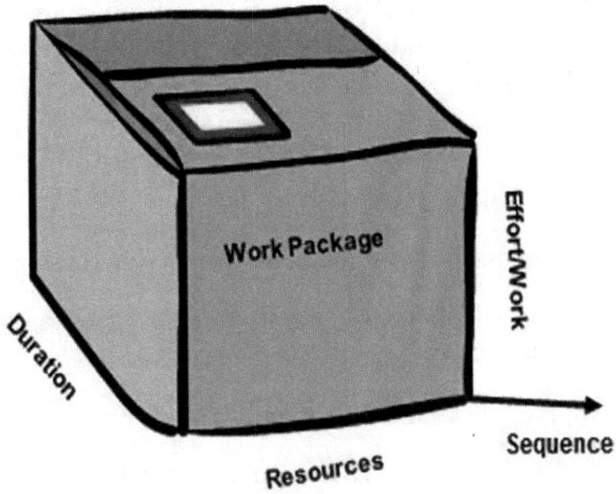

FIGURE 5.5 Work Package (WP).

All boxes in the WBS have a scope-type perspective, but the WP level has more of a multi-dimensional scope perspective. These boxes show a specific product view specification and the time and cost related to that function. Containing the box metaphor, the collection of WP boxes conceptually represents the total project work required. In this view, the WBS is a schematic model of the project. This visual perspective lies at the heart of traditional project planning models, as we will see in later sections.

Let us take a quick side trip to see if you follow this somewhat abstract trail. Assume all the WPs defined are accurate—work, time, and cost. If the estimated cost shown for all of these were added up, that would be a measure of the direct cost of the project. If all the boxes were laid end to end, it would indicate the total duration of the project; however, that is not how we might want to execute the defined work, so more effort needs to be done on task sequencing. With this view of the project, we are now exploring the 1950s view that led to the current method. The WBS was one of the defining artifacts, and linking boxes (work units) was the second. We will see more details regarding this process in upcoming chapters.

5.4.1 Work Package Specification

The WP view has been identified as the formal target for work specification and project management. As these boxes are defined in the WBS, details regarding their composition become the focus. Let us take a simple example to illustrate what is needed in this specification process. Assume a WP scope specification indicates a hole is to be dug. From this logical specification, let us illustrate some mechanics for defining this WP for the project plan. The entry specifications are as follows:

- The required hole size is 10 × 10 × 5 feet.
- This work will be executed manually with a shovel (scope decision).
- An estimator calculates that it will take 80 hours to complete the project (Subject Matter Expert for the project).
- Two resources are allocated to the task (scope decision).
- Resource cost is calculated based on average labor rates ($20/hr.).
- WP summary data is computed as 40 hours or five workdays and $1600.

There is more sophistication required than shown here, but the general goal for the work specification process should be clear. The planned duration and cost for this WP are recorded in the project plan as five days and $1600. The defined deliverable is a hole to be produced by two human resources.

We have described a reasonable but simplified overview of the project deliverables and related work requirements at this stage in the scope definition process. As this data type is produced, it is time to recognize the need to capture and store the information. Documentation activities in the scoping process are vital because many parallel decisions are being made regarding the project, many of which have cross-project implications.

Specific details regarding the defined work units will be needed for future processes. This data is recorded in a retrievable project notebook or database to make it visible to the appropriate stakeholders. This becomes a valuable resource for both current and future projects as it shows decisions made and various technical data. The formal name for this repository is the WBS Dictionary. That may not be a visible name in many projects, but similar data stores exist in all projects. The proper goal for such data is for it to be easily accessible to the appropriate stakeholders and to be viewed as the formal host of data regarding the project. The availability and integrity of this type of data source are key success factors for the project.

One might view the specification of work units as analogous to defining the molecules in a chemical compound (i.e., mix all these just right, and a magic new product emerges). At any rate, the WBS and its dictionary contain specific work elements for the project, and this, in turn, represents the roots of a project work plan. In a waterfall-based project, the work specifications are recorded before execution, while the iterative development methodology proponents would have less scope definition during the planning phase. Regardless of the development method details, some reasonable level of work unit estimates will be needed for any forecast of schedule or budget. The specific WP planning parameters needed to produce the project plan are as follows:

1. ID reference—a code used to identify where the work unit fits into WBS.
2. Labor allocation for the task (e.g., quantity and skills of workers).
3. Estimated duration for the task, given the planned worker allocation.
4. Materials associated with the activity and their cost.
5. The technical supervisor is responsible for the work execution.
6. Organizational unit assigned to do the work.
7. Defined constraints (e.g., the activity must be finished by December 15, 2030).
8. Key assumptions made during the planning activity.
9. Predecessors—linkages of this activity to other project work units.
10. Risk level for the activity—more work will be needed on this aspect later; at this point, a general measure of risk might be summarized as H, M, or L.
11. Work description—a written description of the required output.
12. General comments—free-form statements that help a reader understand the technical aspects of the work required.

In addition to these work definition items, it is also necessary to produce high-level documentation related to an overall view of the project.

One type of documentation often missing in the scope planning phase is the requirements for stakeholder communication, collaboration, and project control aspects. A document titled "Delegation of Authority" is also an important specification for the project's operation in its future decision-making. This defines organizational groups and job titles authority level for expenditures, change control procedures, and operational guidelines for handling schedule and budget variances. Failure to define such management processes leads to future confusion and wasted time. Recognize that the operational specifications regarding how the project will be managed are just as important as defining the project deliverables and work specifications.

5.4.2 Project Dictionary

Various terms are used for stored project data. Even though the model's name is WBS Dictionary, that is a seldom-used term. Regardless of the local name, a significant amount of data is collected during a project life cycle, which has significant value if retrieved. Previous comments have been made regarding the need to document various project decisions. This is a follow-up to that view, except that here the view states that the project files should ideally be standardized across the host organization, which is a rare occurrence. Even though there is no known survey of time wasted in finding project information, it is a significant number. Organizations are now starting to understand the value of well-organized project data. The most logical value for such a data store is internal use by the project and its stakeholders, but the value to other projects is the hidden component. This data store should support the capability of both a traditional data host and an easy-to-access graphical retrieval site as well.

WBS coding is a great tool for showing the broad view of the project, and these schematic hierarchical charts are frequently found hanging on project team walls. However, they are not good for describing the details of work and do not contain the type of detailed data associated with the WPs.

An expansion of the low-level WP data view is needed. The WBS dictionary needs to be created as the source of information needed by the project. The one place to come for such information. PMI (2021) offers the following list of major data groups for the project dictionary:

1. Statement of work (SOW) description
2. Codes to support tracking of organizational resources and project financial details (i.e., WBS or organizational accounting codes)
3. Deliverables status

4. Acceptance criteria
5. Associated activities/tasks (predecessors and successors)
6. Milestone status
7. Responsible organization for each WP
8. Resource estimates for WP
9. Project activities and baseline schedules
10. Status metrics
11. Technical references
12. Contact information
13. Constraints and assumptions
14. High-level risk factors—detailed data would be found in the project risk register

Internet searches reveal sources for "project notebook" type data collections. Because of the data-intensive nature of the project, it is common to find various "islands of data" scattered across various hosts and physical media. A typical target is a local PC spreadsheet that is unavailable to others. The data goal needs to be one of integration in a modern technology data store that approved stakeholders can access. The design of this data store includes not only global accessibility but also proper user interfaces. This data store serves the status-tracking population and the personal planning and control functions. Ease of use is the design mantra.

When reading the design goal of the project data store, it is little wonder that this is not a normal situation. The host organization needs to support such an environment as part of its role in project management. Failure in this role will lead the project to use local options that do not accomplish the strategic goal of having project information for later ventures.

5.5 WBS MECHANICS

Developing a good WBS structure is not an easy task, and the guiding principle is for its design to reflect how the project is viewed by both the project team and external stakeholders. It shows how the project is being managed. To be of maximum value, the project plan structure should be mapped directly from the WBS-defined WPs, which can be traced into the project plan, including ID references.

At the termination of scope planning, a fully decomposed WBS may not be completed as described here. This is often done to speed up the cycle, but it increases the project risk by leaving planned work less specified. However, before work is started for a WP, the schedule and budget information should be documented. Leaving the work definition to be finalized after the planning phase can hurt the schedule, budget, resources, and risk processes. One might translate the situation where work is not adequately defined as "Oh, we will figure it out later." As indicated earlier, there is continual pressure on the project team to move through the life cycle faster. A common method for doing this is creating a new box type for the WBS called a *planning package*. These could represent segments of the overall project to be performed later, maybe even a different phase. When utilized, the WBS no longer has the schedule and cost integrity described to this point. However, just because the work is not yet well defined, the project is likely still required to produce a schedule and budget, even with planning packages included. This is where corporate memory can cause a problem. There is a formal term called Rough *Order of Magnitude (ROM)*, meaning the stated error rates could be +/–100%. So, the project team inserts a ROM value for the planning package and combines all of that into a pseudo-project schedule and budget. Somewhere along the way, it is forgotten that the ROM value meant a rough estimate, and the actual values are double those given earlier. This is just one of many places where the project team's reputation can be damaged by what appears to be clear and accurate data.

As a result of the speedup mentality, the decision could be made to proceed with a higher-level work definition that is not completely defined. This also means a higher potential error in the resulting plan values for schedule and cost.

When involved in projects where the scope definition does not follow the more detailed view outlined earlier, one must communicate forecasts more clearly. Also, the management and stakeholder groups have to be more partner-oriented. It is one thing to overrun a project that has been carefully scoped as described here. It is quite another to "run fast" with the minimal future scope defined and published schedules and budgets with little real substance behind the numbers. If the external "partners" hold the project team responsible for meeting such published goals, that project environment would be very stressful for the project team. Explaining future overruns in a laissez-faire environment is not a comfortable management situation. In many cases, a forecast given by the project team will be remembered beyond the circumstances for which it was created.

One "move fast" technique in the project planning literature is the "rolling wave." Envision a WBS with the left-hand side representing the standard project with multiple "phases" shown on the right-hand side portion. The total project is essentially approved as though all segments were fully elaborated to the WP level. The project commences, and phase 2 is initiated, followed by the other phases. Each phase is a "wave," and the project rolls through each. So long as each wave follows the guidelines outlined here, it is not unusual. A major flaw occurs when ROM values get inserted into the plan and forgotten later, when the project results do not meet those values.

5.5.1 WBS Architectural Principles

Some sponsoring organizational entity and a defined PM should own a WBS. The list below offers some basic WBS design principles to consider:

1. The top-level box defines the project name.
2. The WBS represents the *entire* project scope. Work not defended in the WBS is not considered part of the approved scope.
3. The top two layers of the structure identify major phases or components needed to deliver the project objectives.
4. Decomposing the middle layers of the structure will outline product or process elements related and linked to higher-level units.
5. The decomposition process continues until manageable WPs are defined at the lowest level in the hierarchical tree structure.
6. The WP level will define owner, work, resource, and cost data.

The theory and concepts for a WBS are easy to understand; however, the owner entity must identify the proper structure that reveals how the project will be managed. The list below contains guidance to help define the proper design:

1. Identify logical partitions or major phases in the project.
2. Show major milestones that could represent key management groupings.

3. Reflect any business cycles that need to be considered (e.g., tax period, production, downtime schedule).
4. Evaluate constraints that might dictate a grouping.
5. Evaluate internal or external organizational life cycles that might impact the structure.
6. Evaluate the development methodology process's impact on logical grouping (i.e., construct, test, approve, release).
7. Evaluate risk areas that need to be recognized in the structure grouping (e.g., technical, organizational, political, ethical, user, legal).

Active involvement of the project team and key stakeholders is imperative for the structural design of the WBS. A recommended drafting method is using sticky notes or whiteboards that allow proposed boxes to be moved freely. An actual WBS development exercise is shown in Figure 5.6 as technicians debate the placement of WPs.

This process facilitates discussions and allows boxes to be edited and moved around on the board until a particular work organization is agreed

FIGURE 5.6 Developing the WBS on a whiteboard.

upon. This process should typically work from a top-down view of the project scope. Once a particular WBS level is defined, it may be possible to delegate a segment of the structure to subgroups for detailed discussion and decomposition. Later, each subgroup would make its presentation to the whole team to ensure that other "horizontal across-the-tree" logic is consistent. This view should occur on a level-by-level basis. As the process moves downward, the level of interaction across the structure may be able to decline as the work definition becomes more isolated. For example, if the specification for a physical component is defined, the level of external communication is minimal so long as the component stays within that design envelope.

When the draft process is complete, a final review of the overall structure is required at the team, stakeholder, and management levels. The final WBS design will provide further guidance in understanding the project for all those involved. At this point, the team must buy into the structure and its represented work. From this schematic view, a deeper understanding of the project scope, management roles, responsibilities, potential risk areas, and critical assumptions can be gained. In addition, this overview stimulates team communication and a spirit of collaboration.

A design and review process, as outlined here, can complete a draft WBS structure in a short time. The author once worked with stakeholders to lay out a multi-billion-dollar project direction. At inception, the group could not find a way to envision the effort, whereas in two days, they were able to develop the major segments of the project and produce a ROM level of data for timing and cost. This group had never seen a WBS before the exercise and only understood it as a group of boxes. This type of capability is one of the most notable roles of the WBS.

5.5.2 WBS Box Coding Scheme

This chapter has described the evolving process of defining project delivery requirements. The WBS and its companion dictionary represent a visible artifact from this process. Labeled boxes describe the structure, and there are two aspects to the box names—their title and a related code number.

The visual structure of the WBS can get overrun as a project grows. This is affected by the number of layers and the ability to show the structure schematically. When that happens, the WBS must be "flattened" into a list format, which is also necessary for computer processing. Regardless of the resulting format, a box naming convention is an important design consideration. The standard recommendation is to use nouns in the box titles rather than verbs. This helps focus on what is to be done and not how it will be produced.

The second design view involves the coded numbering scheme used for box labeling. Figure 5.7 shows a WBS with box codes.

As with all usages of the hierarchical structure, there are multiple numbering schemes. One typical approach uses a "1" to depict the project to box, while another alternative approach uses a "0" to depict the top box. In a third view, the organization may code the project name as a prefix to the numbering scheme. In this case, a sample box name for project ABLE would be ABLE 3.1.1. The goal is to identify each box with a code name in all these cases. Note in Figure 5.7 that each layer of the structure inherits a reference value and a layer code. Also, a decimal-type box coding scheme is used to show this. Use box 1.2.2.1 as a coding example. This box is linked to box 1.2.2, the second box (phase 2) under the top box. Also, box 1.2.2.2 is Work Unit 2. Think of this coding structure as a project mailbox of work or delivery units. As project team members and stakeholders become familiar with these numbers, they become a shortened internal communications vocabulary. For example, a team member might say they are having trouble with 1.2.2.1 (or maybe shortened to 1221). That coded source and linkage reference represent a method to clearly understand where the problem occurs.

As the project progresses, these coded box values take on an operational meaning for the participants. Document numbers can also be linked to this numbering scheme. The value of a WBS for communicating project scope during the planning stage has been previously made. As the scoping process is completed and the numbering scheme is finalized, a schematic picture of the project with "mailbox" addresses with named component linkages results. If a project data dictionary is created as described earlier, its contents will be strongly linked to this numbering system. Also, a tangible product-related project, such as house

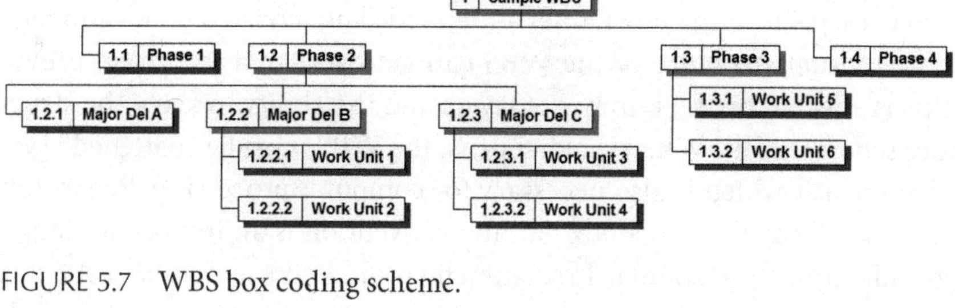

FIGURE 5.7 WBS box coding scheme.

construction, will utilize these codes for design drawings to describe various components.

As with all neat theories in the project world, there is always a caveat. In this case, the scope planned WBS only shows the project as it was initially defined. Scope changes during the execution phase are universal and impact the planned structure. Hopefully, most of the changes would be at the lowest level, but they could occur at any point. If a scope change were to add new work at the lowest level, that would erode the number system described. To counter that, one strategy is to use numbering increments of ten rather than single units. This would leave space for additions without destroying the logic outlined. Maintaining the integrity of the WBS coding scheme will be challenging, but it is important to use a logical coding scheme for internal communication and status reporting.

5.5.3 WBS Management Roles

Much has been made in this chapter regarding the WBS role in project planning, and that value will continue to be shown in future chapters related to schedule, budgeting, risk, control, and other management processes. As the project moves into the execution phase, the WBS becomes visible as a communication and status-tracking tool. However, at this point, there is a new focus on execution and control of the identified WPs. This view often means that groups of WPs are aggregated for management focus and status data collection. These formal aggregations are called *Control Account Packages* (CAPs), representing the project's resource data collection point. This aggregation is sometimes chosen to recognize that actual costs cannot be accurately defined below that level. However, in other cases, it is to focus management responsibility on that group. In more recent times, a new WBS-related management role has emerged in the form of a *Control Account Manager* (CAM). This supervisory-level role, in partnership with the PM, is delegated general responsibility for the successful execution of their defined group. This technique is recommended as a method to improve buy-in and control at the lower levels of the project team. More details related to the cost side of this topic will be explored further in Chapter 8 on cost management.

5.5.4 Other WBS Views

The concept of formal structuring of work has many operational benefits to the project management process. It is typical to re-sort the basic scope-defined WBS data into other views to support different control or analysis needs.

Table 5.1 illustrates a *flattened WBS* with two separate management perspectives. Translating the defined work scope view into a more linear data list is necessary for linkage to computer processing, or a method to show large WBS layouts that do not fit the graphical format. However, in this example, there are two different WBS views of the same project with the same work defined in both. Please note that option A divides the work by technical work area, while more of a life cycle process view is reflected in

TABLE 5.1 WBS Structure Comparison

Software Development A		Software Development B	
1	**Software**	1	**Requirements**
	1.1 Software requirements		1.1 Software requirements
	1.2 Initial software design		1.2 Hardware requirements
	1.3 Final software design		1.3 Hardware requirements
	1.4 Software construction		1.4 Hardware requirements
	1.5 Software complete		
2	**Hardware**	2	**Design**
	2.1 Hardware requirements		2.1 Initial software design
	2.2 Initial hardware design		2.2 Final software design
	2.3 Final hardware design		2.3 Initial hardware design
	2.4 Hardware acquisition		2.4 Final hardware design
	2.5 Hardware installation		
3	**Documentation**	3	**Construction**
	3.1 Hardware requirements		3.1 Software construction
	3.2 Develop documentation		3.2 Hardware installation
	3.3 Approved documentation		3.3 Develop documentation
4	**Training**	4	**Training materials**
	4.1 Hardware requirements		
	4.2 Training materials		
	4.3 Approved training materials		
5	**Testing**	5	**Testing**
	5.1 Develop test plans		5.1 Develop test plans
	5.2 Develop test cases		5.2 Develop test cases
	5.3 Unit test (multiple)		5.3 Unit test (multiple)
	5.5 System test		5.4 System test
	5.6 User acceptance test		5.5 Approved documentation
			5.6 User acceptance test
6	**Project management**	6	**Project management**
7	**Project Close**	7	**Project Close**

option B. This dual example shows that the same work units can then be sequenced in more than one way with an identical related schedule for the defined effort. Which one, right? Which one is best? If the performing organization had a strong departmental culture, option A might reflect that view best. In contrast, option B looks at the effort more as an organizationally coordinated group of activities evolving through life cycle phases, which may be the preferred view. Please also note that the traditional method for software development is to view option B as the most likely structure selected. Regardless of the actual structure chosen, the design of the overall WBS architecture should be one of the early project team conversations.

Conversely, suppose the project target involves a physical product deliverable. There would be a strong bias toward viewing the proper work structure mirroring the physical device components at the higher levels. As an example, Figure 5.1 previously outlined an automobile WBS showing its physical components at the top level. In this case, there is a major focus on the work required to develop the major components. This view makes it somewhat easier to conceptualize the work necessary to produce the defined high-level components.

Another common restructuring occurs when third-party vendors have contractual responsibility for a subset of the project. This substructure is called the *Contract WBS* (CWBS). It is connected to the master WBS but typically stays under the design and administrative control of the third-party vendor to manage without specific details being visible to the project buyer. Assuming that is the case, the contracted "branch" of the WBS would be extracted (cut off) from the internal view and allocated to the contractor as though this piece were the complete project, and it would be for the supplier. In this scenario, the third-party vendor might need to expand the WBS to define their portion of the work internally. However, in more complex situations, the vendor might be fully involved in the original scoping process as if they were internal team members. If the contractual arrangement is a fixed price, that contracted substructure of the plan might be shown on the overall WBS as a single summary box with only summary-level detail. Status reporting for this segment may only require high-level schedule completion status. A fixed price contractual relationship might only be required to report the general schedule status without any other details.

A second common WBS re-sort focuses the work defined into organizational groups. This view is called an *Organizational WBS* (OWBS). The view sorts the WBS box links to the responsible organizational unit.

As work is scheduled, the same view can be converted to dated work plans by the organizational unit.

A third sort of the WBS involves resources allocated by work unit and time frame. This view is called a *Resource WBS* (RWBS). Microsoft Project (MSP) and other project software utilities can produce this view for resource capacity analysis. This view is a time-phased sort by resource categories used to show total resource requirements for each WP and period. Data sorting of this type can produce various total and time-phased cost views. Examples of this would be cost-related data showing project direct cost cash flow, resource type, and organizational groups. In all these examples, one can see the role played by the base WBS in organizing views related to work, time, human resources, cost, and other data captured from planning and operational processes.

5.5.5 Data Migration

It is necessary to link the WBS codes to formal organizational systems to supply actual data into the various WBS formats. For example, the organizational payroll system is most often used to pay the project team members. The team member will enter time spent with a project code similar to a WBS number. That number must be translated to the organization's system code, which can then map the actual cost. This value will be returned to the appropriate WBS for project status reporting. Through this type of linkage, actual resource costs can be mapped to the WBS. Similarly, the inventory or general accounting system will need a translational table to map those resource elements (i.e., material, facilities, and overhead). The WBS numbering system represents the technique of collecting actual data and mapping it back to the project work structure. This is a subtle but important mechanic in the management process.

5.5.6 Project Baseline

The *project baseline* is another WBS view that is often referenced. In model theory, the project WBS data are captured, approved by management, and frozen for the life of the project. As the project unfolds, actual values are collected and compared to the baseline value to quantify project status. This is often described as "plan versus actual," with the plan value being the baseline element. In concept, this is the traditional model status-tracking method for measuring the project's progress.

On the surface, one has a hard time refuting this concept. However, on the reality side, every PM would disagree with it as defined. Every project has a scope change, leading to increased cost and possibly schedule. If the project's actual cost exceeds the earlier gasoline plan because of these changes, is that logically an overrun? Yes, it is an arithmetic overrun of the baseline values. However, it does not necessarily mean the variance should reflect negatively on the project team's performance. This comparison is of questionable status value until organizations understand that the typical baseline plan does not reflect either scope change or risk events. Future textbook sections will describe these gaps in project plans to better understand the issue described here. For now, accept the baseline concept as a model idea with questionable underpinnings.

5.6 CONCLUSION

This chapter has described the management process called scope management. The goal here has been to define how project deliverables evolve in the planning process and then link to the associated resource, schedule, and cost factors related to that view. The underlying design goal of the scoping process is to identify the planned project's functional delivery and, from this, develop a clear work specification. Failure to accomplish this goal will leave gaps in the later processes, which tend to become significant sources of project failure later.

The WBS has been highlighted as the most important management artifact in the life cycle. Examples provided here have illustrated its role in capturing project requirements and its importance as the project moves into execution. There is a clear linkage between work units defined in the scoping process and future management and control tool steps. The numbering scheme of the WBS provides a needed reference scheme. Many more examples of this tool's importance in ongoing project management will surface throughout the rest of the text.

KEY VOCABULARY TERMS

1. WBS ID—a coding scheme used to identify WBS boxes
2. Deliverable—a formal project goal item
3. ROM—a Rough Order of Magnitude estimate (approximately +/−100%)
4. WP—a target-sized work unit called a work package (model size of two weeks and 80 hours of work)

5. Planning package (PP)—a project segment that has not been decomposed
6. Control Account Package (CAP)—a collection of WBS boxes that will be managed as a single group
7. Control Account Manager (CAM)—a manager charged with overseeing the performance of a CAP
8. Constraints—project factors that are formally stated to be obeyed (i.e., schedule, cost, resources)
9. Assumption—a factor that is used to produce planning estimates
10. Ibilities—a set of design measures that characterize the deliverable
11. Flattened WBS—translation of the schematic WBS into table format
12. WBS Dictionary—the project data store, which is a companion to the WBS
13. Contract WBS (CWBS)—contractual portion of the overall WBS
14. Organizational WBS (OWBS)—the WBS sorted into organizational responsibility groups
15. Resource WBS (RWBS)—the WBS sorted into resource groups

REFERENCES

Department of Energy (DOE). (2014). DOE G 415.1-1, Information Technology Project Execution Model Guide. https://www.standards.doe.gov/

Department of Defense (DOD). (2024). Work Breakdown Structure of Denese Materiel Items, Revision F. https://quicksearch.dla.mil/qsdocdetails.aspx?ident_number=36026

Project Management Institute (PMI). (2021). *A Guide to the Project Management Body of Knowledge*, 7th edition. Newtown Square, PA.

Wrike. (2024). What is Scope Management, Wrike.com, https://www.wrike.com/project-management-guide/faq/what-is-scope-in-project-management

CHAPTER 6

Quickstart Example

Project N Motion Videos for This Chapter

6MSP1	Loading MSP	Intro to MSP
6msp3	Pool Project	Pool Project Schedule

Reader Note: This chapter will be more instructive if you first review the Work Breakdown Structures (WBS) translation and MSP keystroke checklist videos [MSP0] [6MSP3].

6.1 INTRODUCTION

This chapter aims to provide a quick introductory demonstration showing how the WBS contains the required data needed to create a first-cut schedule and budget. This approach links the scope definition to the resulting project plan, an often-misunderstood process. Understanding the mechanics regarding the production of a project plan is a critical learning step in that the plan should directly reflect the defined scope. Any tasks added to this suggest it is separate from the official scope. One of the model rules of a WBS is that it contains a 100% representation of the approved scope. Thus, the plan should be built from the scope view. More complexity in this process will evolve in upcoming chapters.

DOI: 10.1201/9781003532989-8

The scoping process produces an approved definition of the work required to meet the deliverables. This work is envisioned as a three-dimensional box with sides modeled as output function, duration, and cost. One of the vocabulary terms for the box is a work package (WP) containing data regarding the duration and sequencing of this unit in the overall structure. These work packages represent delivery activities such as producing a physical product, installing piping, testing software, or producing a user guide. View these simply as a defined set of work associated with a more extensive collection. The formal model vocabulary term introduced in Chapter 5 for this is a WP, although we also need to note that the WBS may contain summary levels that are linked to these. Also, be aware that the usage term for this activity is not standardized. For example, MSP describes all WBS references as "Task Name."

6.2 POOL PROJECT

A swimming pool is installed here to demonstrate the mechanics related to the plan development process. The required task shown gives a somewhat simplistic view of the task required to install a swimming pool. Since this is a small project, it is possible to represent the whole activity set via a single WBS layer (see Figure 6.1).

The staggered layers in the figure are required to fit the visual structure inside the page space, but there is only one layer. The first step in the translation process is to flatten the WBS as previously described in Chapter 5. Also, sequencing predecessor data is attached to the task structure (see Table 6.1).

Note that the flattened activity listing contains the same twelve "Task Names" from the WBS, and all task WBS codes indicate that they are subordinate to the total project. Theoretically, this is an overly simplistic WBS in that all the boxes represent low-level work. The only decomposition required here is to recognize that all tasks are subordinate to the total project, and the WBS coding shows this relationship.

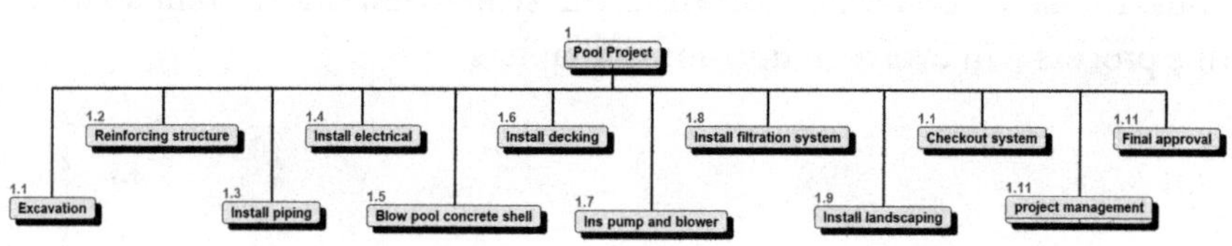

FIGURE 6.1 Pool project WBS.

TABLE 6.1 Pool Activity Listing

ID	WBS	Task Name	Dur.	Pred
1	1	Pool Project		
2	1.1	Excavation	5d	
3	1.2	Reinforcing structure	3d	2
4	1.3	Install piping	4d	3
5	1.4	Install electrical	3d	3
6	1.5	Blow pool concrete shell	5d	5, 4
7	1.6	Install decking	5d	6FS+5d
8	1.7	Install pumps and blower	4d	7
9	1.8	Install filtration system	2d	7
10	1.9	Install landscaping	5d	9, 8
11	1.10	Checkout system	1d	10
12	1.11	Final approval	0d	11

6.2.1 Structure Comments

Six items in the flattened structure deserve explanatory comments. These are as follows:

- Each item has an ID value that will be used as a reference number for predecessor specification.
- The top item is a summary unit accumulating the entire project values (i.e., duration, dates).
- Item 12 is a milestone (coded by zero duration) that represents a key event. These can be inserted anywhere in the project.
- All predecessor values are represented by line numbers. The "3" value shown for line number "5" indicates that line "5" cannot start until line number "3" is finished. This is the mechanism that will be used to produce the schedule.
- All task predecessor codes with numeric values are assumed to be Finish–Start (FS) by default, although line number "7" has a different specification.
- The specification for line number "7" is a gap linkage. This specifies that line number task "7" cannot start until line number "6" is finished; it also must wait for five more days. This is a "gap" specification.

Two of the specification data elements come directly from the WBS. That is the WBS code and related Task Name. Three additional data elements are needed to finish the specification table. These are duration, predecessor, and task cost estimate, which are optional. These planning parameters are defined as part of the scoping process and were previously stored in the WBS Dictionary.

6.2.2 MSP Process

Data from Table 6.1 can be fed directly into a scheduling utility such as MSP or Oracle's Primavera [6MSP3]. Figure 6.2 shows how the software is set up to receive the required data.

Once the raw data is loaded into the software, four additional steps are required. These consist of the following:

1. The WBS structure must be defined. One of the "idiosyncrasies" of the MSP software is that it does not use the WBS codes to define its internal structure, so this must be manually produced. This would be a nice default option for the software to use the WBS codes and create this automatically because the current method is a primary source of process errors.
2. Similar to the WBS structure option, the software default is to manually schedule the tasks. This is another questionable design in the author's view. This option can be changed to signal the software to automatically schedule, which is discussed in more detail in the various support videos.
3. A project start date needs to be set.

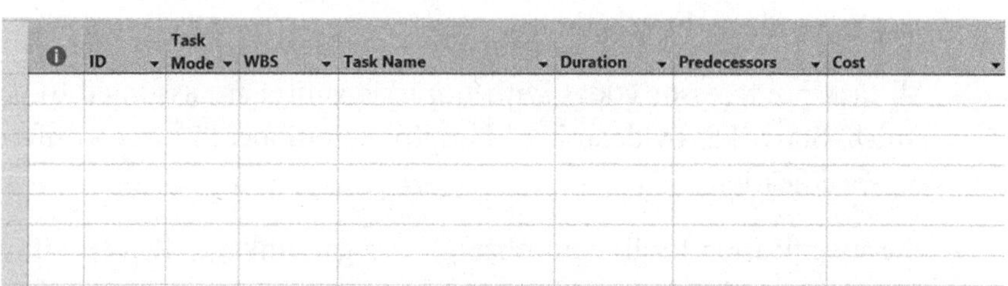

FIGURE 6.2 MSP table view.

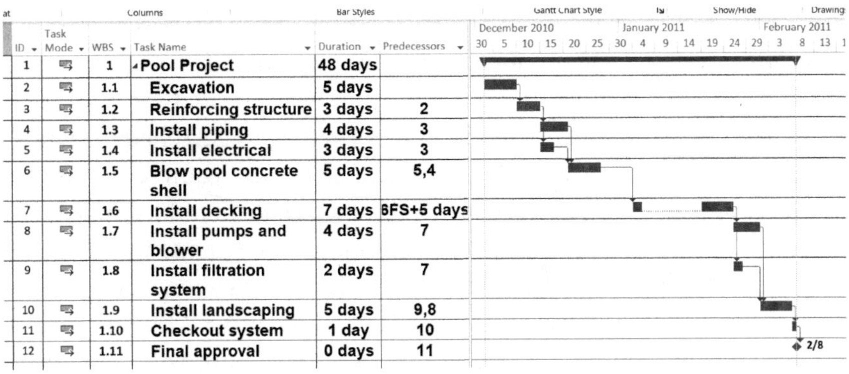

ID	Task Mode	WBS	Task Name	Duration	Predecessors
1		1	Pool Project	48 days	
2		1.1	Excavation	5 days	
3		1.2	Reinforcing structure	3 days	2
4		1.3	Install piping	4 days	3
5		1.4	Install electrical	3 days	3
6		1.5	Blow pool concrete shell	5 days	5,4
7		1.6	Install decking	7 days	6FS+5 days
8		1.7	Install pumps and blower	4 days	7
9		1.8	Install filtration system	2 days	7
10		1.9	Install landscaping	5 days	9,8
11		1.10	Checkout system	1 day	10
12		1.11	Final approval	0 days	11

FIGURE 6.3 Pool project first-cut plan.

4. The internal work calendar needs to be verified because this is used to produce the scheduled dates. The default form is a standard 5 × 8 or 40-hour work week.

Once these execution parameters are set, the result is a first-cut project plan (see Figure 6.3).

Note that the software utility schedules the defined work tasks according to the predecessor codes and durations. Also, note that the total pool project duration has been calculated using the task data. This output format represents one of the primary values for using software for this process. The underlying scheduling logic can be a very time-consuming effort that is saved with this approach. The linkage calculation is laborious, and the software recognizes weekends and other non-work times for the dates shown. Although visible or described here, the software can also compute cost and resource status.

6.2.3 Output Review

The project schedule output contains a data table and an associated Gantt graphical format. Collectively, a large amount of optional status data is calculated and stored. These parameters can be moved into the visible table space as desired (i.e., start date, finish date).

The calculated results indicate that the project would require 37 workdays (elapsed time). Note that the bold black bar ID 1 is a summary bar summarizing the entire project. Various internal data for the project are calculated and stored, but only high-level schedule data is visible here. From this example, these model mechanics resemble Lego blocks being stacked.

This example represents a powerful start to the project schedule modeling process as it illustrates the core data elements required to build a mechanical plan schedule. One of the new data elements added here is the recognition of coding sequence relationships for the defined work units. These are called *predecessor relationships*. The values shown in this column define how the various activities link to each other. Other predecessor relationship coding options can be used, but more will come later. From a conceptual model view, all projects would use this same set of variables to produce a schedule and budget. It is recognized that real-world WBSs are much larger than this and have more layers. Furthermore, the predecessor relationships would be more complex, but the plan generation mechanics would be identical to this example.

Before feeling confident that this is all required to produce a project plan, remember that Murphy's Law dictates that something will go wrong at the most inopportune time, and that is the unrecognized view of this problem. Nevertheless, if the PM can keep the defined WP variables in line, this set of mechanics is an important project management step. It represents a key output process that will be used to communicate goals and status for the project as it evolves through the life cycle.

CHAPTER 7

Developing the Project Schedule

Project N Motion Videos for This Chapter

7MSP0	WBS Translation	Loading Mechanics
7T0	Time Mgt Overview	Time Overview
7T2	Plan Checklist	Plan Creation Steps
7T4	MSP Plan Parameters	MSP Initiation Parameters
7T6	MS Predecessors	MSP Predecessors

**Note that the video lesson 6MSP1 is also relevant to this material as this chapter heavily uses MSP.

This chapter aims to demonstrate the evolution of a project plan from scope definition. Various estimating techniques are covered, and the mechanics of network calculations are shown. Finally, a scheduling specification is translated to a first-cut project plan using the estimated parameters described in the chapter. Details regarding the estimating and sequencing processes are outlined. MSP mechanics will illustrate the process of creating a project plan.

DOI: 10.1201/9781003532989-9

7.1 INTRODUCTION

The entry point for this process is the detailed Work Breakdown Structure (WBS) specifications. From the scoping exercise described in Chapter 5, this is recorded in the approved WBS Dictionary Defined work units from the WBS and will be estimated and sequenced as work components of the project. This estimating and translation process will show how the scope results drive the follow-on schedule and budgeting plans. Work unit specification data represents the core of the scheduling process *[7T0]*.

Figure 7.1 shows the process decision flow linkage model that outlines the two parallel activities that lead to the scheduling parameters needed to produce the initial schedule. The development of a project schedule starts with the scope definition derived from that planning effort. Work units are the primary target for the scheduling process, and each will appear in the evolving project plan. Figure 7.1 describes the decision flow from the WBS to the work unit schedule parameters.

Note that this process shows two essentially parallel paths to produce the required schedule parameters. The three logical groupings of this process are listed below:

1. The incoming work unit will be reviewed to produce scheduling and cost estimates for the unit.
2. The second parallel path will specify the technical work sequence for the work units. This segment will require iteration and communication with key stakeholders to define and approve the work unit scheduling sequence.

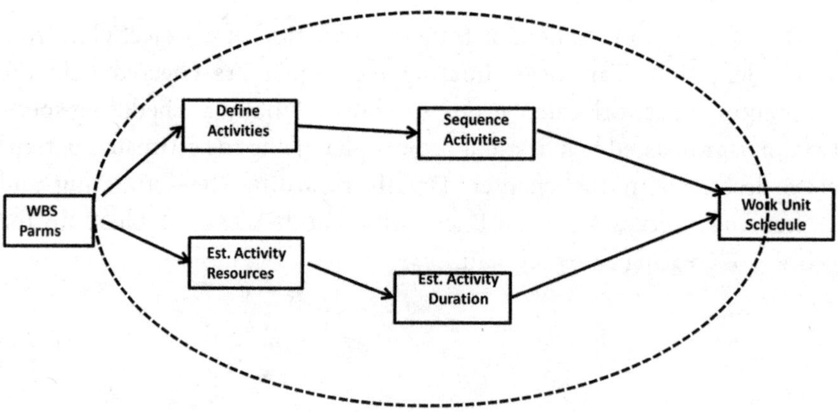

FIGURE 7.1 Plan estimating process.

3. The resulting project plan will combine all the estimated work units and scheduling parameters to produce a delivery schedule that constitutes a significant segment of the project plan. The associated cost plan will be described in Chapter 8.

Figure 7.2 illustrates the logical flow of the basic schedule development process.

This diagram clearly shows the key outputs of the work unit estimating process. Estimates produced by this process will pass through scrutiny from various knowledge areas, stakeholders, and management. Once approved, the parameters are captured in a WBS-like format representing the core planning artifact leading to a project plan. Many adjustments can and likely will be made to the first-cut view, but this is an important planning step for the project. One of the required modifications to a plan is resource shortages. Note that the resources were planned here, but the total requirement and timing for that is unknown until the work units are combined into the plan and the total resource timing and quantity are calculated. This process is called Capacity Management, and the final plan must match the availability of resources (i.e., human, material, or monetary). All of these are physical constraints that the plan has to work through. The worst-case scenario is to find an unworkable situation that affects the Charter parameters for cost, time, or functionality. If there is a compatibility issue with the plan parameters, now is the time to deal with it and not wait to recognize it later. This is just one reason why the term "first-cut" is used. The goal for the resulting plan is for it to be technically and resource viable.

After the plan has been reviewed and verified, model theory says it is baselined (frozen) and used for future status comparison. We will describe this topic later, but it is a suspect theory in practice. Chapter 17 will have a lot to say about this approach. At any rate, this approved plan becomes a visible target for the project team and stakeholders. It will contain

FIGURE 7.2 Work unit estimates.

planned schedule completion dates, as well as total budget values. One can see that this is an important artifact that will be reviewed through the life cycle.

It is also important to point out the dynamics of project work. Multiple factors produce additional work for the project beyond what has been recognized to this point. The two main areas that add new work to the project are scope change requests and unknown risk events. Both of these areas cause significant problems in maintaining a schedule and cost forecast. Our best strategy, for now, is to focus on the initial mechanics and then come back in a later chapter to deal with some of the dynamics. Projects are not static entities; the planning process must recognize this. You will note that the examples in the following chapters give the illusion of stability. The logic for this is to seek the proper goal and then deal with the factors that negatively affect that goal. The current strategy is to work on understanding factors that lead to success.

7.2 DEFINING PROJECT ACTIVITIES

There are two important management considerations for sequencing. First is the notion of *technical sequence*, which is based on the technology inherent in the defined work unit. For example, the technical sequence required to construct a building would define that the foundation must be in place before constructing the walls. Likewise, a software system would have to be coded before being tested. All project activities have some technical or logical sequence generally dictated by their underlying technology or purpose.

A second consideration of sequencing falls beyond the technical issues. Many activities that might be different from the technical view can be performed at management's discretion. In these cases, the PM may sequence the activities to accommodate factors such as availability of resources, weather, physical location, or other factors. In both the technical and discretionary sequencing situations, it is incumbent on the project team's skills to make the appropriate activity sequencing decision. However, it is important to recognize that these decisions often impact the schedule results, so sequencing considerations are usually an iterative activity based on current events. Regardless of how the estimating and sequencing decisions are made, developing a first-cut project schedule is the goal of this process.

7.3 ESTIMATING TECHNIQUES

Several estimating techniques are used in different scenarios. The most common examples are as follows:

- Bottom-up
- Expert judgment
- Analogous
- Heuristic
- Parametric
- Phased
- Effort distribution (top-down)
- Monte Carlo techniques
- Delphi

Each of these techniques is only optimal for some cases, and in most situations, multiple approaches should be used to confirm the derived value. To produce accurate estimates successfully, one must have a good understanding of the capabilities of each technique. Beyond developing a work estimate for the effort, the second half of the equation involves understanding the workers' productivity. Expertise with both of these variables helps with the resulting accuracy. This section describes several standard estimating techniques and the environment where each is most effective.

7.3.1 Bottom-Up Estimating

Experts consider this technique the most accurate of all the techniques described here. However, a fully decomposed WBS to the Work Package (WP) level must be utilized appropriately. In this view, the total project consists of relatively small work units that can be estimated reasonably well by the performing work groups. In most cases, the actual estimating method used for the low-level work units would be provided by the individuals doing the work. Once the entire collection of individual WP estimates is complete, they are rolled up in the WBS structure to generate the higher-level aggregations. From a statistical point of view, each of these estimates would be subject to typical estimating errors; however, over the full WBS range, the errors should somewhat compensate. Given that these estimates are produced by individuals with good knowledge of the work environment, the estimate accuracy should be within reasonable bounds compared to the other methods reviewed. The primary constraint to this technique is the requirement for detailed WP information. Failing that, the accuracy assumption weakens, and other methods become comparable.

It is not hard to imagine a work group's feeling when an external estimator delivers a value to the group and says, "This is how long it should take you." Buy-in to an estimate is part of the accuracy equation, and this external approach defeats that segment. The concept of commitment is lost in this case.

One other disadvantage of this method is the time and effort it takes to produce. A secondary issue with the process is the ability of the local workers to pad the estimate to make themselves look good. External estimators do not have this motive, but the external groups do (Johnson, 2007, p. 224).

Work unit estimating aims to derive estimates that do not have excessive padding and represent reasonable values for the activity. To ensure this, some cross-checking is required. Readers interested in the psychology of this phenomenon should review the Theory of Constraints (Critical Chain) discussion in Chapter 21.

7.3.2 Expert Judgment

Expert judgment is a popular technique for estimating large project sizes and work unit levels. According to a software industry study, 62% of cost estimators in that industry use the expert judgment technique (Snell, 1997). An estimator using this approach relies on his expertise and is often guided by historical information and experience with similar projects. Expert judgment is often used in combination with other estimating techniques for improved accuracy. For example, imagine you wanted to provide an estimate to repair an automobile transmission. One expert opinion could say that a standard repair of this type is 100 hours. However, given the extensive history data with this brand, the inclination might be to accept the value. A lack of experience with that brand product would likely lead to padding the estimate.

There are three main advantages to using expert judgment as an estimating technique. First, it requires little data collection and uses experience from past projects. Second, it has enough flexibility to be adapted to the particular target. Finally, it provides a quick assessment because the expert will have a large stored knowledge set to derive the estimate (Snell, 1997).

There are also drawbacks to consider when using the expert judgment technique. First, the estimate provided by the expert will not be any better than the objectivity and expertise of the expert (Snell, 1997). An accurate estimate requires the estimator to have extensive experience dealing with the type of task or project being estimated. Secondly, estimates made by experts can be biased, which will produce a number with no audit trail to be used to verify logic. Finally, the expert often bases the estimate on personal background, which may not fit the current target.

7.3.3 Analogous Estimating

Analogous estimating has similar traits to expert estimating in that it is based on prior experiences. However, in this case, the comparison is based more on data from the previous experience. This process would typically produce an order of magnitude estimate unless the new project is similar to the comparison one. This estimating technique often uses measures of scale such as size, weight, and complexity from a past situation to make the estimate. When using analogous estimating, the estimator needs to factor in any differences between the new work being estimated versus the previous effort being used for comparison. Examples of complexity factors include new technology being utilized or any changes in the complexity of the task or project.

Analogous estimating is best used in the early phases of a project before significant details are visible. This method provides quick and easy estimates for projects or tasks that are not very complicated; however, the main drawback is that the results are often inaccurate (Baca, 2007, p. 136). To ensure the estimate is as accurate as possible, use past projects that are technically similar and not just similar in appearance. There is also an element of estimator expertise involved in this method (PMI WBS, 2006).

A simple example of an analogous estimate would be determining how long it would take to unload a ship and move it back to its home port. Experience with ships of the same size and cargo offers the base comparison value. From this, a review of variable differences such as resource availability, weather, and other factors would need to be considered. Based on this type of analysis, the estimator should be able to derive a reasonable work or elapsed time estimate.

7.3.4 Heuristic Estimating

A heuristic estimate is based on a "rule of thumb." This process is derived from parameters created from past experiences. In that sense, heuristic estimating has flavors of both expert judgment and analogous. Except in this case, the estimate is translated into a mathematical type expression. Realize that expert judgment estimating must be performed by the expert, whereas heuristics estimating can be transferred to others capable of manipulating the defined relationships. As an example of this technique, a local roofer has been working in a neighborhood for quite some time. All of the roofs are reasonably similar in design. The roofer has noted that a roof estimate is a function of the size of the root multiplied by a constant, such as something like 15 times the roof area, even if that person has never previously been involved in the process. This heuristic equation can be provided to someone else to market roof replacement even if they have never seen the house or done roofing. Therefore, it is a heuristic estimating formula. This class of estimating would typically be found in situations where the same type of project is performed repeatedly such that the relationships are well established (i.e., roofing, suburban house construction, installation of a water heater, auto repair). Whether the vendor is willing to make this a firm fixed price or not is based on the potential variability of the task. A heuristic house construction bid might have several caveats and be only a sizing rough order of magnitude (ROM) estimate, whereas the roofer might feel comfortable making his bid a fixed price. Notice that when you go to a car dealer for a transmission overhaul, they will quote a fixed price on it and may need to do other repairs as well. This is based on heuristic data.

The benefit of using heuristics is that they are easy to use and do not require much detailed research. If multiple estimators were used to make bids, it would be necessary to formalize the estimating parameters so the estimates would be consistent. In many situations, the heuristic estimate is not expected to be highly accurate and should only be used when the inherent risks are acceptable (Mind Tools, 2007). As an offhand conceptual example, if one used the average 25 MPG for his car as a forecasting heuristic to decide whether to drive 100 miles across an isolated area, would it make sense to do this with an estimated four gallons in the gas tank? What if you were flying an airplane across a water route with similar variables to consider? Do you see the risk elements in these two examples?

7.3.5 Parametric Estimating

Parametric estimating is "an estimating method in which an algorithm is used to calculate cost or duration based on historical data and project parameters" (PMI, 2021, p. 244). Parameters based on size, footage, or other scope-related values can be used to produce time and cost estimates for the related work. The process of creating a formal parametric model involves collecting data from thousands of past projects and using this data to produce a regression-type model with multiple variables (more sophisticated and complex than the heuristic models). Parametric techniques are most useful in the early stages of a project when only aggregate sizing data is available. These estimates are considered ROM-level accuracy because they lack detailed requirements precision (Kwak & Watson, 2005). When either heuristic or parametric methods are used, the key question is how well they match the environment where they are applied.

Parametric techniques were first developed by the U.S. Department of Defense (DoD) during World War II to estimate high-technology projects such as weapons systems and space exploration. The technique is most commonly used now in the construction and auto repair industry. Other applications include, but are not limited to, determining the conversion costs associated with new technologies in electronics or manufacturing. The technique can be applied to any situation in which sufficient historical data are available.

The parametric estimating technique needs to take into consideration the delivery methodology that was used in its data collection because this could impact the calculated value. For example, when developing software using the waterfall method, the cost of documenting the requirements before coding needs to be considered because this is a common practice in this method but less so in the environments.

The main drawback of the parametric estimating technique is the accuracy of the estimate it provides. According to one researcher, estimates that are based on the final project design will be roughly in the range of +10% to −5% accurate. If the estimate is based on project designs that are from 1% to 15% complete, the accuracy will range from +30% to −25% (Kwak & Watson, 2005). Accuracy can also be affected by factors such as the experience of the individual making the estimate, the level of scope change, and other incorrect assumptions. Accuracy is especially problematic for construction companies that use parametric estimating to produce bids for projects. They will not win the project if the bid is too high, and the project will not be as profitable if the bid is too low. So, low accuracy makes the provider vulnerable to adverse results. With the increased amount of historical information available in databases, innovations in statistical applications, and the availability of expert data over the Internet, improved accuracy of parametric estimating techniques seems likely.

7.3.6 Phased Estimating

Phased estimating is based on a life cycle-oriented rule of thumb and is most frequently used when insufficient detail-level data supports the overall estimate. In this situation, the estimating process is broken into a series of sequential decisions linked to project phases. At the defined decision points, called phase gates, a management review process will determine whether it is appropriate to continue with the project, redefine its scope, or terminate it. According to Verzuh (2005), the performance baseline should be considered reset at each phase gate, and the project should be re-evaluated. The following series of steps are used to create a phased estimate:

1. Break the project life cycle into phases, which will be considered a subproject.
2. For the first phase, a detailed estimate is made for the cost and duration of that phase. At this point, the estimator should create a ROM-level planning estimate for the entire development life cycle.
3. If the first phase is satisfactory, the project is approved to move to the next phase. Marginal performance of the first phase opens up a critical review to cancel or adjust the project plan.
4. If the project is approved to continue to the next gate, a detailed estimate is created for that phase, and the phase estimating equation for all subsequent phases is updated based on current information.
5. The cycle of phased estimating continues until the project is completed or terminated (p. 174).

When using the phased estimating or *rolling wave* technique, it is important to understand what is required at each phase gate. Each active cycle should contain three main components. First, the phase must specify its required deliverables. Each gate will have a different deliverable or set of deliverables. Secondly, each gate should define a set of success criteria. These criteria determine whether the project should proceed to the next phase or be terminated. The final component is the specific outputs. These outputs help answer the question, "What is the purpose of the gate?" The number of phase gates that will be required depends on the size and complexity of the effort. When the project scope is well-defined, fewer phase gates are required. Project teams should try to use the same gates at consistent points for similar projects, as this will help improve the review process.

The main benefit of the rolling wave estimating approach is that it allows the effort to move forward quicker and recognizes that early planning in an uncertain world may be a waste of time. However, this approach requires much more monitoring from management, and the level of predictability regarding future budget and resource requirements is low. This approach does not help significantly in comparing the two options in that the decision metrics are known to have high ranges.

In this model, the uncertainty regarding the project will decrease as the project progresses so that each subsequent phase estimate will be more accurate. Another subtle issue in this approach is that the project team has more potential to meet the short-term phase goals, whereas the longer-term full estimate approach could lead to situations where the team would have a lesser chance of success.

One of the main criticisms of the use of phased estimating is that many experts feel that the project team loses accountability. They argue that the team cannot be accountable when the baseline is reset after each phase gate. One method to hold the team more accountable is to require justification for the cost and schedule estimates. When estimates change, the project team should be able to justify these changes with evidence of changes in difficulty or scope (Verzuh, 2005, p. 4).

Organizations whose phase reviews are performed in this matter utilize phased estimating in their estimation methodology. Those organizations that do not meet these criteria typically use milestone reviews instead of phase gates. This latter approach tends to hold the original baseline plan intact through the life cycle and measure variances from that point.

7.3.7 Effort Distribution Estimating (Top-Down)

This method looks at the project as a whole and apportions the total estimate into high-level groupings. An approach like this may be necessary in early planning when sufficient details for lower-level approaches are not yet available. Once the high-level value is derived, a WBS is used to assign percentage values to the main levels in the structure. Normally, this type of allocation would only go to one or two levels below the top of the structure. Historical data for similar past projects are typically used to determine the percentages for each phase and/or the summary activities within a phase (Horine, 2005).

This technique works reasonably well in organizations, using a standard methodology for similar types of projects and good historical data. This method also works well with the phased estimating technique in that the historical data provide reasonable estimates for future phases without extensive definition. For example, when a phase gate is completed, the actual amounts from the previous stage can be used to project future values.

One hazard in using this approach is that an aggregate estimating error for the project is proliferated through all segments by the apportioning method. Another drawback is that the technique uses historical data to define the apportioning formula (Verzuh, 2005, p. 176). If the comparative projects are not technically similar, these values can provide erroneous results that are once again proliferated through the life cycle. For example, if the organization's projects do not have the same number of phases or the phases are different for every project, applying a consistent apportioning formula to new projects will be difficult.

To illustrate a top-down approach, imagine that a project goal is to design and deploy a new car model. The organization has defined five phases for this effort and an estimated overall budget of $250,000. These defined phases include initiation, planning, design, construction, and deployment. The construction phase is further decomposed into frame, exterior, and interior activities. Past projects indicate the following main phase resource cost distribution:

Initiation = 10%

Planning = 15%

Design = 15%

Construction = 40%

Deployment = 20%

Furthermore, at a lower level, the activities within the construction phase typically consume resources in the ratio:

Frame = 14%

Exterior = 13%

Interior = 13%

**Note that this latter group allocations add to the 40% value shown for the construction level.

Using these allocation values, the construction phase would receive 40% of the estimated budget, which would compute to $100,000 (0.40 × $250,000). Similarly, within the construction phase, the frame activity would be estimated to cost $35,000 (0.14 × $250,000). Assuming that the effort distribution and the phased estimating techniques were applied, these estimates could be updated at each phase review point. The original allocation estimate for the initiation phase was $25,000 (0.10 × $250,000), but the actual cost was $30,000, which is 20% higher than the plan. From this result, the overall cost estimate would be revised by 20% to $300,000, so the new estimate for the construction phase will increase from $100,000 to $120,000, and the estimate for building the frame will increase from $35,000 to $42,000. The key issue in these comparison methods is whether the project team can modify the planned values dynamically. Some organizations will allow this, and others will hold the original value constant and pressure the project team to take corrective action elsewhere to stay on the plan. These are management philosophy issues beyond the basic estimating methods.

7.3.8 Monte Carlo Simulation

This technique's name was popularized by movies and similarities between its roulette wheel gambling at the Monte Carlo casino. In this model, individual work estimates are recognized as having a range of values rather than a single deterministic one. This range of values is then "simulated" using random numbers similar to the casino roulette wheel to produce a time estimate. The overall project is "simulated" hundreds of times from this process to produce a probabilistic probability distribution view of its outcome. Because of the volume of calculations related to this method, it requires automation support, and software is available to support this type of analysis. Specific details of simulation modeling contain a somewhat complex set of related details that are too broad to cover here. However, this is the focus of Chapter 20, so further details on this method will be deferred to that chapter.

7.3.9 Delphi Technique

The Delphi technique is another variety of expert judgment. It is most effective when making top-down estimates in the early stages of projects where there are many unknowns and a single expert is not sufficient because of the breadth or uncertainty of the goal. A radical example is "How long will it take to develop a product to go to Mars?" In complex situations such as this, the Delphi approach gathers the estimates from a group in the space industry. This technique involves the following steps:

1. Experts are given the specifications of the project and an estimation form.
2. The group meets to discuss any estimation or product issues.
3. Each expert provides their estimate without collaboration with others.
4. Estimates are returned to the group, indicating the median group estimate and the individual's estimate.
5. The group meets again to discuss the results and individual logic for their estimate.
6. Each expert again provides their estimate.
7. Steps 3–6 will be repeated until the group of experts reaches a consensus (Snell, 1997).

Table 7.1 shows a sample estimation form that the members of the Delphi panel can use to record their estimations. In this example, an estimate is given for each of the six tasks in the project and totaled at the bottom. Once the first iteration is complete, the panel meets to discuss the results; then, estimators review their estimations in the Delta column. The Deltas are totaled, and changes to the total estimation are made at the bottom. This process will continue until a consensus is reached.

TABLE 7.1 Delphi Technique Estimation Form
Project Name: Gary's Super Mansion

Date: 6/30/2026 Est. Units: Days

WBS	Task Name	Est. 0	Δ1	Δ2	Δ3	Key Notes
1	Foundation	10	+3	+2		Question soil composition
2	Walls	15	+3	+2		Drawings not clear
3	Plumbing	5	+1	+1		
4	Electrical	5	+1	+1		
5	Landscape	9	+6	+5		
6	Ext. paving	12	+4	+4		The scope is not defined well.
	Est. Delta		**+18**	**+16**		
	Total	**56**	**74**	**72**		

Normally, the Delphi technique is used for highly complex estimates with little historical background; however, the concept works in situations where a nonbiased estimate is needed. The example in Table 7.1 shows that the original time estimate was 56 days. After two iterations, the Delphi experts refined that, as shown by D1 and D2. The first round was summarized as having the "expert" estimates at +18 above the original estimate. The second round shared the logic of D1 estimates, and the total variance was +16. From this, a consensus estimate would be made as 72 days.

Because this technique seeks estimates from multiple participants, it tends to remove bias and politics that can occur when an estimate is based on only one expert's judgment. Group meetings also allow experts to discuss issues or assumptions impacting the estimate. One of the main drawbacks of this type of technique is the significant amount of time it can take for the panel of experts to reach a consensus. Larger numbers of experts will increase the number of iterations required and, therefore, the time it takes to reach a final estimate. Another potential issue is the panel's experience level. If the panel is made up of individuals who are not very experienced, the estimate will likely not be so accurate. It is also important to ensure that strong facilitators are available to guide the group and keep them focused on the topic. The central idea of Delphi is to anonymously share the estimates made by others along with their basic rationale. The next iteration would allow the group to think about other views and possibly adjust their estimates. This is essentially a consensus-building process. A contemporary example of a modern Delphi process can be found in the *Digital Office Complex* by Jackson and Richardson (2024).

7.4 TIPS FOR ACCURATE ESTIMATING

The first tip involves selecting the proper method to produce an estimate. The models outlined above have highlighted both pros and cons for each. That suggests that there is no single approach to developing an estimate for a work unit or a project. Deterministic forecast estimates should be avoided since they are essentially wrong by definition. Range values or probability-type distributions better reflect the complexity of estimating. When single-value estimates are given, some indication of accuracy should also be provided.

Organizations have different methods, cultures, and tolerance levels when dealing with data accuracy. Two common terms can help keep a proper focus. These are rough order of magnitude (ROM) and Budget ROM estimates. The ROM estimate reflects a low confidence level with a variance of up to +/–100%, while a budget level variance is more in the range of +/–10%. Beyond these two views, the real question when provided an estimate is its level of accuracy. At different points in the project life cycle, an estimate provided might be at the ROM level, but later, it could be more like a budget level. Sometimes, the early ROM estimate gets recorded as a project budget value, and the PM's life becomes more stressful when trying to stay within this boundary. When providing estimates to external groups, it is important to point out that the data has some variance and attempt to represent that range.

The accuracy and use of estimates need to be disciplined in terms of how they are used. One can understand why budgets and task times are padded when the local culture penalizes the project for overruns. An often-unanswered question is what variance is acceptable as noise level and what is considered a hard constraint. Projects are not noted for accurate results, so unplanned overruns are a fact of life to be dealt with.

7.5 ACTIVITY SEQUENCING MODELS

The modern history of project task sequencing can be tracked to the mid-1950s, with two independent project groups identifying a better method for developing a project plan than the historic Gantt chart that did little for this purpose. One source for this was the DoD Polaris missile project, characterized as a high-technology weapons system with difficult-to-estimate task times. This scheduling approach used an arrow diagram to show the task sequence. Each task was represented by a variable time using three estimates to define the range. In addition to the sequencing logic, a broader development methodology titled the Program Evaluation and Review Technique (PERT) was defined with a tool called WBS as part of it. This is one of the most important events in the classic history of project management. More of this background can be recalled from Chapter 1, and Wikipedia offers a good technical background reference list.

At approximately the same time the PERT model was evolving, a similar view of task scheduling occurred in a commercial project environment. This group was interested in the project's cost and schedule. Their task sequencing network was similar, but the tasks were deterministic, and boxes were used to define the tasks with arrows to connect.

The idea of viewing a set of project tasks as a network was novel in that period, and the logic for doing this caught on quickly as a scheduling model. Unfortunately, technology gaps upset the process that followed. Access to computers in the 1960s was only available to large organizations, and software to process the networks was crude. The computation mechanics to produce a network were laborious for networks with more than 50 tasks. Also, the calculation mechanics for the PERT three-time estimate model further complicated this problem. As a result, the three-time estimate idea was never popular, and the expansion of the idea lingered until the 1980s when small computing matured along with inexpensive software. Since this time, the use of network models has become the standard approach to project scheduling. There are occasional arguments about which network model is best, but the computer format has covered this argument.

So, that is a brief historical background to project scheduling. There are more details to be covered, but this is a good starting place for the next step. For the sake of history, we will describe the two network options and then leave that argument behind.

To make this history lesson coherent is impossible because the evolution of these two ideas is not very coherent. First, the Activity on Arrow (AOA) side was interested in high-technology projects where accurate estimating was difficult, while the Activity on Node (AON) side was more interested in the project's cost, and their task estimate errors were much lower.

If this scenario had occurred in the 1980s, the field of project management would be much different from what happened in the 1960s. At this point, the new approach of project planning required computers, given the increased work needed to process the wither of the networks. This means that initially, only large organizations could use either method. While this technology gap blocked access to processing tools, the two methods got intertwined, so most do not know the difference today. On top of this, looking at project plans, our networks did not seem to appeal to the masses, and the Gantt schedule bars remain fixed in the culture. Then, in the mid-1980s, inexpensive, small computing resources hit the market along with software to process the network. Today, anyone involved in project management understands the sequencing network idea, but by this time, the Gantt culture rejected it. To Microsoft's credit, they found a way to display a network resembling a Gantt bar chart. That is the current view of project scheduling. There is so much more that could have been said about this evolving, but the theme, for now, is to show how the historic network roots morphed into the modern view of scheduling. The examples described will cover the mechanics of this method and the new vocabulary that emerged from this model.

At about the same time, a consulting organization conceptualized a similar network. In this case, the network used boxes to show tasks and arrows to connect the boxes. This network was named AON. Purists will argue that one of these is better, but we avoid that argument. Both can produce the same results and use the same inputs.

7.5.1 Activity on Arrow

The AOA network design approach constructs the project plan as a set of nodes connected by arrows to show how the various activities are sequenced. This design method was introduced in the late 1950s as a component of the PERT development methodology, which is the historical root of the waterfall method.

This scheduling network was named an AOA. Arrows represent tasks, and circles (nodes) depict the start or stop of that task. A project schedule can be produced by working through the task paths defined. This view of project scheduling would be more theory than fact until computers became available to produce the calculation involved. Also, the three-time task estimates fell out of usage quickly, and that loss of practice is now considered a regression in project scheduling practice. Many consider this view of scheduling one of the most important tools in project management, along with the WBS that was also part of the PERT model. The inventors recognized the schedule deficiencies inherent in the Gantt chart, and this was supposed to kill the bar chart and make networks with variable time estimates the new planning tool. Once again, technology gaps allowed the original view to erode, keeping the Gantt chart alive and in

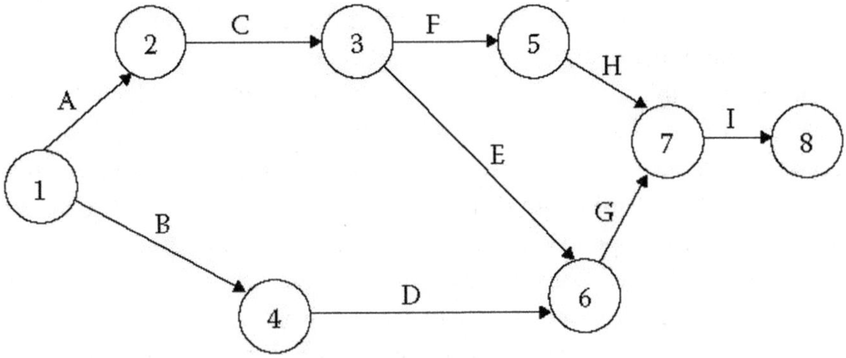

FIGURE 7.3 AOA network.

use today. Regardless, this method ushered in the modern network view of project scheduling.

A sample AOA network is shown in Figure 7.3. Depending on the scheduling role, the arrow can represent a high-level or very detailed work specification.

This diagram structure shows an orderly, systematic series of tasks or work units that must be completed in the defined order (as directed by the sequencing arrows) to reach a specific definable objective (node 8). The AOA network diagram has two primary elements: arrows and work units (event). The work units are sequenced by an arrow, and the node represents both the end of one task and the start of the next. In this view, it is important to understand that an activity (arrow) consumes time (Stires & Murphy, 1962). Without exception, every activity except for those that start and stop the project must have both a predecessor and successor event node. Every task has a predecessor and successor node. These paired relationships provide the specifications necessary to draw the network diagram. In many ways, this diagram takes little time to understand what it represents, which was one reason it caught on so quickly. The problem with the diagram starts when one wants to know the specific schedule represented by the task values.

7.5.2 Activity on Node Model

The use of the AON model for project scheduling is generally credited to DuPont in the mid-1950s, although other organizations are also mentioned in the literature as using similar techniques. Regardless of the source idea, a network-type view of task scheduling emerged while the PERT methodology evolved.

Figure 7.4 shows a sample AON network format with boxes defining task information and arrows to show sequence.

The only visible difference between the two networks is the layout of the tasks and the amount of data shown on the diagram. However, there is a declining concern over the diagram format as computer technology has taken over the role of computation and display. It is important to credit this source with popularizing the term Critical Path Method (CPM) for describing the longest path through the network (e.g., Critical Path). This is just one of the crossover cases between the two model views.

The evolution of these two network models since the 1960s has morphed them into one view. The term PERT/CPM is still recognized, but few know the origin of it, or what each acronym represents. However, the project industry has absorbed the concept, and related core variables for project scheduling remain. As described here, these two models represent founding ideas for scheduling. History has already begun to wash away the tidbits representing what each one brought to the solution. Computer technology has taken over the calculation mechanics and hidden the laborious mechanics away. However, this is a tools textbook, and a serious project professional should understand how the network solves the network schedule. The method illustrated here will be mechanical, and the software algorithms are much more sophisticated than the method shown. Nevertheless, the computed parameters will be as described here.

7.5.3 Sequence Codes

The decision for task sequencing is a combination of technical factors and a host of internal issues related to weather, resource availability, unplanned issues, and management discretion. Figure 7.5 illustrates a finish – start sequencing decision.

This type of relationship specifies that the predecessor activity has to be finished before the successor can start. This is the most common relationship and is the default in computer software. In addition to the default relationship, there are three other common codes:

Start – Start (SS)—Both activities start at the same time.

Finish – Finish (FF)—Both activities finish at the same time.

Start – Finish (SF)—A little-used relationship signifying the successor cannot be completed until the predecessor starts. This fits the Just-in-Time process model.

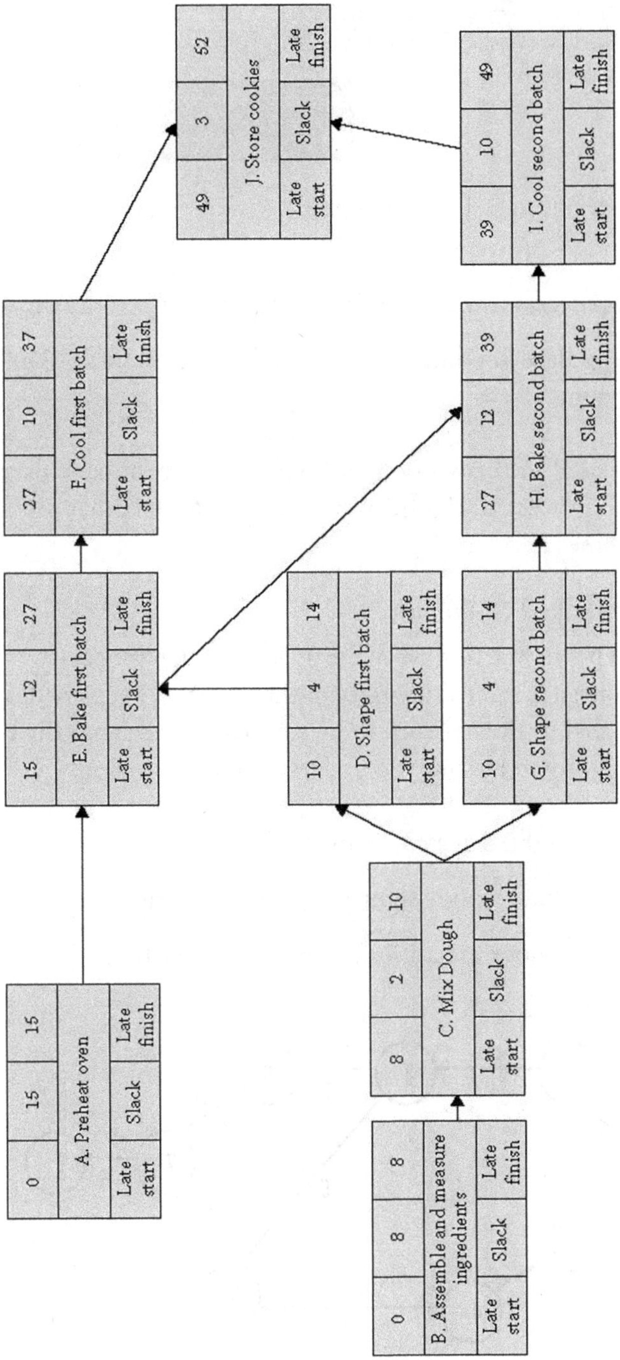

FIGURE 7.4 AON network.

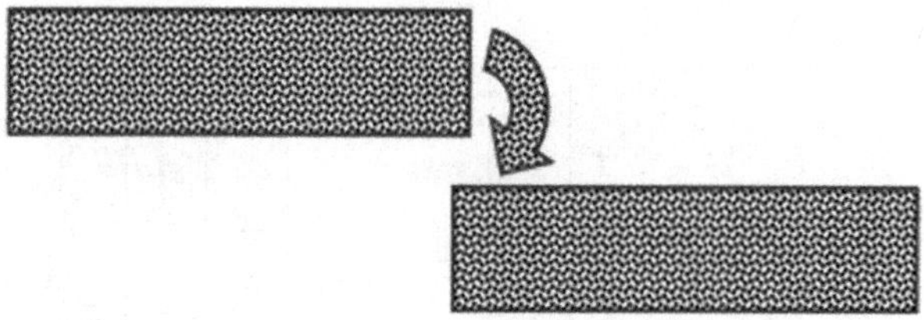

FIGURE 7.5 Finish – Start (FS) relationship.

Lag—The predecessor must wait by the lag amount before starting.

Lead—The predecessor and start before the successor is finished by the Lead amount.

The examples used in this section will only use FS relationships to keep the calculation process straightforward. The use of the other sequencing options will come later.

7.5.4 Sample Project Definition

Figure 7.6 will be the starting point for the calculation process.

The AOA format is arbitrarily used for the example. Let us review some mechanics related to the network to be sure you understand the coding. Review the diagram to be sure you understand the following:

1. Activities A and B can start at any time.
2. All activities have a node signifying a start and stop.
3. Activity C cannot start until activity A is complete.

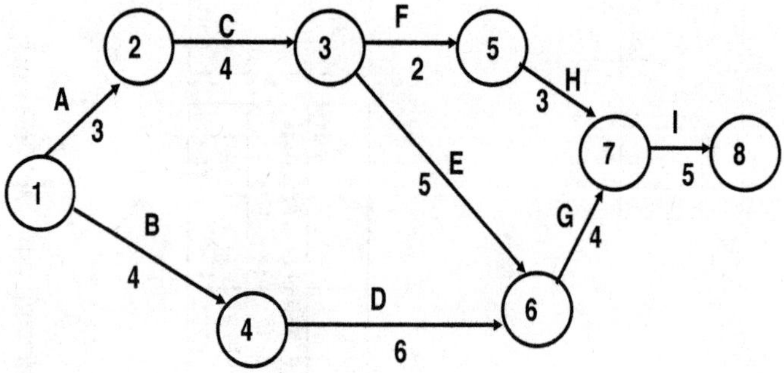

FIGURE 7.6 AOA network.

4. Numeric values adjacent to each task represent the estimated duration.

One common misunderstanding at this stage is the calculation of total project duration. Some would add the duration times in the table and define the project as 36 time units or the sum of all durations. However, because some activities occur in parallel, this is not a correct view, as we see in the calculation process. Producing this type of schedule insight was the event that made this technique recognized as a valuable tool.

The sections below will take this model through its steps to show how the result is produced, and then computer software will be used to duplicate the calculation.

7.6 FORWARD PASS CALCULATION

From the initial network setup, step two of the mechanical plan formulation is to perform a *forward pass*. This means we need to calculate the time through the network while obeying the predecessors. These early start times are recorded on the nodes in the format of "T(E)/." Figure 7.7 shows the calculated T(E) values for each node. The node "8" tag value calculation says that the project will take 21 time periods (workdays). A T(E) value signifies the quickest this point can be reached.

The example calculations are straightforward except for two nodes: 6 and 7. For node 6, two activity paths must be considered for the parallel activities D and E. Node 4 has a T(E) value of 4 and a following activity D time of 6; therefore, the resulting value for node 6 would be 10. However, the other parallel path coming into node 6 starts at node 3 with a value of 7 and an activity E time of 5, so the path value is 12. For forward path

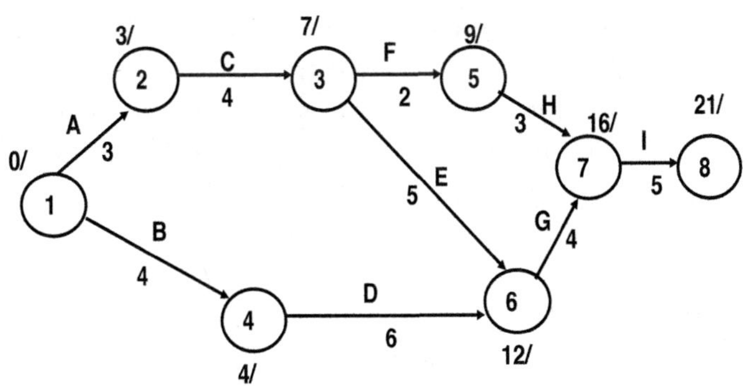

FIGURE 7.7 AOA Network with T(E) calculations.

calculations, the highest value is selected for such parallel paths. Thus, we assign the value 12 for T(E) at node 6. This means that the earliest we can claim completion of project activities at this point is 12. The same logic applies to node 7, where a value for T(E) of 16 is calculated. The key calculation mechanic for the forward pass is remembering to take the *highest* value for multiple input paths. T(E) values at each node represent the earliest time that these points in the project can be reached. This calculation also shows that the final node (8) can be reached at period 21. This defines the project duration as 21, given this set of parameters. We have a crude schedule resulting from the duration and sequencing parameters (see Figure 7.7). However, there is more that is needed beyond just the total duration. We will get that out of the backward pass. Review the node values to ensure you understand the forward pass calculation process.

For this basic network, the assumption is made that the planned time at termination is equal to the forward pass calculated time shown (e.g., 21 time units). Another way to say this is that the forward look specified that the project would take 21 time units, and we accept that value for the backward pass.

7.7 BACKWARD PASS CALCULATION

The backward pass inverts the previous step by working from the endpoint back to the beginning. The rationale for doing this is likely unclear yet, but let us first look at the unexplained calculation values in Figure 7.8 resulting from the backward pass.

It is important to understand the logic for doing the node calculation twice. Backward pass calculations generate a second node value that represents how long it will take to reach the end of the project. These values are defined as T(L) and formatted on the network nodes as "T(E)/T(L)." Each node in the network now shows two data values. For example, node "5" has a calculation score of 9/13. That translated to the interpretation, *The earliest time the project can reach this note is 9 time units, and the latest it can leave from that point and get to the end on time is 13 time units.* Another scheduling parameter just emerged. There are four days of idle or slack time at this note. More of this interpretation will follow the calculation mechanics.

To calculate T(L) values, we start at the completion node (8) and insert the same value for T(L) as is shown for T(E). That yields 21/21. From this value, we start working backward through the network at the next available node, node 7. The T(L) value would be 21 - 5 (for activity I), yielding

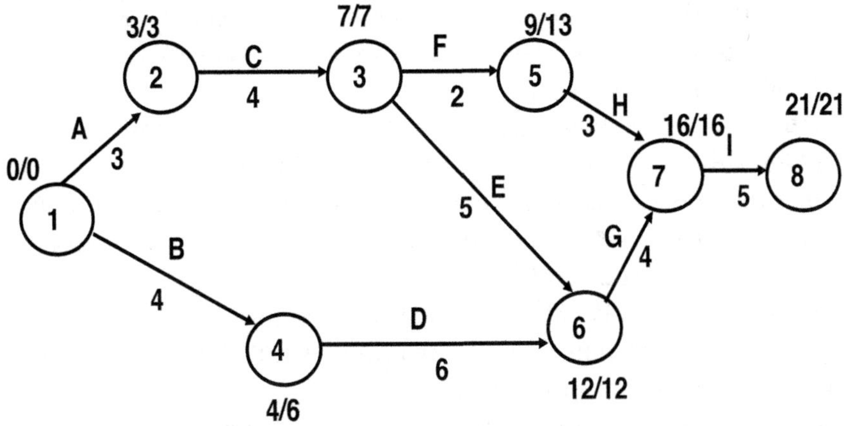

FIGURE 7.8 AOA network with backward pass calculations.

16 for node 7. Node 6 is next in the schedule chain, and its T(L) would be calculated as 16 – 4 or 12 (for Node 6). First-time users of this logic often have trouble with the backward arithmetic, which can be confusing, especially at branch points.

Node 3 is a diverging branch point for tasks E and F. For the F path, the calculated T(L) is 13 – 2 or 11. For the E path, the calculated T(L) is 12 – 5, or 7. There are two T(L) values for nodes (11 and 7). For the backward path, choose the smallest number. Remember, for the forward path, the most significant number was selected for a converging node—largest for forward and smallest for backward. Review the node values shown in Figure 7.8 to ensure you understand the basic idea of what T(E) and T(L) represent.

7.8 DEFINING CRITICAL PATH

Once the values for T(E) and T(L) are computed, it is possible to analyze two important timemanagement factors (i.e., the longest path through the network and slack time details). The simplest item to calculate is nodal slack time. This is simply the difference between T(L) and T(E). So, node 3 would have a slack value of zero (7 – 7), whereas node 5 has a slack value of 4 time units (13 – 9). Each value defines the amount of time this node can stay idle without affecting the project completion date. Also, note that the computed slack for the start and finish of the network will be zero using the assumption of 21/21—that is, we are happy with the original computed project duration. You might be thinking about what you would do if this was not the case. Changing the T(L) for nodes 8 – 25 would mean you are planning for at least four days of slack through the network. The calculation mechanics are the same.

Three other slack-type status parameters can be derived from this view. These are total float, free float, and late finish. Each relates to activity views rather than node calculations but can be derived from the node values. Total float relates to the amount of activity slack before project completion is impacted, while free float deals with the same view only for the successor activity. Late finish describes the latest time that the activity can be completed without impacting the schedule. All of these values will be available with the software results.

Calculation of the critical path requires a little more logic. Step one is to mark all of the nodes with zero slack. This means the critical path passes through nodes 1–2–3–6–7–7. If there is only one path between the critical notes, that task is on the critical path. Tasks A and C fit that definition, so they are marked. Task F fails, so it is not on the path. The same is true for task H. D is not on the path because there is slack at its endpoint. Tasks E, G, and I all fit the definition, so these are marked as critical. This means task path A–C–E–G–I contains the lists of tasks containing the schedule, thus the critical and longest paths. Adding the durations for each adds to 21, which validates the forward path calculation. Specifically, the critical path should be considered a vector containing both the activity list and the total duration, so it would be more proper to state the critical path as A–C–E–G–I and 21 time units. Knowing where the critical path resides in the project plan is considered a critical management issue for the project (Brown, 2002).

Figure 7.9 shows the network calculations with the critical path marked with bold arrows.

This example had a simple critical path. For more complex networks with parallel paths shown between the same two nodes, both would appear to be on the critical path according to the rule given, whereas only one is. Each path had to be tested to verify that it was critical.

Management importance of the critical path lies in the fact that any delay in these activities will delay project completion. Thus, it is important to focus management attention on this set of activities. Other activities with Slack generally require less rigorous monitoring because a minor slippage in these will not affect the completion date as long as Slack remains. For example, node calculations for the start of activity H indicate that it could start as early as period 9 or as late as period 13 without

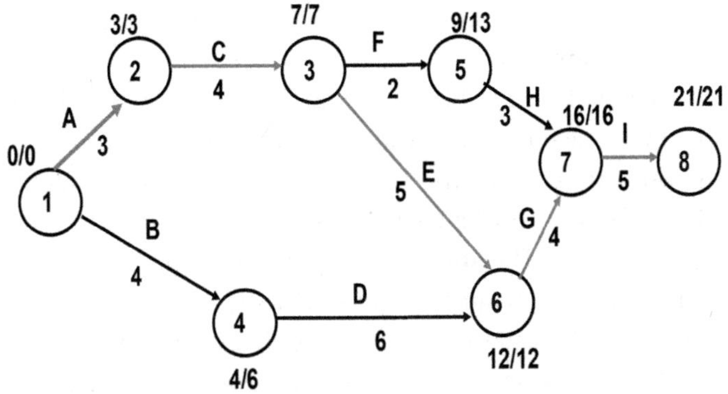

FIGURE 7.9 AOA network with critical path defined.

affecting the completion schedule. Similarly, activity F could start as early as period 7 or wait as late as period 11 with no adverse impact. Task H has four units of slack and could wait until period 13 (it is a late start).

Understanding and interpreting Slack values is vital to time management. Also, recognize that errors in duration estimates can invalidate the status interpretation of the plan. Duration values should not be considered deterministic, which should be kept in mind as the plan is revised. Understanding activity slack or float gives the PM flexibility in scheduling the start of a particular activity within the early start and late start time ranges (Uher, 2003). This also allows PMs to have flexibility in allocating resources across the project plan. For example, if two activities occur within the same time range and compete for the same resource, one is critical, and one is noncritical. Knowing this, the PM can allocate the resources to the critical activity first and then use the float time to delay the start of the noncritical activity.

7.9 AUTOMATED PLAN CALCULATION TOOLS

One of the project team's most important activities is generating the base task work schedule. The following three planning parameters are required to produce the first-cut view:

1. *Work activities* in the project (WPs and planning packages)
2. *Duration* estimates for all activities
3. *Sequence relationships* (predecessors)

Table 7.2 defines a sample project activity list using the network activities and sequence relationships.

We first see a flattened WBS. Note that the same list of tasks and durations is shown in the table view. The sequencing logic should also be readable. Compare this data to the network to ensure you understand the tabular view. The premise is that you will never have to return to the graphical view again. The software is going to do the same calculation described above. As previously mentioned, the maturation of computer software for this task has removed the need for manual or graphical views of this model. The logic is the same, but it is now data-driven.

Feeding the data table shown in Table 7.2 into MSP generates the result in Figure 7.10, which duplicates the network schedule parameters.

Note that the software accurately converts the graphical AOA model into an equivalent Gantt bar format; however, observe that the Gantt bars

TABLE 7.2 Sample Activity List

Activity	Duration	Predecessor
A	3	—
B	4	—
C	4	A
D	6	B
E	5	C
F	2	C
G	4	D.E
H	3	F

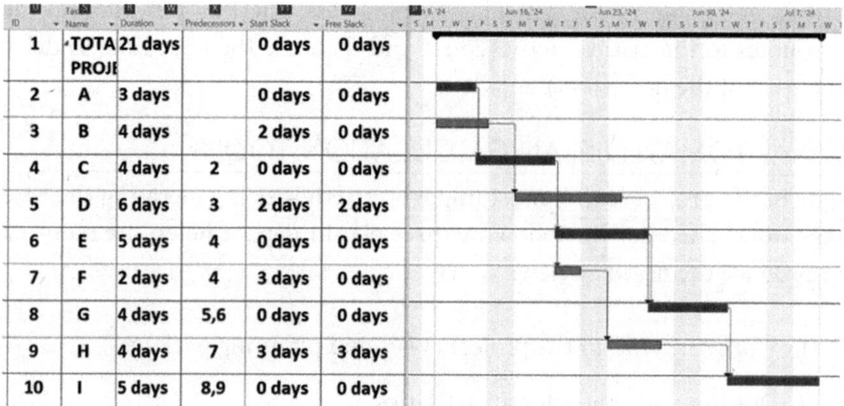

FIGURE 7.10 Automated schedule calculation.

have connectors, unlike a standard version. Recall that the industry was reluctant to give up the Gantt bar view, so when the software became available, the quandary was how to make the computer version acceptable to the audience. Microsoft deserves credit for its creativity in finding a suitable solution. The Gantt bars in the computer output are networks with bars for tasks and connector lines that duplicate the network. So, the graphical view is merged into one. There is still a challenge in showing the tasks as fixed duration and fixed calendar date, so this is a concept battle to fight in the future.

Note that the software can also produce various supporting data about the task. Starting Slack and Free Slack values are included in the output, but other data is also available. The key scheduling lesson learned here is that software can save much time in project planning. In addition, changes are easy to make and evaluate, which would not be the case if this were still a manual process.

Even though hundreds of software titles purport to process this form of project scheduling, the two most popular commercial software utilities are MSP and Oracle's Primavera. Microsoft dominates small- to medium-sized projects, while Primavera is prevalent in larger projects.

7.10 MANIPULATING THE SCHEDULE

If all project work unit activity estimates were perfect, all required resource skills would be readily available, and all defined materials would be available as needed. In such a pollyannish world, a project schedule like the sample shown here would fit reality as tasks would remain static along with their duration estimates. That is the view represented by both the AOA and AON networks, and it is an important perspective from a theoretical standpoint; however, it is also important to recognize that all such static and risk-free simplifying assumptions related to projects are suspect. Real-world projects are very complex regarding both the initial planning estimates and the ongoing predictability of the outcome. Thus, the job of project time management becomes one of dynamically tweaking and guiding the variables involved in an attempt to match the target goals.

The examples presented in the textbook up to this point have been defined as the first-cut *view*. This is intended to mean there will be other issues to consider in refining this view into a final picture. For example, the plan may not yet be approved or may not have been evaluated for available resources. It simply means it has gone through a step in the process leading to a final version.

Project plans need to be viewed as dynamic artifacts designed to reflect the current view of the situation. As the level of work definition evolves, it is common to find that the original estimates contained sufficient errors that necessitate major rework—both in task duration and sequencing. In addition, the first-cut view may be rejected by management or key users and recycling the process may be necessary.

Another view of plan manipulation occurs when a plan exceeds the desired schedule, and the goal is to shorten it without changing the scope of work to be included. The question then becomes how to manipulate the plan variables to decrease the calendar cycle time. All of the issues outlined here make the planning process viewed as iterative until a final version is approved by management. Techniques to manipulate plan variables involve tradeoffs and strategy changes between scope, time, cost, and resources. If the current view of the plan exceeds the desired cost or time, it may be necessary to cut the scope. Likewise, if the scope expands with newly approved work, the schedule will likely expand, or the cost might otherwise be affected. Dynamics related to these areas trigger the variable manipulation actions.

Two classic plan manipulation vocabulary terms must be understood and used appropriately. These are as follows:

Crashing—allocating extra resources to the critical path to shorten the schedule but likely increasing the budget.

Fast-tracking—rearranging the predecessor relationships to shorten the schedule.

To decide which options to pursue, evaluating the value of recovering time versus the cost is necessary. Also, consideration must be given to risk and other factors involved in the chosen approach.

7.11 BIKE WBS

Previous chapters have illustrated how the WBS provided a detailed work definition for the project and activity estimating produced time estimates for those work units. Once these preliminary decisions are made and a technical work sequence is defined for each work unit, the fundamental building blocks necessary to produce the first-cut plan are in place. Beyond this core view, it will be necessary later to add other supplemental non-direct work activities to this structure. However, this example offers the core steps related to project plan construction. The planning challenge is to produce a plan such as this that mirrors the structure of the WBS and represents the total work required to execute the defined scope.

As a starting point for this example, assume that Subject Matter Experts (SMEs) associated with the project have collectively worked to create the project WBS (see Figure 7.11). This represents the scope of the bike construction deliverables and related work units. This will be the driving specification for constructing the plan.

To translate the WBS graphic view into something more compatible with the computer input format, it is necessary to reformat the graphic view into a flattened WBS. This means reformatting the definition data into table rows and columns. This modified WBS view is shown in Table 7.3 [*7MSP0*].

Study this translation process until you understand the logic of the mechanics related to this translation. The data for the right-hand column will be explained below. This artifact becomes the data set driver for plan development. From a logical view, there is one not-so-obvious point regarding this data set. *Note that only WP-level tasks have duration estimates.* Higher-level summary groupings will inherit data from the lower level. This subtlety means that the internal computer network scheduling mechanics should use the work unit data (bottom of the WBS hierarchy) to calculate summary-level details. Also, recognize that schedule times for a summary task may not be the arithmetic sum of the lower-level

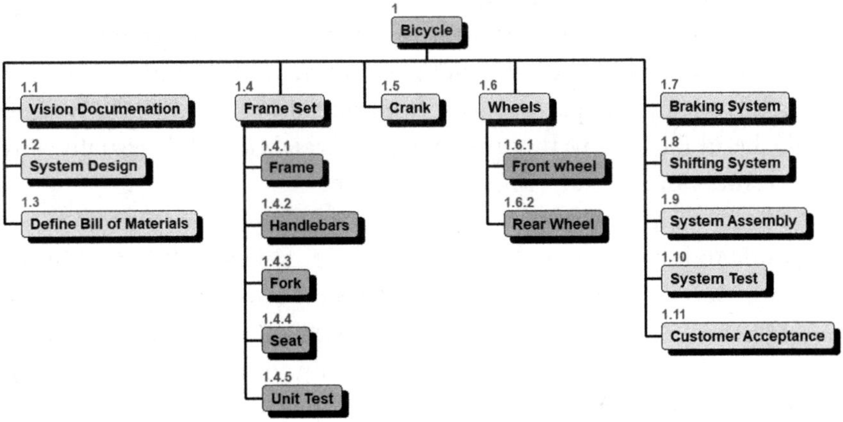

FIGURE 7.11 Bike project WBS.

produced by WBS schedule Pro.

TABLE 7.3 Flattened Bike WBS

1	**1**	**Bicycle**		
2	1.1	Vision Documentation	10 days	
3	1.2	System design	10 days	2
4	1.3	Define bill of materials	3 days	3
5	**1.4**	**Frame Set**		
6	1.4.1	Frame	5 days	4
7	1.4.2	Handlebars	5 days	6
8	1.4.3	Fork	5 days	7
9	1.4.4	Seat	5 days	8
10	1.4.5	Unit test	5 days	9
11	1.4.6	Project Mgt Review	0 days	10
12	1.5	Crank	10 days	11
13	**1.6**	**Wheels**		
14	1.6.1	Front wheel	5 days	12
15	1.6.2	Rear wheel	5 days	14
16	1.7	Braking system	10 days	15
17	1.8	Shifting system	10 days	16
18	1.9	System assembly	10 days	17
19	1.1	System test	10 days	18
20	1.11	Customer acceptance	0 days	19

components. For example, if two lower-level tasks were to be performed simultaneously, the summary duration would not be the sum of these durations. This somewhat complex task-related arithmetic process is one of the valuable aspects of using software for the process. One only has to do the schedule logic manually once to understand this statement.

The movement of the data from the flattened WBS into the software is essentially a copy-and-paste process from spreadsheet to MSP. However, it would be unfair to say this migration is straightforward. Usually, any setup errors can be fixed within the software.

To understand how this process evolves, it will be necessary first to see the result and then work on more of the underlying details buried in the software. Several layers of complexity will be demonstrated as we move through the chapters. Adding the support elements is required to represent the additional work required. Conceptually, the process shown here is simple. Tasks can be represented by boxes of time, cost, and duration. The boxes are sequenced according to predecessor codes (work order). From this, a schedule can be calculated for the network on tasks *[7T6]*.

7.11.1 Bike Project Mechanics

The first step in the translation process is to migrate the flattened WBS data into an MSP Table view with the same column titles. The *Project N Motion* Basic Keystrokes video will walk you through the setup steps described here *[6MSP*1].

Column headings can be added to the default table view by right-clicking the desired column location and selecting insert "column name." Hiding an existing column can be accomplished by right-clicking and selecting "hide." The following four setup data items need to be moved from the table to MSP *[7T4]*:

WBS

Task

Duration

Predecessor

Review Table 7.2 to see what the MSP table view should look like before moving data from the WBS.

It is time to explain how the tasks will be linked in the model to produce a schedule. Note in Table 7.3 that the first column contains row numbers, and the right-hand column contains numbers. Sequencing the scheduling task is simply a matter of coding the sequence. For example, row 6, Frame, shows a linkage of row 4, which is the Define Bill of Materials. This means that in the scheduling process, the row 6 task will be scheduled after row 4. This is called a predecessor relationship. Other leak linkage options are available, but for this example, we will stick with the simple FS, which just requires a simple numeric code, as shown. Specifications of this type are assumed by default. If the predecessor code for line 6 is 4FS, it is assumed that you can leave off the FS. More predecessor complexity will be shown in later examples. Regardless of the coding, the predecessor codes aim to specify how the task work sequence will be performed. The software translates these linkage codes into a time sequence to produce the project schedule.

As indicated, the data migration process is a cut-and-paste exercise. However, experience shows that using the Paste Special/data option is better because this eliminates some internal MSP logic editing that is not needed at this point. If the loading process fails, it often suggests a logic error in the WBS or something less obvious. The Paste Special option is usually good enough to force a starting position. As the data is moved into MSP, the result is shown in Figure 7.12.

This view uncovers two MSP model idiosyncrasies to correct (one hates to call this a bug in popular software, but we are tempted to call it this). First, the WBS structuring indicated by the WBS codes is not obeyed by MSP automatically nor is the desire to automatically schedule the resulting Gantt bars. There are three manual steps required to correct three related items. First, go to the Task menu and identify a small green right arrow within MSP. This icon is used to indent tasks manually and tells the system which tasks (lines) are subordinate to specific lines. In this case, there are two subordinate levels to define. Also, the system default is set to schedule tasks manually. This needs to be changed to automatically schedule based on predecessor values. The following three mechanical steps are required to resolve the hierarchy and scheduling issues:

1. The first subordination level indicates that all tasks below line 1 are subordinate to the top line (i.e., total project). Use the cursor to paint Tasks 2 through 20 and click the right arrow to indent this group subordinate to the top line.

2. A second-level indentation is needed to show two summary groups, 1.4 and 1.6. Indent the six tasks under WBS 1.4 and two tasks under WBS 1.6 using the same logic as above.

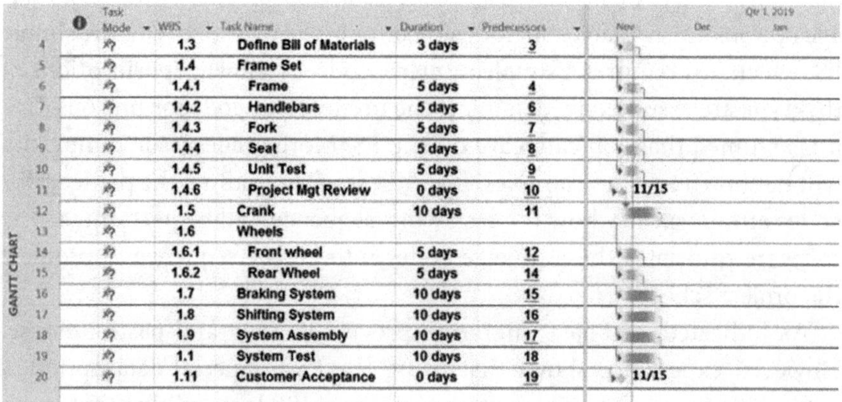

	Task Mode	WBS	Task Name	Duration	Predecessors
4		1.3	Define Bill of Materials	3 days	3
5		1.4	Frame Set		
6		1.4.1	Frame	5 days	4
7		1.4.2	Handlebars	5 days	6
8		1.4.3	Fork	5 days	7
9		1.4.4	Seat	5 days	8
10		1.4.5	Unit Test	5 days	9
11		1.4.6	Project Mgt Review	0 days	10
12		1.5	Crank	10 days	11
13		1.6	Wheels		
14		1.6.1	Front wheel	5 days	12
15		1.6.2	Rear Wheel	5 days	14
16		1.7	Braking System	10 days	15
17		1.8	Shifting System	10 days	16
18		1.9	System Assembly	10 days	17
19		1.1	System Test	10 days	18
20		1.11	Customer Acceptance	0 days	19

FIGURE 7.12 Initial bike project load.

3. To signal the system that it should obey the predecessor links and automatically schedule the task, paint the entire task group with the cursor and then go to the Task menu section and select "Auto Schedule." All of the bars in the Gantt section should now show the linkages defined. The Gantt bars shown in Figure 7.12 reveal that manual scheduling is set. You can see the changed view from auto-scheduling in Figure 7.13.

Notice the Task Mode column and the shape of the icon there. This is the auto Schedule icon. Compare this to the scheduling column icons in Figure 7.12. Use the File/Options section and change the schedule default to Auto Schedule.

The steps described have now created a first-cut raw plan; admittedly, it is not very complete or formatted nicely yet. To improve the initial view, four frequently used keystrokes can be exercised. These are as follows:

1. *Zoom.* To move the Gantt bars into viewable space, place the cursor on the Gantt bar timeline and right-click. Select Zoom/Entire project. This will format the bars into the available display space.

2. *Critical Path.* The default option does not show the critical path. To change this, go to the Format menu and check the box titled Critical Tasks. This will color code all of the project's tasks (i.e., the software will identify the critical path tasks in red, noncritical blue (slack time), and black as a summary task).

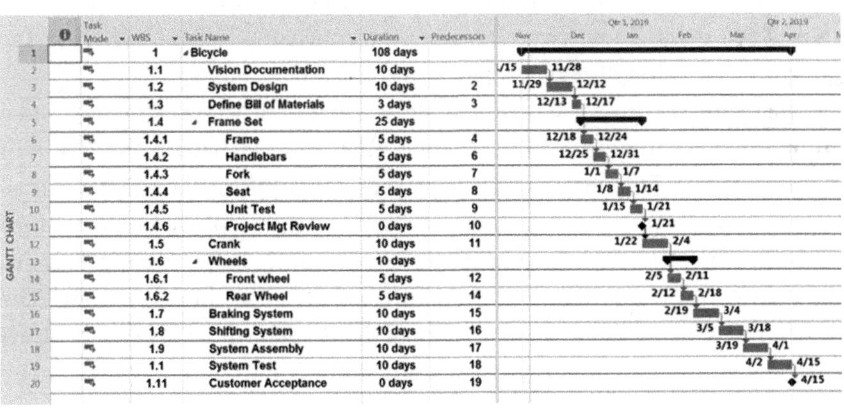

FIGURE 7.13 Formatted first-cut plan.

3. *Gantt Chart Wizard.* Adding this command to your menu will save much time-consuming formatting of Gantt bars. Use the help command to load the Gantt Chart Wizard.

4. *Recurring Task (not shown in text example).* A recurring project management task (number 21) was not transferred from the WBS. This is a special level of effort task type that regularly recurs, rather than the typical one-cycle task. This will be ignored for now as it adds more complexity to the setup mechanics, but you need to understand that these exist in a real plan. Instructions to create this can be found in the keystroke sheet, or you can add it using the instructions Task/Task/Recurring task.

The results from the initial setup and keystrokes described are shown in Figure 7.13. As these introductory examples have demonstrated, learning how to manipulate MSP to produce this form of output is an important core literacy step in handling many other examples in later chapters. MSP is not toy software; it can be challenging to emulate some project requirements.

At this point, a start date for the project has not been set. That value is defined in the Project/Project Information menu options. The sample plan has a calculated project duration of 108 workdays starting on 11/15 and ending on 4/15, with the final Customer Acceptance milestone task specified by a black diamond icon (and zero duration). The cycle time between the start and completion dates is called *elapsed time* and differs from task duration, which estimates actual working days. Here, the project elapsed time would be five months or close to 150 calendar days, while the duration is 108 days. These two terms can be confusing when answering the question "How long is this project going to take?"

The resulting project schedule is shown in a table (data elements) and a companion Gantt graphical bar format. Also, realize that an abundance of status data is calculated but not displayed. Any of these data items can be opened and displayed in the table structure by right-clicking on a column and selecting the chosen data name. Some samples of missing column data in this example are start date, finish date, cost, total slack, etc. The challenge is to find the available space to display desired items, which requires hiding and inserting to customize the view.

Later, as resource types or specific worker names are allocated to tasks, the system will calculate the cost of the task and the total project based on those allocations. This is yet another value gained through the automation process. The software will also show tasks that cannot be executed because of inadequate resources. This is called Capacity Management or Resource Leveling and is the sole topic of Chapter 15. These are future topics, but this description provides insights into some of the software's additional capabilities.

Real-world WBSs would undoubtedly be much larger than this sample program. More boxes, more layers, and the predecessor relationships would be more complex, but the plan creation mechanics would be identical to this example. Think of this process with 1,000 boxes in the WBS. Automation is the only way to deal with such ventures, and even small projects can exhibit calculation complexities.

Before feeling overly confident and too good about this nice-looking plan, remember that Murphy's Law dictates that things will go wrong at the most inopportune time. This means that an initial plan will undergo significant changes over time. In most cases, the first-cut view will not be what the final plan looks like. Nevertheless, the goal is to keep the operational plan reflecting the current view of the required work. Much more needs to be said about this process, but this represents a good jumping-off point into other topics. The reader must not get lost in the details of the lower-level discussions and forget that the overall goal of project management is to *influence the desired outcome as defined by the approved plan, which minimally involves specified deliverables, schedule, and cost. All of this should align with organizational objectives as approved by management.*

7.12 WBS SCHEDULE PRO

Several utility add-ins are helpful in various MSP structures, formats, and status processes. WBS Schedule Pro is an excellent utility for creating the coded WBS view as an add-in to MSP. Common user errors occur in the task indentions, and this utility can help graphically see where the error occurs. It was stated previously that MSP does not automatically obey the input WBS codes. They are just reference values with no related hierarchy assumed. Manual task indenting is required to match the WBS. This is one of the most error-prone areas of the plan setup, and WBS Schedule Pro graphically shows what is coded. This utility generated all of the WBS figures shown in the text. A 30-day free evaluation copy is available from the vendor at www.criticaltools.com. Once loaded, this tool is embedded into MSP and reads the internal plan code to produce the equivalent graphical WBS structure. Figure 7.11 shows a sample version of this format. It is recommended that the initial WBS load be tested with this utility before moving forward. This will help avoid schedule calculation errors, which can be difficult to fix later.

7.13 CONCLUSION

This chapter describes how time-related aspects of the project evolve from scope definition and are then translated into a first-cut schedule for the project. There are many more management issues related to producing a final schedule, but for now, take this introductory deterministic model view as the conceptual process by which scope is translated into an activity list. Then, that view evolves into duration and sequence specifications. We call this version a first-cut plan, meaning that more massaging is needed before it is finished.

The network models illustrated here are imperfect for showing how the project will be executed. However, they represent the underlying scheduling model and best illustrate the method to introduce the fundamental aspects of time management. Embedded in the network model is quantifying the scope, work activities, duration estimates, sequencing, resource allocations, and budget. For this reason, network models need to be understood as an essential tool in the PM tool kit, even though modern computer software will handle the arithmetic calculations to generate this result.

The simple bike project example was chosen to demonstrate the mechanics of creating a first-cut project plan and to illustrate the fundamentals of MSP keystrokes. Also, becoming familiar with the output format of this software is an important step in understanding how it formats the project's graphical and data schedule. The mechanics described here are fundamental to the project plan setup. From a learning self-test view, spawning a new project WBS and going through the loading flattening and loading processes is recommended. This will help reinforce the common step. Also, the pathway from the visual scope to the initial schedule is represented by these basic steps. Recognize the complexity of going through this process manually and commit yourself to becoming proficient in the automated approach. Projects with over 30 or so tasks would be challenging to process manually. With an automated calculation utility such as MSP, it is easy to evaluate and repair setup errors with incremental effort. Similarly, changes in task duration, predecessor links, and work calendars do not fit well with a manual process *[7T0]*.

At this stage, we focused on fixed-time tasks and project estimates but recognized that this view does not accurately map to the dynamic nature of tasks during execution. Subsequent chapters will open up these limitations for further review. Let us proceed to the next chapter and see how project cost emerges from the schedule model.

KEY VOCABULARY TERMS

The following list of ten basic terms involved in the input and output parameters of a project plan is important to understand:

1. Activity/Name/title—a short description of the work [Input]
2. Activity duration—number of working periods (days) [Input]
3. Predecessor—defining the order of execution [Input]
4. ES or T(E)—early start; earliest time the activity can commence [Output]
5. LS or T(L)—late start; the latest time the activity can start and not affect the project schedule [Output]
6. Slack—measure of idle time in the node or activity [Output]
7. EF—earliest time that an activity can finish [Output]
8. LF—the latest time the activity can finish and not affect the project schedule [Output]
9. Predecessor—a defined task that must be completed before the current task can start
10. Total float—the difference between the *earliest* date that the activity can start and the latest date the activity can start before delaying the completion date (LS – ES) [Output]
11. Free float—the amount of time an activity can be delayed before the successor activity will be delayed (EF – ES) [Output]
12. Slack—a value that represents extra time to execute a task, often called float
13. Critical path—the longest task path through the network; zero slack

14. Summary tasks—a collection of tasks that represent a higher WBS level

15. Modified Gantt (connecting links)—MSP method of showing the calculated network

PROBLEM SECTION

1. Use the information provided below to build the network plan defined. Show the critical path and node slack. All activity duration times are five.

Task	Predecessor
A	—
B	—
C	—
D	*A*
E	*B, D*
F	*C, E*
G	*B, D*
H	*B, D*
I	*A*
J	*I*
K	*H*
L	*K*
M	*I*
N	*G, F*

Using this activity description above, draw a valid precedence network that satisfies the predecessors (use proper format). Using this specification, manually draw the network using AOA or AON format.

2. Use the network created from problem #1 and produce the schedule calculations.

   A. Compute the node values for all nodes on the network and mark the CP (show all calculations on the network diagram).

   B. How long can you wait at activity J before making the project late? ______________________

C. How much slack is in activity B?

D. How long could you wait to complete activity A without affecting the completion? _______

3. Use the sample network diagram and duration data below and calculate the CP and node schedule values.

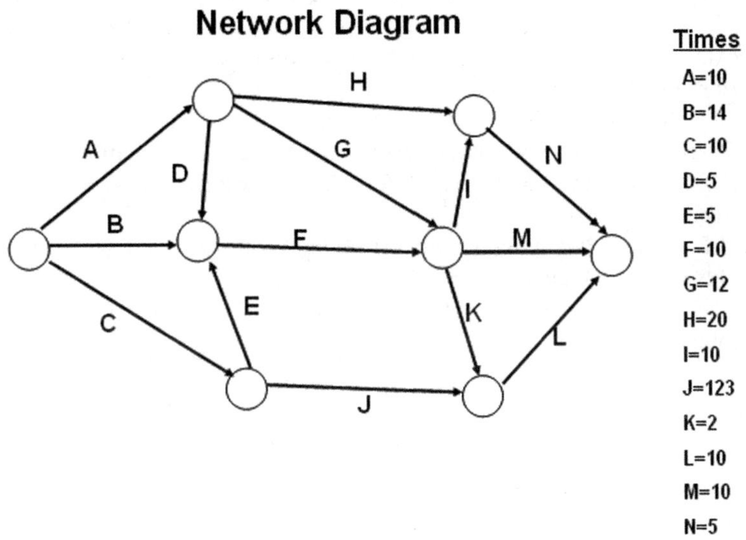

REFERENCES

Baca, C.M. (2007). *Project Management for Mere Mortals.* Boston: Addison-Wesley.

Brown, K.L. (2002). Program Evaluation and Review Technique and Critical Path Method—Background. Reference for Business: http://www.referenceforbusiness.com/management/Pr-Sa/Program-Evaluation-and-Review-Technique-and-Critical-Path-Method.html (accessed March 17, 2008).

Horine, G. (2005). *Absolute Beginner's Guide to Project Management.* Toronto, Ontario, Canada: Que Publishing.

Jackson, B.M. & G.L. Richardson. (2024). *Digital Office Complex.* CRC Press, Boca Raton, FL.

Johnson, T. (2007). *PMP Exam Success Series: Certification Exam Manual.* Carrollton: Crosswind Project Management, Inc.

Kwak, Y. & Watson R. (2005). Conceptual Estimating Tool for Technology-Driven Projects: Exploring Parametric Estimating Technique. http://home.gwu.edu/~kwak/Para_Est_Kwak_Watson.pdf (accessed April 10, 2008).

Mind Tools. (2007). Heuristic Methods: Using Rules of Thumb. http://www.mindtools.com/pages/article/newTMC_77.htm (accessed April 12, 2008).

Monsey, A. April 1988, Estimating-Fact or Fiction, *ASC Proceedings of the 24th Annual Conference*, California Polytechnic State University - San Luis Obispo, California, pp. 107–115.

PMI WBS. (2006). *Project Management Institute Practice Standard for Work Breakdown Structures.* 2nd edition. Newtown Square, PA: Project Management Institute.

PMI. (2021). *A Guide to the Project Management Body of Knowledge (PMBOK® Guide)*, 7th edition. Newtown Square, PA: Project Management Institute.

Snell, D. (1997). Expert Judgment. https://natlawreview.com/article/using-chatgpt-interpret-insurance-policies-eleventh-circuit-opens-door-ais-role#google_vignette (accessed April 10, 2008).

Stellma A.N. and Greene, J. (2006). Applied Software Project Management, O'Reilly. https://www.stellman-greene.com/ (accessed September 2, 2025).

Stires, D.M. & Murphy M.M. (1962). *PERT/CPM*. Boston: Materials Management Institute.

Uher, T.E. (2003). *Programming and Scheduling Techniques.* Sidney: University of New South Wales.

Verzuh, E. (2005). *The Fast Forward MBA in Project Management*. Hoboken, NJ: John Wiley.

CHAPTER 8

Developing the Project Budget

Project N Motion Videos for This Chapter		
8C1	Cost Management	Cost overview

This chapter weaves through the layers necessary to produce the project budget and check it against resource availability. The direct cost segment of this process dollarizes WBS specifications based on work estimates and labor rates [8C1]. However, beyond this step, there are multiple other process layers required to reach an approved budget. The following four steps are needed to produce an approved cost and human resource budget:

1. Direct cost reflected in the first cut plan
2. Incorporation of non-direct resource items
3. Evaluation of resource adequacy
4. Organizational approval process

DOI: 10.1201/9781003532989-10

8.1 INTRODUCTION

As in most project management areas, the process of budget construction is both simple and complex. Yes, it is possible to mechanically calculate dollar costs for the defined WBS work units using reasonably straightforward techniques, but that simple view will only define some costs related to the project. One of the complicating factors in this process is how the financial community looks at resources (people, money, and equipment). Internal team members view the budget similar to their household account with a checking account and bills to pay. This view would be called direct expenditures since these elements represent visible cost items under their control. The PM must understand the broader perspective defined by the financial side. From the budget structure viewpoint, this translates into various line items and terms that are not familiar. The one area that both the technical and financial sides agree on is the role that financial resources play in the execution process. That is, the project runs on resources and without these, it will die. Recognizing this fact means that the technical side must align with the financial group's view. Where the technical side of the project would like to view the budget as consisting of a pile of dollars that could be used as necessary to execute the project, the financial group focuses on categories of funds, sources, types, and other vocabulary terms that are hard to relate.

Fortunately, the first step in this process can ignore the complexities and just quantify the direct budget component based on the approved requirements definition parameters previously described in the scoping process and recorded in the WBS Dictionary.

8.2 ALLOCATING DIRECT RESOURCES

Previous discussions have taken a *peel-the-onion* view of a target topic (i.e., pull away one layer at a time and inspect it, then move to the next level for another view). That approach seems even more appropriate here. Additional financial components can be attached to that view once the work unit's direct labor cost is determined.

Direct costs are linked to specific project work units or larger planning packages. Each of these now contains some combination of human, physical, or monetary resources. So, the stated cost at this point would be the combined dollar amount calculated. The Estimating process was described in Chapter 5. We know that the goal of the work unit estimate is hours of work, along with details regarding the quantity and skill of the human resource. When using MS Project, it is easy to feel that this calculation produces the only cost for a work unit. When producing a full cost view of the project, offloading the direct labor cost calculation to a separate file for further manipulation may be easier. The essential data elements to be left in the Microsoft Project (MSP) are the resource allocations to work units. This layer produces cost estimates derived from named resources required for each work unit. The following four-step process describes the process:

1. Each skill group defined will have a corresponding charging rate factor.
2. The related direct labor cost for the work unit is computed by multiplying the charge rate by the number of estimated hours.
3. Add all the individual resource costs defined to produce a total direct labor cost for each WBS work unit.
4. Add other cost elements, such as material and travel charges, to produce a total direct cost for each work unit.

This collection of computed cost items is called the work unit *direct cost.*

Concern for computational accuracy is universal in project management, and this area is no exception. One has to view work units as floating through time with unknown work resources. Mechanically, the labor rate assumed can be wrong, and the duration of the work unit can vary. Both variables make the direct cost calculation also variable. Many PMs need help with handling these difficult-to-quantify areas, and there is no simple solution to this reality. The planning question here is the degree of variability. If the estimate is good, the assumed labor rate is reasonably close, and the execution productivity is as planned, the result can be very good. Project plans are targets. Deviations from the target energize management to move the result back closer to the desired goal. There are computational accuracy issues related to this process that must be addressed later.

8.2.1 Direct Cost Calculation

The mechanics involved in computing the human resource cost component outlined above will be further illustrated using the following work unit example:

1. Two carpenters are assigned; their work schedule is a standard 5-day X 40 hours/week calendar.

2. The average resource rate is $30/hr.

3. The estimated work unit duration is two days (16 hours), assuming a standard work calendar.

So, the total resource requirement is two carpenters working for two days at eight hours per day, which constitutes 32 total hours of work at $30 per hour. The calculated direct labor resource cost for this work unit would be $960 (32 × $30). The hanging question is whether this rate represents the actual base salary for the human resource or a fully burdened organizational overhead cost view. For now, let's assume that this is only the direct pay rate.

As one can see, this arithmetic work unit calculation process is simple, but the accuracy of the results should always be questioned. One potential error source is related to the rate value used for the allocated resource. During the earlier estimating process, some assumptions were made as to the skill and experience level of the individual. Realize that the actual name of the person and their specific rate is unknown at the original time of estimation; therefore, by necessity, an average "generic" rate is commonly used. At some future point during the execution of this plan, the actual rate variance will be an item of interest. Even though we recognize this process has potential accuracy errors, it is a necessary part of the planning process. As a project unfolds and actual resource names become known, it is possible to replace the generic resource name with a specific name and rate and then recalculate the initial plan value to achieve a more accurate estimate. However, even though this may improve arithmetic accuracy, that extra step is not typical because of the administrative overhead. This process would be confusing as named individuals could be added frequently through the life cycle. For this reason, the examples shown here use generic names and rates, but the process described would be the same in either scenario.

8.3 MSP RESOURCE COSTING PROCESS

Cost values have been previously shown in the project plan, but the source of these numbers has yet to be explained before this chapter. Indeed, you can enter a task's total cost value with no supporting background, and MS Project will properly compile the total budget reflected by those values; however, the problem with this is that there is no tracking as to the estimating logic behind the number. So, if a future goal is to evaluate the work involved, that can only be done by allocating estimated resources to the work unit. For this reason, this area of the costing model must be understood. Failure to go through this resource allocation step leaves the plan in a limbo state regarding the resource status for these tasks. That omission will leave the PM somewhat in the dark regarding his resource status through the life cycle.

8.3.1 Resource Sheet

Figure 8.1 shows the direct resources available for allocation to project tasks.

The project resource pool is called the *Resource Sheet* in MS Project. This topic is also discussed in Chapter 15. A brief explanation for each data element in the example sheet follows:

- *Resource Name* is an identifier for the resource.
- *Type* represents one of the three direct resource types—Work (human), material, and Cost (dollar expense).
- *Material Label* defines the unit of measure for the resource—square feet for sheet metal, per day for equipment, etc.
- *Initials* represent an abbreviated name for the resource.
- *Group* can be used in multiple ways—team name, organization, shift, etc.
- *Max Units* is the assumed maximum quantity of the resources—200% represents two resources and 50% represents one resource for half the time.

	Resource Name	Type	Material Label	Initials	Group	Max Units	Std Rate	Ovt Rate	Cost/Use	Accrue At	Base Calendar	Code
1	Electrician	Work		ELEC		200%	$40.00/hr	$0.00/hr	$0.00	Prorated	Standard	Generic
2	Plumber	Work		PLUMBER		200%	$40.00/hr	$0.00/hr	$0.00	Prorated	Standard	Generic
3	Concrete laborer	Work		CONLB		200%	$35.00/hr	$0.00/hr	$0.00	Prorated	Standard	Generic
4	General laborer	Work		GENLB		200%	$25.00/hr	$0.00/hr	$0.00	Prorated	Standard	Generic
5	Rebar labor	Work		REBAR		200%	$40.00/hr	$0.00/hr	$0.00	Prorated	Standard	Generic
6	Supervisor	Work		SUPER		100%	$50.00/hr	$0.00/hr	$0.00	Prorated	Standard	Generic
7	Front loader	Material	.d	FL			$1,000.00		$500.00	Start		MATL
8	Backhoe	Material	.d	BH			$1,000.00		$500.00	Start		MATL
9	Concrete blower	Material	.d	CB			$1,000.00		$500.00	Start		MATL
10	Other Expense	Cost		O						Prorated		Expense

FIGURE 8.1 Project resource sheet.

- *Std* and *Ovt Rates* are the charge rates for the resources in either regular time or overtime. This can be in hourly, daily, or annual format.
- *Cost/Use* represents a one-time setup cost for the resource.
- *Accrue At* signifies when the cost is to be recognized—start, end, or prorated through the work time. In the example, the setup time is specified at the beginning of the work unit. All other resources are prorated.
- *Base Calendar* indicates the work schedule for the resource. The standard work schedule is eight hours daily, five days a week. The organization or project will define holidays or other non-working days in their calendar.
- *Code* can be used to sort the resources. In the example, the resources are shown as generic to signify the type being allocated. This does not affect the calculation process.

The total available supply source of Labor skills is defined in the resource pool by the defined Max percent for each named resource. From this data, the planning process estimates will allocate a resource type and quantity to specific work units, as illustrated above for the bike example.

8.3.2 Assigning Resources to a Work Unit

Resource allocations for each activity are shown in the table view by opening a *Resource Names* column. To illustrate the allocation process, right-click on a WBS line item and select *Assign Resources* from the popup box. Figure 8.2 shows the resulting allocations for that activity. The Cost data field should reflect the total allocated resource value of $56,400, the sum of the three-line items computed from these allocations.

Let's examine the allocation of the Mechanical Engineer resource. Note that 300% was allocated, and the associated cost was calculated to be $28,800. The 300% value indicates that three engineers have been allocated to this activity. The results of this allocation would now appear in the Cost column of the table view, and it would obey the calculation rule described above (i.e., Rate times Hours). What is less evident from this view is the fact that the software knows how many hours this resource would have worked for that duration. If the planned duration was 10 days and three engineers were involved, the software would know from its

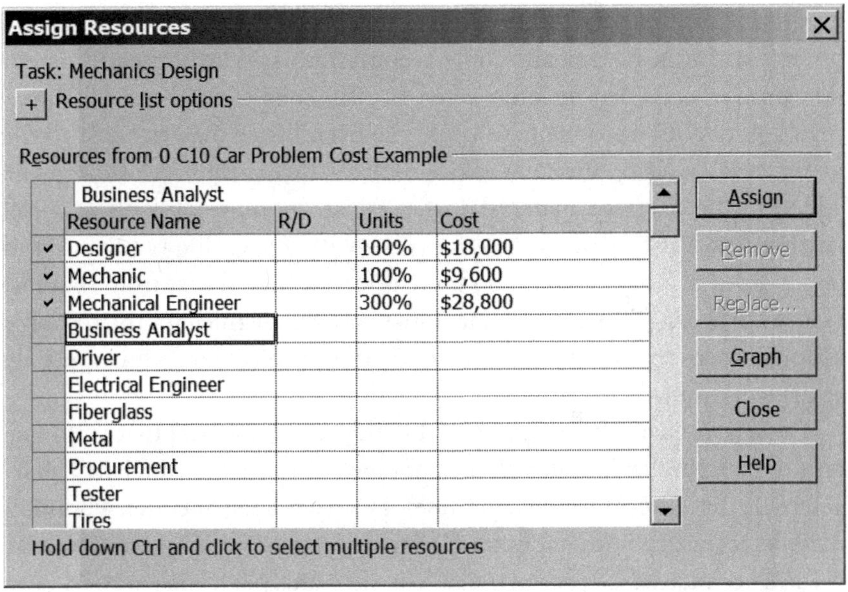

FIGURE 8.2 Resource allocation for a single WBS.

internal calendar that they would have worked for 80 hours each (8 hr/day times 10 days). One can see both the simple formula mechanics for this and the massive amount of related arithmetic performed with these calculations. This again illustrates the value of using a software approach to handle repetitive mechanics of this type.

8.3.3 Material Cost

Direct material costs are more complex to accurately calculate since this class of resource is often purchased in large batches and then piecemeal allocated out of inventory to the project. As a result, these costs are usually kept at the total project level and are not considered part of a direct work unit. Large inventory items related to a specific work unit can be linked for analysis purposes, and this suggests access to a separate inventory and/or analysis file external to MSP. At the work unit level, many items fall into a bulk purchase category (i.e., nails, common materials, etc.). These would be impossible to link to individual work units. As an example for mapping expensive inventory to a work unit, assume that stainless-steel sheets cost an average of $10/sq. Foot and the work unit require 100 square feet, then $1000 could be allocated for the work unit for "high cost" material expense. However, from an enterprise view, there may have been several orders for stainless steel, and some amount may still be housed in inventory or even scrapped in some cases. In situations of this type, the organization must decide how to allocate the overall raw material cost to the using entities. Also, line-item allocations for inventory overhead may be added to the prorated material cost. One of the more difficult material costing issues is related to material types with a shelf life requiring it to be scrapped at a defined date, so there is an additional cost from that lost value source as well. Many such factors make this segment of the resource costing model complex to calculate both at the project level and especially at the work unit.

There is a second factor associated with project material that is important to mention here. Material does not have the same productivity characteristic that direct human labor does. From a management viewpoint, it may be necessary to extract material costs from the total project view to get a truer picture of productivity. The first analysis consideration is to view the ratio of material cost to human resource cost. If material cost is a significant component, it must be separated from the productivity review process. For this discussion, assume that some process will be used to value the material according to an approved internal standard. Once again, it is up to the PM to understand what this allocation represents. It is often possible to buy an item outside of normal procurement channels at a lesser cost, but the project team should resist the idea of setting up a project procurement department to "save" that increment. The project must live within the organization and its infrastructure; accepting the overhead allocations is part of that reality.

8.3.4 Direct Dollars

A third direct expense item involves many non-labor and non-material activities that consume dollars. Typical examples of this third resource group are plane tickets, car expenses, food, and other miscellaneous cash flows related to project activity. This category is labeled "dollars" and is any resource expenditure unrelated to internal labor or material.

Third-party expenditures sit on the fence of resource categorization and may be defined as a fourth resource category. As in the case of material, it may be important to segregate contractual expenditure items when performing a project teamwork productivity analysis. A clear understanding of how these cost groups are captured in the accounting system is important when looking at overall project status analysis.

8.4 BIKE PLAN EXAMPLE

To illustrate the mechanics of resource allocation, a slightly modified version of the bike plan introduced in Chapter 7 will be used here. Figure 8.3 shows the modified schedule for the project.

This view would be considered a standard first-cut version of the plan, indicating that the direct project cost is $34,560 (review the Cost column). This plan is highly simplified and uses a single resource for cost calculations. That resource is named "generic" and is shown in the Resource Names column. This allocation implies a single person for each activity.

To illustrate the resource allocation, right-click on WBS 1.4.5 and the window shown in Figure 8.4.

Note that the worker allocation shows as 100%, which means one resource is allocated. It is also possible to show a dollar allocation here as well. Material resources could also be defined in the resource pool and allocated here, although that is becoming a less popular approach. A more common practice now is to handle material costs external to MS Project.

Creation of the work unit cost estimate will be produced by allocating a resource type and quantity to specific work units, as illustrated above for the bike example. This allocation will then produce a direct cost for that item as shown in Figure 8.4. When the work unit cost values are created and linked to the predecessor data, one should be able to envision a time-phased view of the direct cost for the defined schedule. Note that the various WBS units are arrayed across the schedule timeline. Each of these will be populated with resource allocations and related costs. Collectively, this is a time-phased cost plan for the project.

8.5 INDIRECT BUDGET ITEMS

At this juncture, we have described the process for creating a budget plan that has been evaluated for available resources. Recognize that this still is not the total budget that would be presented to management for approval. Many other components need to be included in the overall cost of this venture from the enterprise viewpoint since various less visible and direct cost items are not yet included. The following is a summary of the major cost elements not yet specifically dealt with:

1. Various support resources that may be utilized by the project (i.e., computer tech support, equipment, facilities, etc.)
2. Material costs not charged directly to a work unit

	ℹ	Task Mode	WBS	Task Name	Duration	Cost	Predecessors	Resource Names
1			1	◢ Bicycle	106 days	$34,560		
2			1.1	Vision Documenation	10 days	$3,200		Generic
3			1.2	System Design	10 days	$3,200	2	Generic
4			1.3	Define Bill of Materials	3 days	$960	3	Generic
5			1.4	◢ Frame Set	25 days	$8,000		
6			1.4.1	Frame	5 days	$1,600	4	Generic
7			1.4.2	Handlebars	5 days	$1,600	6	Generic
8			1.4.3	Fork	5 days	$1,600	7	Generic
9			1.4.4	Seat	5 days	$1,600	8	Generic
10			1.4.5	Unit Test	5 days	$1,600	9	Generic
11			1.5	Crank	10 days	$3,200	10FS-2 days	Generic
12			1.6	◢ Wheels	10 days	$3,200		
13			1.6.1	Front wheel	5 days	$1,600	11	Generic
14			1.6.2	Rear Wheel	5 days	$1,600	13	Generic
15			1.7	Braking System	10 days	$3,200	14	Generic
16			1.8	Shifting System	10 days	$3,200	15	Generic
17			1.9	System Assembly	10 days	$3,200	16	Generic
18			1.10	System Test	10 days	$3,200	17	Generic
19			1.11	Customer Acceptance	0 days	$0	18	

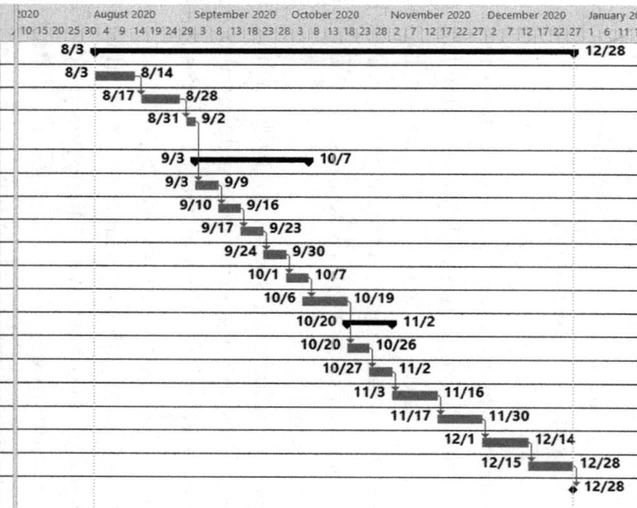

FIGURE 8.3 Modified bike project plan.

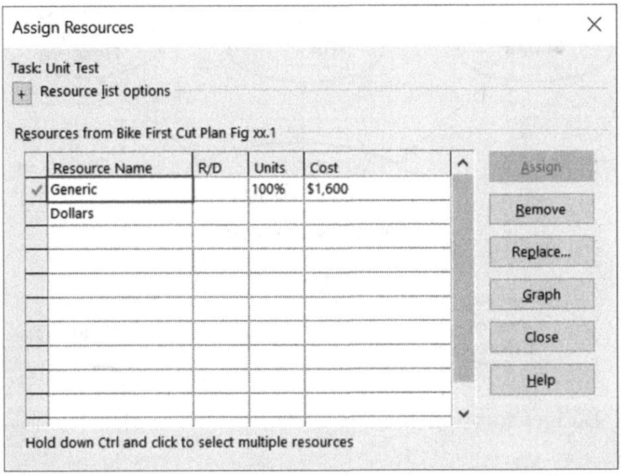

FIGURE 8.4 Resource allocation window for WBS 1.4.5.

3. Various overhead charges related to the human segment (medical, retirement, etc.)
4. Management reserves—unknown/unknowns
5. Risk contingency reserves—known/unknowns
6. Scope reserve—budget process to handle approved scope changes
7. Organizational overhead (often called G&A for General and Administrative)
8. Profit—if this project is being performed under contract

The items shown in this category arise from three basic sources. First, there are direct resource elements used by the project that are often not easy to see or allocate at the WBS level. As an example, various support services can be charged to the project. One of the most obvious would be the charge for computer support or the organization's computing infrastructure. These could be in the form of technical human or hardware-oriented charges. The second category comes from the need to allocate some form of reserve for risk, management, estimating, or other undefined items. Third, overhead allocations come from the enterprise as the project's host. Figure 8.5 shows these external items schematically.

The role of a project resource budget plan is to define all project work elements into which resource charges would be incurred. It is now

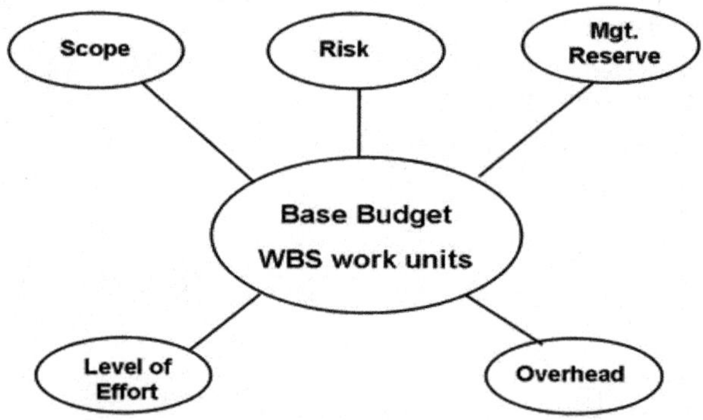

FIGURE 8.5 Budget components.

important to categorize the major groupings for this goal. Recognize that these other non-direct resource needs exist to support the project life cycle.

Young technical PMs often have difficulty understanding why all of these additional items are needed in the budget. As a personal example, when the author executed his first project along the lines outlined here and was ready to make a significant profit on his project, these additional allocation items shocked him. They took away most of the anticipated profit. While reviewing the draft budget with the local financial support person, one after another, these strange new items to the skeleton budget began to emerge. The need to understand that experience is the key message for this section. Simply stated, a project budget must reflect its full resource impact on the organization (positive and negative). To accomplish that, it is essential to understand the logic of these additional charges. Another important aspect of this broader resource view is that some represent controllable items, while others are less so. That management perspective will emerge as the project moves into the execution phase. Our goal now is to explore these groups in a little more detail. Note that the ranking of these additional costs moves from relatively easy to those that are much less visible. One common term used in this area is showing all the *fully burdened* budgetary items. The items outlined here represent much of that concept.

8.5.1 Direct Resource Support

This cost category can arise in many ways. A common example of this comes from various support services that may be allocated to the project.

For example, an agreement could be reached with the information technology (IT) department to supply the needed technical support for the project. That agreement might be for $10,000 per month. The actual work performed is not linked to specific work units so this amount would be a monthly charge. This class of allocation is the *level of effort (LOE)*. Similarly, various equipment could be "loaned" to the project and charged as LOE. Realize that there are many other similar examples of this type. These charges differ from overhead in that the support resource is direct but hard to isolate to a particular work unit or a summary WBS level. For example, expenses related to a vehicle used by the entire project team would be challenging to track in a work unit. In that case, the vehicle costs might be collected at the project level. Whenever possible, it is desirable to evaluate resource consumption at the point of WBS level usage and thus be able to compare plan versus actual values later during the execution phase. Failure to be able to do this makes the actual cost of lower-level items more difficult and error-prone.

Another example that should be described in the non-direct category would be third-party contractual expenses. The decision to contractually procure an item externally versus produce it internally will create a different cost link to the project. Those efforts can be linked to a WBS, but the associated costs fall into a different review category. The final test of this category would be any charges considered direct support of the work effort but not easily charged to a specific work unit.

8.5.2 Reserves

The area of reserves opens a topic that we have yet to explore in any detail and an area that is quite variable across organizations. The concepts outlined here are technically valid, but the methods of recognizing these are quite varied and may differ from these statements. Failure to understand how these reserve-type items impact the project will majorly affect future outcomes. There are three yet-to-be-incurred resource areas for which some reserve strategy is needed. These are risk, scope change, and management (undefined).

8.5.2.1 Risk

The formal definition of project risk is *known/unknown*. This says that we have identified various events that have a probability of occurrence and, if they occur, would require extra resources to resolve. None of these events would be shown in the plan since they may or may not happen. However, when the risk event does occur after the plan is finalized, this *contingency reserve* fund would be used to cover the cost of dealing with the event. The proper vocabulary term here is to say a risk event *triggers*. There is much more to say about the related mechanics for risk assessment, but for now, let's assume that we can quantify these events and allocate a fund to protect the project. The essentialcritical item to note is that if this is not provided, the event's occurrence would create a budget overrun. Chapter 16 will delve deeper into risk management practices related to this.

From a categorization point of view, risk is defined as the group of events that logically could occur but may or may not. From a management viewpoint, we can't set aside budget funds for all of these, but if one does occur, where does the resource come from to deal with it? The answer to this is to establish a risk contingency reserve. In an ideal world, a budget pot of resources would be set aside to handle such events. For this reason, some such external funds should exist and it should be reflected in the budget. It also should be evident that visibly showing such a fund might be rejected by management in a low-maturity environment. Many would think that this was padding the budget for something that would not happen. Resolving this is one of the Achilles' heels of budgeting. If the external environment does not understand this item, the tendency is to hide it under other categories (i.e., task padding). We have already stated that this must not be the proper approach to project management. Our motto is to show the resource categories logically, which is a critical element of that philosophy. Risk events are real, and the best method to handle them is through a separate risk contingency reserve.

8.5.2.2 Scope Change

Projects pursued under contract have a direct mechanism to handle changes in scope. In this environment, approved changes in scope often come with additional time and budget as part of the approval process. In this case, a scope reserve would not be necessary. However, projects pursued internally often need a change mechanism. In these environments, changes may be approved without additional time or resources added to the budget. One can quickly see the flaw in that logic. The original budget, as currently outlined here, has been carefully constructed around the original requirements. It should be obvious that adding more requirements without adding time and resources to the plan creates an imbalance. Once again, immature organizations need to recognize this, so the project teams often pad their estimates to anticipate what may occur.

The effect of this strategy is to distort the planned values and make future comparisons reflect something quite different than what is estimated. For the process to work correctly, the direct budget should reflect the original work specified in the requirements. Academically, the way to handle future unspecified changes to the requirements is to set aside a reserve fund for scope change and then track that fund balance as new changes are approved. Funds would be extracted from the fund as new work is approved. Look at the new project scope in a timely manner to ensure the control process is smooth.

8.5.2.3 Management Reserve

Of all theoretical reserve options, this one is the most difficult to rationalize from a logical viewpoint. However, some level of undefined reserve is needed in the budget to cover a general category of undefined events that resides with senior management to allocate. Proper project management suggests that such reserves be held separate from the internal project control. A myriad of things happen in a project that are unknown and can't be reasonably planned, and these are the elements dealt with by a *management reserve.*

Given the above non-direct budget variability categories, the proposal for completing the total budget view is establishing three separate reserves—risk, scope, and management. One way of looking at this family of reserves is to recognize that estimates and future events will be missed, and a reserve fund is set aside to handle the resulting associated budget plan variance. The goal of making these events more visible is to improve the management aspects of each. Further discussion of budget mechanics is beyond the scope of this book, but it should be recognized that these three events impact the final project cost and, therefore, should be explicitly identified in the approved budget. As stated above, all reserve accounts are viewed with suspicion, and it takes a mature organization to handle them properly.

8.5.2.4 Enterprise Overhead

The concept of organizational overhead is much like taxes. In this vein, the project receives charges that it cannot verify. Rules for of various enterprise overhead amounts are formally defined in most organizations, and these line items are typically standardized in budget formats. Also, the mechanics for calculating these multiple categories are formalized. Regardless of how the overhead values and categories are reflected in the project budget, the PM needs to ensure they are included in the final cost structure. Some of the typical organizational overhead allocations originate from:

Procurement—support for the purchasing and inventory activity

HR—employee-related factors (i.e., pension, medical, retirement, bonus program, etc.)

Facilities—charges based on usage of organizational facilities and infrastructure

Organization—general enterprise administrative factors

All of this overhead charge will be outside the project's direct control, and it is essential to recognize that these allocations can easily double the direct budget amount.

Many higher-level overhead charges are allocated to the project's direct budget on a percentage basis. For example, direct labor estimates might have an additional 100% added, 5% added to material costs, and 50% added to the total direct project cost for administrative overhead. Additional fees such as this significantly increase the initial budget estimate, yet they are an organizational reality that must be dealt with along with the direct cost items. Recognize that these charges arise due to the project being housed inside a larger organizational structure. In many cases, the best way to minimize the indirect allocation is to finish the project as quickly and efficiently as possible since some administrative overhead fees are time-based. The PM must understand these various categories to avoid unpleasant surprises later.

8.6 BUDGET LAYERS

Several cost groups have been derived to this point. The question is, "Which of these budget elements is the PM responsible for?" The most accurate answer to this is "everything," but that is not precisely the case. The model rule is that the PM would theoretically be held responsible for all of the layers shown in Figure 8.6 up to the *Performance Measurement Baseline* (PMB), and, in some cases, might be assigned a portion of the management reserve (MR).Scope changes would be under the purview of the formal change control process. Responsibility for handling the contingency reserve is varied; technically it should be handled by the PM with oversight from the Project Board.

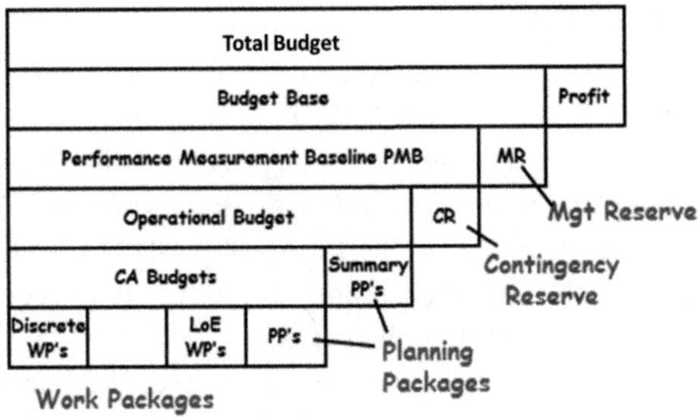

FIGURE 8.6 Total project budget components.

This overview diagram summarizes the various budget components discussed throughout the chapter. The additional concept of a *Performance Management Baseline* (PMB) is introduced here. This budget-level term is beginning to be used as a control point for the project from a management perspective. In other words, the PM would be held accountable for all layers below the PMB. In some cases, the PM might be delegated some portion of the management reserve, representing an attempt to recognize the existence of unknown elements. There is no formal industry process around such delegations, but there is reasonable logic to the above. Each organization handles this situation somewhat differently, so the local practice must be understood. The hidden question is, "How much overrun is considered excusable?" In some cases, that value is zero, so one must not assume that any level of overrun is acceptable without formal confirmation. From a project management viewpoint, not having some known degree of unknown reserve overrun would be stressful and would likely lead to hidden padding. If the organization wants to get away from that practice, some part of the management reserve needs to be shown under control of the project.

Note that no cost element is shown in the budget diagram for scope changes. In the authors' opinion, this is a gap in the model process. Where are the funds going to come from for this anticipated event? Once again, it is essential to understand how scope change will be funded. The diagram is accurate if each change is funded from new external sources. However, if all changes must be covered by the approved budget, there is a gap in required resources. That issue must be resolved before the final budget version is approved. If one wants the published budget to represent the end reality of a project, all of these factors will come into play.

8.7 COST BASELINE

A critical component of project control is the project Cost Baseline. A sample view is shown in Figure 8.7. The monthly bars represent budgeted cost estimates or planned cash flow for the project. Collectively, these constitute the aggregate project cost baseline. This term means the project cost arrayed across its life cycle. This form of project cash flow is a vital project control variable. Also, the report might need to be segmented into actual cash flow versus a total budget view.

This discussion has moved away from the direct WBS activity scope- and time-oriented structure to a more complete budget view. From this, we can see more clearly how the overall project costs are defined. Status reports will be generated for various stakeholder groups using these cost components. Example cost views one might see are:

- Project planned (Baseline) versus actual expenditures
- HR costs and staffing level
- Procurement costs
- Contract status
- Project total cost forecasts

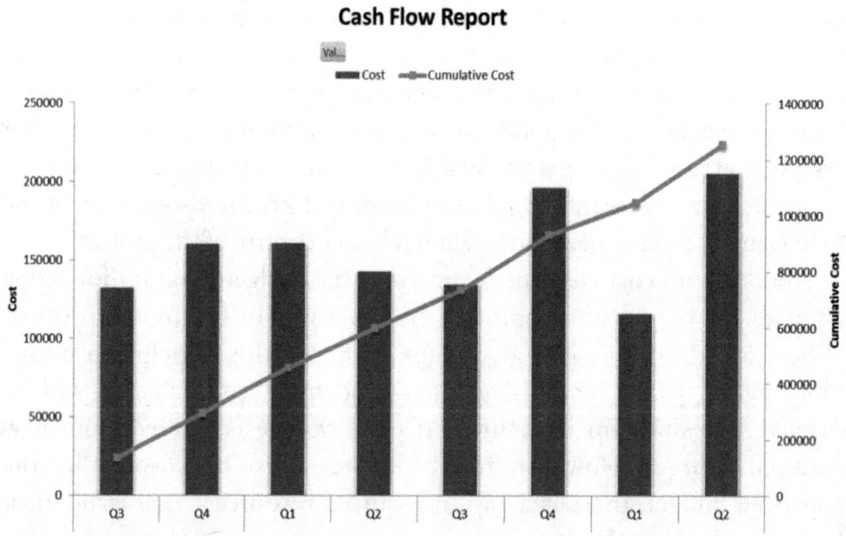

FIGURE 8.7 Cost baseline.

The goal of these various resource displays will be to show current and future projections in both table and graphical formats.

8.8 BUDGET MANAGEMENT GOAL AND PHILOSOPHY

The scope, schedule, cost, and resource mechanics described to this point are viewed as the core planning processes in project management. More activities are required to evaluate the viability of a project plan, but much of the basic mechanics to assess these other considerations are now in place. One way to conceptualize what has been shown thus far is the mechanics required to translate requirements definition, work estimates, and resource decisions into an equivalent descriptive view of the project. In this view, we have combined the project's scope and corresponding work requirements into an integrated work plan. After views of this sort are constructed, potential ambiguity remains regarding the related management processes that deal with these elements. The following list contains some of the related resource philosophy questions that need to be resolved and understood:

- If the project plan time or budget overruns, what is the best way to deal with the variance issues? How much variance is tolerable?
- Where do additional funds come from for approved scope change?
- Who controls the contingency and management reserve funds, and what is the process for moving resources from reserve to operational status as these categories of events trigger?

Let's review what the final budget contains. First, project requirements are converted into work through the WBS that outlines the project's scope. From that, defined work is translated into a schedule and budget by estimating the work and allocating appropriate resources to those work units. This process created what was called a first-cut schedule and cost plan. Evaluating available resources then leads to various adjustments to the plan such that the interlocking variables of scope, time, cost, and resources are synchronized. Finally, additional cost items are recognized above the direct plan elements. These extra cost elements relate to factors such as indirect support, reserves, scope change, and enterprise overhead, which are additive to the direct project cost budget. Each element is then accumulated into a final schedule and budget view that constitutes the approved baseline plan.

8.8.1 Project Profit

The final layer in the budget schematic is Profit. This implies that the project is being performed under contract for a third party, and the internal organization wishes to earn a profit profit from the venture. Recognize that if overruns occur in the lower budget layers, the profit layer becomes the last reserve buffer, and without careful management, it erodes into a financial loss. In many cases, this might be a significantly adverse career event.

8.9 CONTROL ACCOUNTS

As the project begins to move into the execution phase, another aspect of management becomes more relevant. That is a strategy for project control. Once again, the WBS is the focal point for defining this activity. The arbitrary definition of a Control Account represents a modified level of responsibility in two ways. First, this is a physical collection of work units, and the specifications that they are a group may imply that all costs will be collected at that level rather than a lower work until or Work Package as defined earlier. This means that lower-level costs will not be available, and the control level will now be at the Control Account (CA) point. Second, contemporary organizations are starting to assign a level of responsibility to this grouping in the title of a Control Account Manager (CAM). This person may be delegated specific authority or responsibility shared with the PM. Regardless, the idea is to improve focus on selected areas of the WBS. This is both an accounting and a managerial concept. The WBS remains the focal point for this, but control is focused on the group level rather than the lower work units. Individuals who have had to charge time to a WBS understand the confusion that can result from excessively granular levels of control for no added value. The alternative approach to this is to establish a group of WPs that are linked as control points. To help make the management of the WBS cost collection more flexible from a control viewpoint, the concept of CA has been introduced. Think of a CA as a formal management and data collection point within the WBS. This can be composed of a Work Package, a Planning Package. or a collection of WPs. CAs become focal points for both financial and operational management. Figure 8.8 illustrates how CAs might be defined.

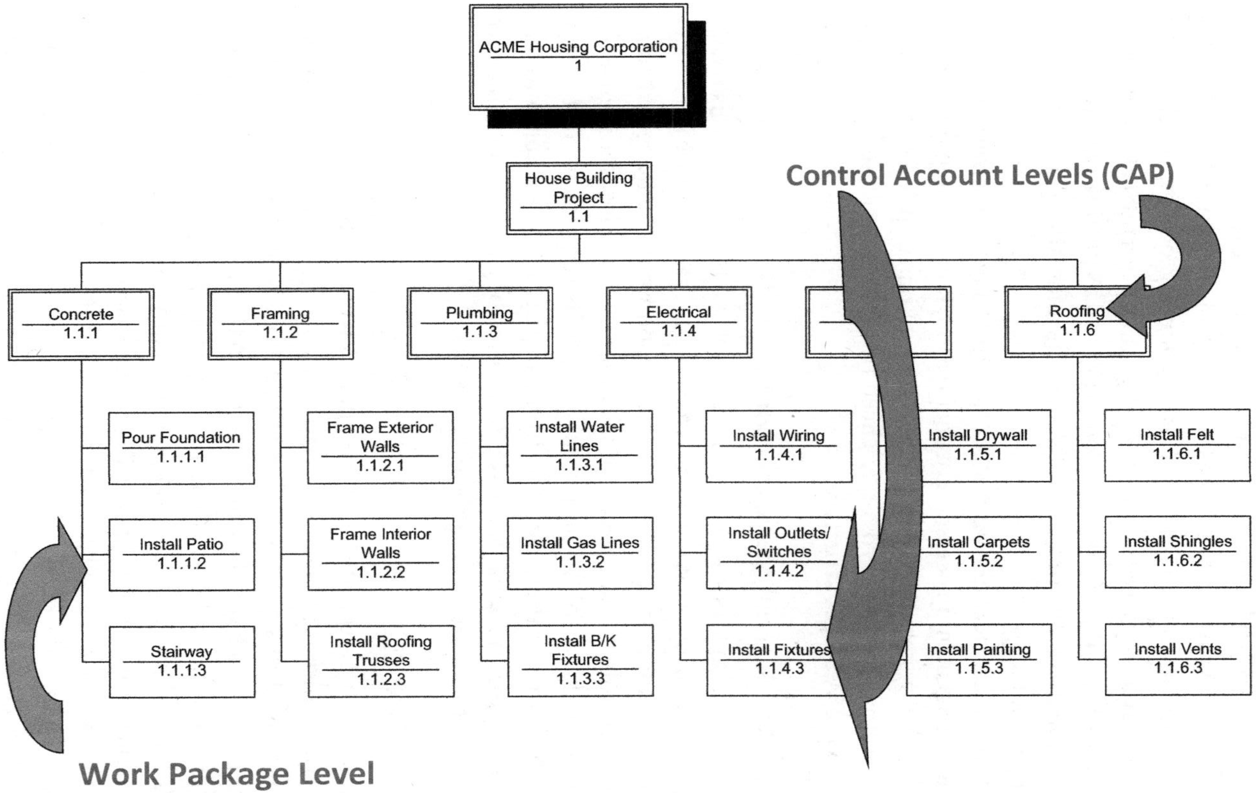

FIGURE 8.8 Control account packages (CAPs).

In this example, note that the summary WBS 1.1.6 is labeled a CA, including all lower-level roofing WBS units below that level. Also, note that WBS 1.1.4.3 is a WP labeled as a CA. It is up to the PM to specify the level of control granularity that best fits the project needs. Related to this decision is the linkage to the enterprise accounting system, which will be the normal *system of record* for collecting and delivering actual resource costs back to specified CAs. This collection of CA groups represents the control strategy for the project and defines the official target locations for tracking resource status.

The enterprise accounting system is one of any organization's most mature management processes. For that reason, it is common for projects to link to this system to obtain actual resource consumption status information. Actual costs are cross-referenced from the enterprise accounting system to the WBS CA structure.

8.9.1 Control Account Managers (CAMs)

As the companion process to financial control, some organizations recognize an emerging operational management role. That role's title is a formal Control Account Manager (CAM). This person would be delegated some level of responsibility for the execution of the CA. There is value in making the operational level of the project more formally responsible for the successful execution of these elements in the increased level of commitment it brings to the operational team. A CAM would be a management partner to the PM for that segment of the overall work.

8.10 BUFFERING

A term somewhat associated with overallocation or variability is the concept of buffering which is used to protect overruns of schedule and cost variability. One view of this concept occurs in situations related to task overruns. Several comments have been made in this section regarding the potential variability of time and cost. There is a stigma against plan variances, and most organizations try to find ways not to incur visible variations. As indicated earlier, the common method is to increase task time and cost values in hopes to avoid an overrun. One would intuitively think padding an estimate would make overruns uncommon. Somewhat surprisingly, that is not the case. Section V of the text focuses on the concept of variability and various ways to deal with this. For now, we introduce the simple process of buffering. To illustrate this idea, a modified version of the earlier bike plan is shown with a project schedule buffer inserted at the end to protect a defined completion date. The buffer is shown at WBS 1.11 in Figure 8.9.

	Ta Mc	WBS	Task Name	Duration	Cost	Predecessors	Resource Names
1		1	◂ Bicycle	118 days	$39,560		
2		1.1	Vision Documentation	10 days	$3,200		Generic
3		1.2	System Design	10 days	$3,200	2	Generic
4		1.3	Define Bill of Materials	3 days	$960	3	Generic
5		1.4	◂ Frame Set	25 days	$8,000		
6		1.4.1	Frame	5 days	$1,600	4	Generic
7		1.4.2	Handlebars	5 days	$1,600	6	Generic
8		1.4.3	Fork	5 days	$1,600	7	Generic
9		1.4.4	Seat	5 days	$1,600	8	Generic
10		1.4.5	Unit Test	5 days	$1,600	9	Generic
11		1.5	Crank	10 days	$3,200	10	Generic
12		1.6	◂ Wheels	10 days	$3,200		
13		1.6.1	Front Wheel	5 days	$1,600	11	Generic
14		1.6.2	Rear Wheel	5 days	$1,600	13	Generic
15		1.7	Braking System	10 days	$3,200	14	Generic
16		1.8	Shifting System	10 days	$3,200	15	Generic
17		1.9	System Assembly	10 days	$3,200	16	Generic
18		1.10	System Test	10 days	$3,200	17	Generic
19		1.11	Project Buffer	10 days	$5,000	18	Dollars[$5,000]
20		1.12	Customer Acceptance	0 days	$0	19	

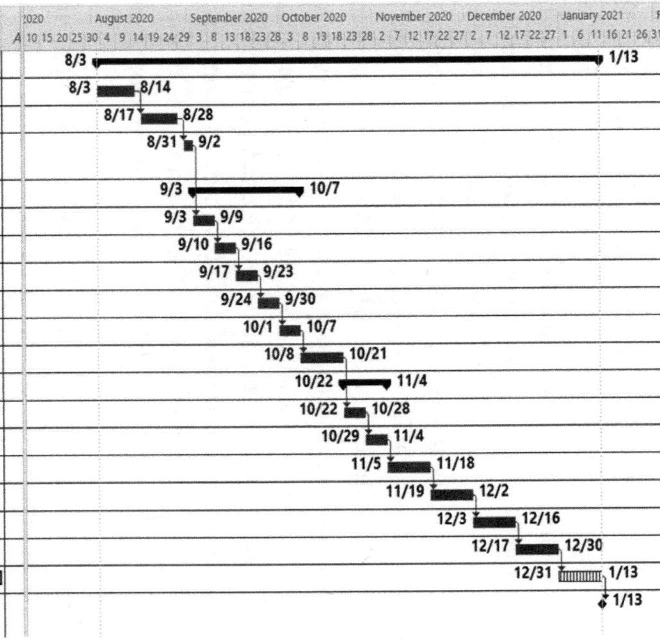

FIGURE 8.9 Bike project plan buffered.

In the buffered plan task, overruns could still occur, and the resulting task variations could be covered through the buffer. This can be done without padding individual tasks with the same desired result. That is, protecting the completion date. The critical question here is how big the buffer should be. There is no scientific answer to that question, and the correct answer depends on the type of project and the maturity of the organization's processes. For this example, let's say that our historical project task time variability is about 9%, and cost variability averages 13%. Using these historical data values as a guideline, we insert a project buffer of 10 days and $5,000 at the end of the project (WBS 1.11). Assuming our historical performance data is accurate and the project progresses according to historical averages, we statistically should finish on the date, and cost indicated. Failure to recognize this variance characteristic aspect of time and cost will likely result in poor completion results compared to a non-buffered plan. The key management point of using this non-padded task approach is to see what parts of the plan varied from our real estimates.

There are many ways to use buffering techniques. As shown in the bike example, they can be added to project phases, major task groupings, or the entire project. From a logical and stylistic point of view, the buffer bar should have a different format or color from other bars since it does not represent a real project task. One of the difficult behavioral parts of this process will be explaining to stakeholders, management, and others why a buffer is needed. On the surface, it looks like padding and it is, but there is valid logic behind it if created with the logic described here. The concept of buffering, if managed properly, will result in lower overall costs and improved schedule performance as the local culture related to procrastination is modified. More discussion of buffering strategies will be scattered throughout the text. The key point to remember is the fundamental logic behind why estimate padding is a cultural process that needs to be curtailed, but one that is behaviorally hard to implement.

8.11 ANALYZING RESOURCE STATUS

The major activity required for resource analysis is matching the defined resource pool against the project schedule with resources allocated to specific tasks. It is not uncommon at this point for the planned resource allocation to not match available resources. This situation is called *overallocation* and is a condition that must be addressed before moving forward. Chapter 15 will delve into this problem in more detail later.

The following step in the resource plan quantification is to define various non-direct resource items that add to the project's cost. From these various sub-activities, a total project plan budget can be compiled outlining the target outcomes and resource strategies. Finally, the project plan is presented to appropriate decision-makers for final approval. At the end of this process, all parties should have a common understanding regarding what is to be produced and a formal agreement to support the effort as defined by elements represented in the plan. At this point, the project team commits that they can execute the defined collection of work units. Equally important, various stakeholders say they will support the plan, and the management decision-makers agree that it fits the organizational goals. Think of this final step as an intra-organizational contract among these key stakeholders. The ongoing control of the project will be based on this agreement called the project-*approved baseline plan.*

8.11.1 Resource Capacity

All previous examples have been defined as though there was no resource shortage in the project resources pool. However, it is now time to peel away another management layer. How do we know that a resource gap exists in a plan? Analyzing this problem is subtler and more complex than what has been described thus far. Estimating the resources resource requirement and obtaining a commitment to supply are two different and often mismatched events in the project life cycle. We have now come to that cloudy intersection. Think of the early planning phase as a wish list. From that vision base, an initial first-cut schedule and budget are calculated. The challenge now is to assess the available resource's ability to execute the plan. One of the status variables that must be monitored is the evaluation of planned versus available resources. It is also essential to understand that this picture is not a constant. Plans change, estimates may have errors, and the planned resource pool can also change. Figure 8.10 graphically illustrates the mismatch between planned and actual resources. Note that the area under the plan curve and above the capacity dotted line represents a *resource overallocation*. During the gap period, the defined plan is only viable and must be adjusted once there is consistency between the time-phased resource plan and actual availability. Also, recognize that this type of analysis is needed for all skill types.

The Resource Sheet has been shown as the method of defining a pool of resources that can be utilized by the project to execute defined tasks. However, since these estimates are defined one task at a time it is not easy

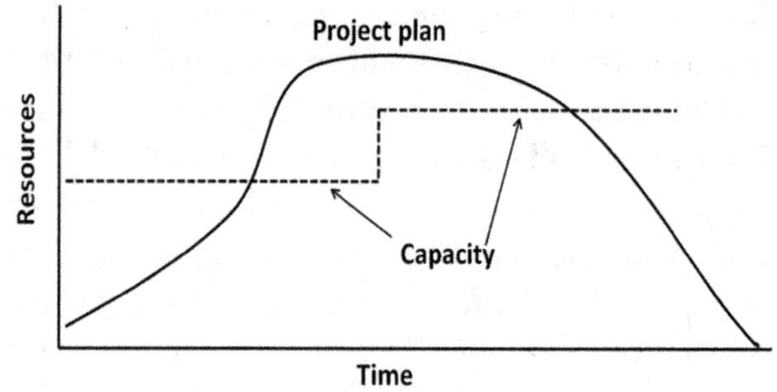

FIGURE 8.10 Resource capacity analysis.

to see the global impact of the allocation. At any specific point, we might find that more resources were allocated in a calendar period than was available. Identifying and resolving this condition requires new utility views of the plan.

One warning on the outside: resolving the overallocation issue is not a simple cookbook process, and multiple decision strategies can be used to alleviate this issue. When a resource mismatch occurs, the PM will have to decide how to bring the overall scope-time-cost-resource linked equation into balance. The following options are typical strategies for this:

1. From the PM's view, the easiest mechanical answer is simply to obtain extra resources; however, that is often the most difficult organizational option to get approved. Also, bringing in a new resource may cause team disruption as they require time to obtain competence regarding what needs to be done.

MS Project attempts to help analyze this situation by first highlighting allocations with a red man icon in the Table view information column. Please note that the textbook images are in black and white, but MSP will display red men. Review the earlier bike plan in Figure 8.9 and note the small red human-looking icons in the left column near WBS tasks 1.4.5 and 1.5. These icons indicate insufficient resources available to support the plan for those tasks. The challenge for resource analysis is to decide what is the best strategy to resolve the problem. This introductory problem has been kept very simple to highlight key elements related to this issue and the role of these basic utility tools. Chapter 15 will discuss this issue and the related decision options process in greater detail.

8.12 APPROVING THE PROJECT RESOURCE PLAN

Much of what has been described thus far falls into the mechanics category. The term *first cut* has been used frequently in earlier sections to describe basic processes for developing plans related to scope, time, and resources. This set of mechanics could often result in your ideal project plan. Or, a first-cut estimate might simply be a rough order of magnitude (ROM) estimate of potential functionality, schedule, or budget.

As the planning process moves into the final stages, there are other unknown considerations to resolve. For instance, estimates should improve in accuracy as the scope becomes better defined. For example, a ROM estimate might be off by 100%, whereas a finished decomposed estimate should be within 10% of the actual value.

This chapter has tried to emphasize the improved accuracy level of a plan as it iterates through the process. First cut, second cut, and final approval level plans should each offer improved levels of accuracy.

8.13 PROJECT PLAN SUBSIDIARY COMPONENTS

In previous discussions, the project plan was based on WBS-defined specifications for scope, schedule, and resources. That is the traditional core of project planning, but at least nine views need to be addressed in the final plan. These are:

- Scope—deliverables (approved)
- Resources required (human, material, dollars)
- HR—timing and skills required
- Risk management process
- Procurement plan—third-party process
- Communications plan—how will stakeholder communications be carried out
- Stakeholder plan—plan for dealing with this segment
- Integration—change control, project close

When the final project plan package is presented to the appropriate decision-makers, these selected areas may be required as part of the discussion. Each can be considered a subsidiary plan to the base overall plan.

8.14 DATA ANALYSIS

Analysis requirements for the financial side of project management require a more robust data access environment than the earlier view discussed. To accomplish this, offloading the core project financial view represented by the planned and actual data contained in the MS Project is recommended. Specifically, data related to duration, labor allocation, and rate information. These data items will be needed to evaluate the accuracy of the estimating process and labor variances. This process, called performance analysis or measurement, is contained in the overall project status view. This will be covered in more detail in the Earned Value and Status Tracking chapters. MS Project is not designed to be a flexible data manipulation tool; this is the underlying logic of moving data externally.

In addition to the performance analysis role, having flexible access to the financial data better supports the external financial analysis needs. Also, this format helps deal with additional data related to non-direct financial items, such as third-party roles. Crating the organizational overhead allocation view can best be performed external to the more fixed project work plan structure. Finally, external status reporting formats can be best handled using this approach.

In addition to these financially oriented needs, multiple process areas will be described in the text where specialized data stores are needed. Each of these can be viewed as a separate data file, but the more appropriate design is to consider each as having a supporting view with potential links to the overall management processes. A data store such as the one described here could serve as a common information source for various data items moving into or out of the project plan for process areas such as scope change, risk analysis, work unit estimating, performance analysis, status reporting, lessons learned, or the budget support process.

8.15 CONCLUSION

The basic project budget and associated resource plan development processes have been described in this chapter. Model theory dictates that project requirements outlined in the WBS drive the scheduling, financial, and support resource aspects of project planning. The overall plan documentation also defines other resources and potential plan category details. Beyond describing the basic project work requirements, the project plan will include many other factors, such as non-direct, reserve, and overhead components. Finally, the planning process needs to cover various other operational questions. Moving the model theory into the real-world stage is now set, and from here, the art side of project management takes center stage. At this juncture, the PM will need salesmanship and negotiation skills to work through the various trade-off negotiations related to the upcoming dynamics of the execution process.

QUESTIONS

Review Questions

1. How can scope creep affect the project plan regarding time and resources? Comment on strategies to control this.
2. Why are indirect costs included in the project plan?
3. If project management is considered a project resource cost, where would you show this?
4. What is the role of a Project Buffer? What is the advantage of a buffer over just padding individual tasks to achieve the same completion date?
5. Resource overallocations were described in the chapter. Give some examples of management techniques to resolve this class of problem.
6. Assuming a work unit is accurately estimated, what is the most likely reason for the actual labor cost to vary from its estimated value?
7. Describe the basic arithmetic in generating a work package cost.
8. Where is the project's actual resource cost information stored?
9. Where in MS Project can you find functions related to resource overallocation?
10. How would LOE-type costs be allocated to the project?
11. When a scope change occurs, where does the additional resource to support this change come from?
12. Why do you need a management reserve?
13. Why do you need a contingency reserve?

14. Why does the organization add an overhead to the direct cost allocation for labor?
15. How does external contracting affect the internal examination of project status?

EXERCISE

Use the library file *10 Car Problem Cost Example* as a practice overallocation example. Identify all tasks containing a resource overallocation and then document each time range, resource name, and overallocation level. Use MS Project tools and attempt to resolve each of the overallocations (don't be too concerned if you fail at this process as it only illustrates the complexity of the problem. Realize that removing tasks or adding resources will always work but may not be a viable strategy).

CHAPTER 9

Creating a Viable Project Plan

Insert Work Files Note Here

9PLAN	Finalize Plan	Viable Plan Concepts
9PLAN2	Reschedule the Plan	MSP Rescheduling

Previous chapters have outlined the basic mechanics for constructing a first-cut scope, time, and cost project plan. Earlier comments warned that different stakeholders would have diverse areas of interest that could affect the plan produced by the formal process. These different views manifest themselves in multiple ways that collectively put pressure on the project to tweak the plan in an attempt to satisfy what are often countervailing views. The focus of this chapter is to illustrate various strategies that the PM must embrace in customizing the initial plan to fit these various constraints or views while keeping the core structure somewhat intact.

9.1 PLAN TWEAKING STRATEGIES

One of the frequent decision scenarios in the project, especially after the end of the first-cut plan, is the need to change some aspect of the approved plan to fit new views or constraints defined by key stakeholders. Based on these pressures, the initial plan will need to be modified. Techniques to

DOI: 10.1201/9781003532989-11

resolve such issues without destroying the integrity of the original plan represent a management art. Each tweaking technique takes some aspect of the project work units and moves them around to eliminate or minimize the issue. We'll define these processes as the PM's magic toolbox of tricks to tweak the plan in an attempt to satisfy various scope, time, and cost constraint issues. Seven common scenarios will be discussed as examples of this process. They are as follows:

1. The plan is perfect, except you have asked for more resources than those available—this problem is labeled *resource capacity.* Chapter 15 will describe these decision mechanics in more detail.
2. Resource commitment issues are the Achilles' heel of a project matrix organization structure. In this scenario, resources are being supplied by a functional organization (say engineering), and you are informed that the promised quantities cannot be provided because of some internal issue in that organization. Without this, the current plan is not viable.
3. The plan consumes more time and cost than management will allow. This is a case where the original Charter does not match the detailed planning result.
4. The project sponsor has informed the PM that the project must be delivered quicker than the current plan indicates, but more resources are not available.
5. The project is overrunning the planned budget and schedule. You have been told to get these parameters back on plan and complete the project on schedule.
6. An internal organizational project is encountering significant levels of change requests. The PM brings this to the sponsor and is told that the project must be delivered on schedule and budget without changes to the original plan.
7. Project requirements are okay, but the technical staff has a new proposal to execute the work better than reflected in the current plan (i.e., latest technology).

These items may appear hypothetical, but any PM in the industry for a few years will testify that each is an example of what the real world brings.

Each event above represents different project issues and strategies required to "tweak" the plan. In each situation, the goal is to preserve as much of the initial base plan as possible while dealing with the stated constraint. Some of these options are more mechanical than others, but a fundamental management process theme is running throughout this set that needs to be understood. The sections below will discuss techniques for dealing with each of the scenarios.

9.1.1 Resource Capacity

Simply stated, resource capacity is a situation where the plan specifies resources that are not available. This topic was previously introduced but will be covered in more detail in Chapter 15. Resource capacity is easy to comprehend at the project macro level but much more difficult to resolve at the task level. Figure 9.1 shows the project resource plan. Note that the required resource level for months six through eight is 15 engineers, while only ten are available. This gap for five engineers means that the existing plan is not viable.

Recognize that a resource shortfall can occur for various skill groups, and quantities can exist for factors other than human resources, for example, contract support, equipment, physical spaces, or even a broader collection of resources. Surprisingly, many project planning efforts do not

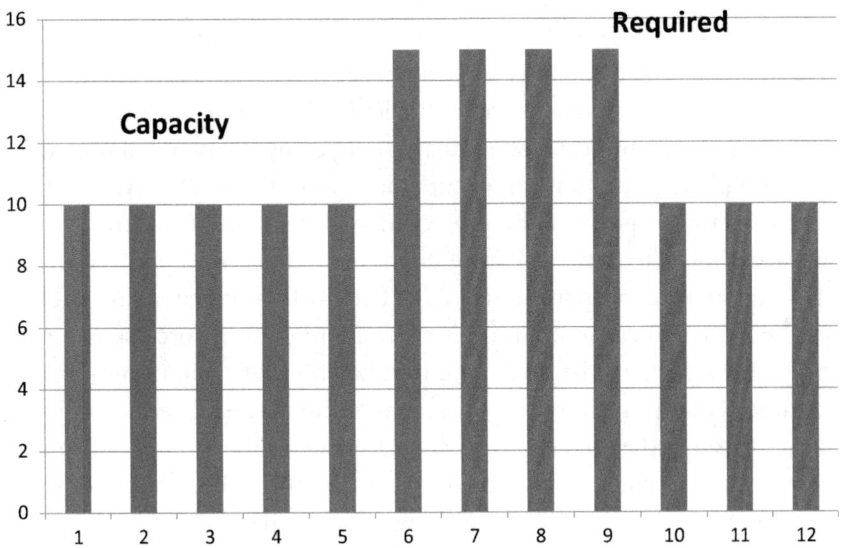

FIGURE 9.1 Resource capacity.

match task estimates with related resources. Let's say that is a better way. Individual work units will specify quantities and skills in the estimate, but that is often not entered into the project plan. If the parts of cost estimating data are not entered into the software, it is not able to assess the resource capacity issues for the project and no other process is likely to duplicate that. This means that the need and location of resource gaps are not visible until they occur in execution. One should understand the chaos that can be created if a critical skill is not available to execute a critical path task and that information is not known until a few days before the need. If the task sequence or existing estimate is incorrect, the resource staffing plan will be significantly impacted. PMs earn their reputation in these types of situations. The goal at this point is not to wade into the decision options that need to be made when this class of problems emerges. That exercise will be saved for later. The message for now is to recognize that this is a significant management problem when it occurs, and there are techniques to mitigate it. However, there are also techniques to warn that a resource gap condition is likely. All resource management practices do not have to be as automated and sophisticated as shown in Chapter 15, but some warning system is needed with sufficient advance notice. Lack of reaction time to find a solution can make this type of problem one of the most difficult to resolve. Human labor and related materials are the key to timely execution.

9.1.2 Resource Commitment Issues

The resource commitment issue arises when a plan has been created, but the supplier is not supporting the plan levels. Remember that in a matrix organization, the project resources are supplied by a functional organization group. This situation can occur at any point in the life cycle. If there is a common problem in the project management environment, it is in this area. Too often, the skills and quantities defined are approved in the plan and unavailable per the schedule. Even with a perfect plan, a deficit of resources will cause the project schedule to drift or force some other mitigation strategy to be used. To minimize this issue, obtaining a formal commitment for a resource supply plan is essential, and then tracking that closely through the life cycle. This is an area where the PM can be blamed for an overrun when, in fact, the cause is directly linked to the resource supplier. Staffing data must be carefully monitored, and corrective action must be taken as gaps arise. Maintaining a match of actual resources per the plan is a significant success factor for the project.

9.1.3 Reducing Scope

Returning to an earlier metaphor, recall that a scope-related work unit was characterized as a box with dimensions of scope (deliverable), time, and cost. These work boxes are represented in the Work Breakdown Structure (WBS) hierarchy and the project plan as part of the defined work units. If there is a need to cut project time or cost, cutting scope is one of the most straightforward ways of accomplishing that. Envision this through the WBS boxes. Scope cutting is represented by cutting boxes out of the structure or reducing the box size to reflect fewer deliverables. Once again, we see a need from the requirements definition process to indicate a priority for each requirement to help guide these rationalization activities. If all deliverables are prioritized, the scope-cutting process can use this data to yield the least negative impact on the project output. In other words, keep as much of the project's high priority intact as possible while achieving the required resource reduction.

9.1.4 Fast-Tracking

Every project has some level of pressure to shorten its overall cycle time. Sometimes, this has to do with market delivery requirements, while in other situations, it may be more weather-related (winter is coming) or some other delivery need. There are many instances where being first to market significantly impacts goal success, and the organization often presses the PM to accomplish faster output delivery. Fast-tracking is one of the most popular because of its apparent zero incremental cost, although that may be more of an illusion than fact.

The term *Fast-tracking* has multiple meanings in practice. At the core, it means reducing the project's overall life cycle time. The fast-tracking process involves selecting activities that are ideally worked on in sequence and moving some parts in parallel. To shorten the project cycle time, the process focuses on the critical path by taking two sequential tasks and overlapping them to some degree to decrease the duration sum. That is the mechanic, but there is more to this technique than just overlapping a work unit.

The first point that must be realized is that overlapping tasks that are not on the critical path will not impact the overall schedule. From another perspective, overlapping causes other tasks to become the critical path; therefore, the original mechanic did not accomplish the desired compression result. The more iterations of this process, the more slack will be taken out of the entire project network, so multiple critical paths will likely begin to emerge and must be handled.

Beyond the described mechanical steps, this process often increases the potential for adverse risk, quality, and management events that negatively impact the outcome. Joel Kohler offers a straightforward and realistic example (Kohler, 2009). If we have two sequential tasks to lay the carpet and paint a room, we might allocate these resources simultaneously by moving them in parallel, thereby cutting the cycle time. The optimum technical sequence is to paint first and then lay the carpet. By doing these tasks in parallel, we risk having paint on the carpet, resulting in lower quality, rework, management issues, etc. In most cases, the original plan should have been constructed with an optimal technical sequence, so fast-tracking can create challenging operational issues that may produce less than desired results.

Before deciding to fast-track a project, it is necessary to evaluate the technical characteristics of the project. The risk level is further elevated if it involves a new type of initiative or high interaction between the selected tasks. If critical path time can be cut by overlapping, evaluate the value of that decrease compared to the increased risk and management effort involved. Recognize that many factors are involved in the decision to fast-track such that it is not a simple cookbook approach. Some of the major considerations to review are as follows:

1. Are the tasks involved located on the critical path? If not, there is no value in doing this.
2. Can the two tasks overlap? What happens to productivity if you do this?
3. Do the associated skill groups work together well in parallel?
4. Would fast-tracking create a resource capacity imbalance?
5. Can the management process handle these tasks in this modified format?

Carroll et al. (2004) describe a related tactic of reducing cycle times by decentralizing decision-making to improve expertise in finding alternate fast-tracking methods. His study shows simulation results from this strategy, indicating that decentralizing the decision-making to speed up the cycle time did not work. There is also some belief that fast-tracking itself is more complex than it appears, with the result being that cycle time is increased and not decreased. Regardless of the underlying complexities in selecting tasks for fast-tracking and the potential negatives surrounding the action, it is a popular time compression technique. This is primarily because it appears free from requiring additional resources, although that is not true in the overlap period.

9.1.5 Crashing

The process of reducing task time by crashing means adding more resources to a critical path task to finish the task quickly. Here is a simple example to illustrate this. Suppose we wished to dig a hole 10 × 10 × 10 feet manually. The estimated work required is 200 hours, and one laborer is allocated to the task. Logic suggests that allocating two laborers to this task would shorten the task. How about three? What would happen if we allocated 50 laborers to the task? Would the same incremental task time reduction occur with each new additional resource? Clearly no! To execute this process in practice, it is necessary to establish a minimum time, called the *crash time* (Uher, 2003). The incremental cost of adding resources is called *crash cost*. The mechanics of crashing involve adding *crash cost* to the work estimate in return for incremental time compression. This process can be repeated until the crash limit is reached.

Two variables must be defined to perform the crashing process—crash time and crash cost. We assume that the original plan value is considered to be the optimum time and cost for that work unit. Whereas the crash time is considered the absolute minimum time for the job, and crash cost would be the incremental cost to reduce a one-time unit. The primary concern for the crash estimate is the decrease gained with each additional resource.

A time-cost tradeoff graph shows the relationship between an additional resource and time reduction. Eventually, crashing will result in diminishing or even negative returns. At the minimum crash point, trying to shorten the activity any further is not feasible. At the extreme crash point, an attempt to crash beyond the minimum crash point will increase both the time and the cost of the effort as overall productivity falls. Review the previous hole-digging example from the estimating discussion with this concept in mind. Then, envision the situation where too many diggers have been assigned and are now in each other's way. This is a visual example of adding incremental resources beyond the minimum crash point.

9.1.6 Simple Crashing Example

For this example, assume that crashing data has been derived for each activity on the critical path. It is also common to assume that crash costs are linear, even though that is a questionable assumption. Figure 9.2 shows a simple example to illustrate the basic mechanics of network crashing. Because all tasks are on the critical path, any task duration reduction will cut the overall project cycle time. Figure 9.2 shows a simple plan that needs to be "crashed."

Table 9.1 contains the crash data necessary to execute the process.

All tasks allow only one day of crashing, and all except one have the same crash cost (Δ$/ΔT). Given this data, cut one day out of the project for $3 by selecting any of the tasks except G. This will decrease the plan duration but add $3 to the budget. Continue to crash the schedule and record the budget increases. Note that a task cannot be decreased beyond the CT value.

In a more complex network, each duration crash iteration would select the minimum critical path task crash cost, add that additional cost to the budget, recalculate the critical path, and continue to iterate the process. The stop point would occur when the desired result had been reached (i.e., schedule target or added budget limit). In other words, the stop point would be reached when the cost to cut another unit of time outweighed the benefit. Also, the network cannot technically be reduced further at some point, regardless of how much money or resources are available. At that juncture, all tasks are at the crash point.

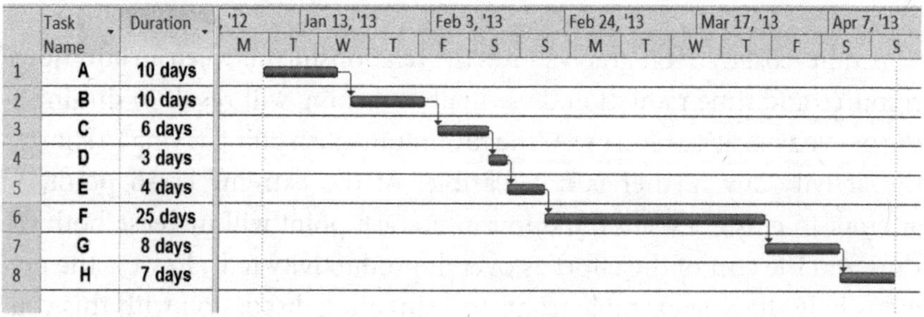

FIGURE 9.2 Crashing plan.

TABLE 9.1 Crashing Task Definition

Activity	NT	CT	Δ$/ΔT
A	10	9	3
B	10	9	3
C	6	5	3
D	3	2	3
E	4	3	3
F	25	24	3
G	8	7	6
H	7	6	3

Notation: NT = Normal task activity time

CT = Minimum time that the activity can be crashed

Δ$/ΔT = Incremental crash cost per time unit

9.1.7 Value Engineering

The concept of value engineering was conceived at General Electric in the early 1940s as a technique to produce products in a situation of scarcity of materials (Save International, 2007). In other words, the goal is to engineer equivalent output using more readily available or cheaper materials. A key part of this early process was a function analysis concept to help measure the value of alternatives. This technique was later called *value analysis*. In the 1950s, the title changed again to *value engineering (VE)* as the design role of the process changed. As viewed today, it is the systematic application of recognized techniques by multidisciplinary team(s) to identify the function of a product or service, establish a worth for that function, generate alternatives using creative thinking, and provide the needed functions, reliably, at the lowest overall cost. As stated in simpler terms, this product or process analysis technique creates the same functional outcome at a lower price. It can be applied to both product and process-type projects.

The standard view of VE is a cost reduction tool; however, it is more appropriate to consider it a value improvement tool, which may or may not include a cost reduction. The Department of Defense (DoD) uses this approach widely and reports savings of $1 billion annually (IDA, 2006). A synonym for VE is design-to-cost. This process attempts to evaluate various requirements regarding cost/value. In the situation of tweaking project cost, we can see how this could be used to cut scope or restate a requirement with less cost than the original specification. The DoD considers this process as a principal in establishing and providing tools for reducing costs and enhancing system performance (IDA, 2006). The following seven basic steps are defined in this process (Save International, 2007):

1. Gather information—current status
2. Measurement—defining how the alternatives will be measured
3. Analysis—what is the requirement, and what does it cost
4. Alternatives—what are the various ways this can be accomplished
5. Evaluate—select the best options to evaluate further
6. Quantify options—develop cost/impact status for each option
7. Presentation—describe results to decision-makers

The first part of the VE process is similar to the creative aspects of brainstorming, and the latter part exhibits many of the quantification traits of the scientific method. Realize there are many dialects of this process, and interested readers can gain more details on this from www.value-eng.org and several DoD references.

9.1.8 Court of Last Resort Strategies

Once the logical plan correction options described earlier have been used or eliminated, the remaining choices fall into the Court of Last Resort category. None of the choices in this group is optimal, but it may be necessary to invoke for overriding reasons.

One typically undesirable management choice could be to halt all changes to the project and focus on delivering the currently defined version. That decision essentially says that no more schedule changes or additional resources will be allocated beyond what has been currently defined. Of course, another option could be to cancel any further execution efforts and stop all work—terminate the project.

A second example would be to decide to increase the risk level of the deliverable. A risk-increasing example of this would be to eliminate some part of the testing program. This has major risk implications and omits the opportunity to identify operational problems by finding them with less negative impact.

A third example is to cut the active project into smaller phases and focus on delivering higher value, focused groups quicker. This is a "stretching out" strategy and may create more total cost, so this is viewed as a tactical rather than a strategic strategy. If this option is used during the execution phase, it could cause a significant disruption for the project team to restructure the work plan into different groupings.

9.2 PLAN ACCURACY

The process of developing an accurate project plan is often more of a wish list than an actual goal. Does accuracy mean clearer specifications and higher confidence in schedule or budget? Or does it relate to some other plan factor? As mentioned before, the operative reaction to a project question is typically formulated by the source of the question, and that applies here. During early planning periods, requests are made for estimates regarding future outcomes.

Too often, these rough estimates in the project initiation phase get locked in as an approved project target before detailed planning is completed. Remember the previous example of Project Titanic. It is important to understand that task estimation accuracy should improve as more details about the project are known. Mature organizations realize this and make their decisions accordingly. For instance, early estimates can easily be +/–100% for a large, complex project (maybe higher). This range might be much smaller for smaller, repetitive projects with historical experience. Regardless, in general terms, the estimate accuracy generally improves through the following three basic stages:

1. Rough order of magnitude (ROM)—an early guess based on minimal data (+/–100%)
2. Definitive—an estimate based on formal WBS requirement definition, but an incomplete analysis related to areas such as risk and resource analysis (+/–25% to 50%)
3. Budget—the normal goal for a project budget after a formal planning phase (+/–5% to 10%)

The terms used above are not necessarily standard industry labels but are used here to represent levels of accuracy. This evolutionary accuracy would be well understood in a mature organization and is an important concept in the overall project management plan development process.

One mature project organization color-coded its project plan documents by phase level. So, a Charter-level proposal might be presented on red paper to fit the ROM-level category, while a budget created during the definitive phase would be presented on yellow paper. Finally, the formal, approved plan would be called a green estimate and distributed on green paper. This approach lets everyone know how to adjust their view of the plan. Unfortunately, this is not a common model, and budget estimates are often not viewed in the manner shown above. One can see the potential miscommunication resulting from not knowing the state of development for such estimates.

9.3 PLAN VARIANCES

Schedule and budget variances occur in essentially all projects. Various studies have shown this to be a common occurrence, and plan versus actual variance levels of 50% and higher are not uncommon. Many factors contribute to this, and one could argue that poor initial estimating is a major actor, but there is more to the issue than this. It is important to understand that many other causal factors exist as well with other primary variability sources listed below:

- Poor initial requirements definition.
- Poor scope control during execution.
- Inadequate risk assessment during planning creates significant problems in execution.
- Inadequate support from stakeholders and management.

Beyond these generally well-known items are a myriad of environmental items related to factors such as the following:

- Technical management factors are based on the defined work process.
- Factors related to human productivity. These can be related to organizational culture, worker training, etc.
- Political-economic factors. For instance, planners may assume the host environment to be the same as found locally but later find that this assumption was incorrect.

From these root causes and others, the project plan variances emerge in the following visible ways:

- The cost of the entire project is higher than planned.
- Tasks expand as new requirements emerge, which creates a variance of observed costs.
- Poor human resource management causes even a good plan to deviate.
- Poor overall management creates an environment that allows the project to drift, which then causes all plan elements to suffer variances.
- Risk events occur beyond the planned level, which can affect all plan elements.

9.3.1 Managing Variability

In a deterministic world, there is no variability, and the project management job would be a checklist checker with much spare time. Reality observation does not support this view, so the root cause of observed variability is assumed to come from the millions of things that can go wrong at any particular time. That list is too long to summarize here, but we now know some of the primary likely targets for variability. One thesis for minimizing variability is to produce a well-thought-out plan. In some project situations, this is an achievable objective, while in others, it is less so. Most projects have sufficient variability to suggest that the prudent PM must constantly be on the lookout for the emergence of some unexpected variance or issue to resolve. Summarized below are some of the main variance driver areas for which to watch:

1. Changes to the original requirements used to create the plan. This is often represented by excessive scope changes during execution. Regardless of the original cause of these changes, failure to control scope changes will most assuredly doom any early completion prediction. Some industries quote an average scope change rate of 2% per month, but this value can vary. The PM needs a general guideline estimate for this in their project. Simple arithmetic says that if a project is experiencing scope creep of 2% per month, for a one-year

project, the new project scope would be 24% more than the original plan. The key management question is, "Where does an additional budget to cover these additions come from?"

2. Task estimating errors will occur; original estimates tend to be lower than the actual result. These variations tend to create overruns in both duration and cost. Once again, what will you use to handle these variations? Previous chapters have offered various techniques for this issue, so this is a remedial question with multiple potential answers.
3. Resource pools give the impression that appropriate skills and quantities are standing by to execute the defined tasks. That is usually not the case regarding time, skill quality, or quantity. Even if the work estimate is correct, the appropriate resources are often unavailable as planned. Variations created from this source can affect not only the schedule but also the timing and magnitude of cost. Managing the flow of required resources is an important process element. Once again, even with a perfect plan, failure to supply the defined resources on schedule will create a variance.
4. Various "bumps in the night" will occur during the project life cycle. For example, the material will be scrapped and have to be reworked. Tests will have to be rerun. Various other unanticipated events will collectively add variability to the actual outcome. Much of this variance will need to be dealt with by a variability reserve.
5. One of the variability sources that has been increasingly recognized recently is a segment called "known/unknowns." Our formal name for this is a risk! This means that some portion of the plan will be impacted by probabilistic events that may or may not occur; because of this, they would not be shown in the base plan but would affect the outcome if they did occur. This category of events is difficult to quantify fully but certainly cannot be ignored. More specifics regarding this topic will be covered in Chapter 16.

The prudent PM should attempt to forecast these classes of variance and deal with their impact through buffers or reserves and an active tracking process. One unclear point from the above discussion is that you cannot put something in a plan that may not happen. However, you can establish a reserve for that event and use it when it occurs. That is a key management point in the area of variability. The one management philosophy previously emphasized is that padding task estimates to cover such variations is not the appropriate mechanism and hides the problem. The key to understanding variation is to expose the various causes through *lean* estimates and ongoing root cause analysis.

Each of the various sources described above is often difficult to identify. When looking at the historical performance of complex projects, it is not uncommon to find large variations. Previous chapters in this textbook have highlighted historical data regarding these sources and project outcomes. Historically, projects have been characterized by significant variation in planned outcomes, and a design goal of this textbook is to outline a management methodology to improve this result. Many textbook chapters isolate a source of variation and offer guidance regarding planning and controlling it. The ultimate goal of project management is to improve outcome predictability. In some ways, one could view project management as managing variation. Assessing variances is part of the learning process, and knowing how to do it better next time is important. This is the essence of a learning organization (i.e., not hiding variances but surfacing them and looking for their root cause). There are techniques to help deal with the various factors described above, but for now, recognize that they exist, and each potentially adds time and cost to the project plan. One final comment to consider is if your plan says that you will be finished on June 5, is this a valid statement? All the chapters in Section V will deal with various questions of reality and project variability. Each of these represents some aspect of the viable plan concept.

9.4 CONCLUSION

A 1960s IBM software development project offers a classic example of plan tweaking gone awry. As IBM struggled to complete its Model 360 computer system, which was a strategic project designed to capture the world computer market, it found that the software side of the project was running behind the hardware segment, so the decision was made to crash the software portion by adding resources to speed up delivery. At the initial stage, about 50 developers were working on the software operating system. Multiple cycles of this resource-crashing strategy were used, and in each case, the result was not as desired. Before the software was finished, some two years later, the software team size had grown to nearly one thousand developers, and later analysis led to the conclusion that this process did not speed up the development and certainly did not help the final quality of the product. In essence, this was a massive crashing exercise created by management's desire to complete the product quickly, but the elements of the job requirements were not understood. Frederick Brooks (1974), the PM, later reviewed this process and surmised that it would have been better to leave the original 50-person team with core knowledge in place and not create confusion by continually introducing new raw resources into the team. This real-world example shows that the process of project tweaking from what was considered the optimal arrangement is not that simple to execute or straightforward in its predicted results. Nevertheless, PMs will often have to utilize this tool kit of options as they are pressured to move their projects back into some desired scope, time, or cost state.

9.5 STUDENT EXERCISE

This exercise aims to test your understanding of basic project compression techniques. This file can be accessed from the student resources on the publisher's website. The sample exercise file is named *9 Crashing Exercise.* To derive full value from the exercise, performing this as a group discussion is recommended to simulate how a project team would go about this. At the end of the exercise, describe what you would tell management and your key stakeholders about this revised plan. Assume that there are constraints for both time and cost (Table 9.2).

TABLE 9.2 Project Compression Exercise

WBS	Task Name	Duration	Pred	Cost ($)
1	Office complex	71 days		549,000
1.1	Determine budget	2 days		6,000
1.2	Select architect	4 days	2	12,000
1.3	Interview architects	4 days		4,000
1.3.1	Architect 1	8 hrs	3	1,000
1.3.2	Architect 2	8 hrs	5	1,000
1.3.3	Architect 3	8 hrs	6	1,000
1.3.4	Select an architect	8 hrs	7	1,000
1.4	Prepare first draft plan	1 day	4,8	4,000
1.5	Review the plan	0.5 days	9	500
1.6	Finalize the plan	0.5 days	10	500
1.7	Obtain construction permit	0 days	11	0
1.8	Plan office layout	4 days		16,000
1.8.1	Prepare layout plan	2 days	12	8,000
1.8.2	Estimate costs	2 days	14	8,000

(Continues)

TABLE 9.2 (Continued) Project Compression Exercise

WBS	Task Name	Duration	Pred	Cost ($)
1.9	Buy materials	5 days		16,000
1.9.1	Rent tools and equipment	3 days	13	6,000
1.9.2	Purchase materials	5 days	13	10,000
1.10	Prepare the site	20 days		140,000
1.10.1	Excavate foundation	8 days	16	56,000
1.10.2	Build the foundation	12 days	20	84,000
1.11	Begin construction	18 days		224,000
1.9.1	Build the pillars	10 days	19	70,000
1.9.2	Lay the roof	8 days	23	56,000
1.9.3	Build the walls	8 days	23	56,000
1.9.4	Flooring	6 days	23	42,000
1.12	Miscellaneous	12 days		126,000
1.9.1	Install plumbing items	6 days	22,26	42,000
1.9.2	Install wires and cables	3 days	22	21,000
1.9.3	Plastering	4 days	29	28,000
1.9.4	Doors & windows	3 days	30	21,000
1.9.5	Furnishing	2 days	31	14,000
1.13	Office complex complete	0 days	32	0

The corresponding crash definition table outlines the time-cost tradeoff data for this plan. All tasks shown in Table 9.3 are on the critical path.

TABLE 9.3 Activity Crashing Data

Task	Task Name	DUR	CT	Δ$/ΔT
3	Select architect	4 days	3	0
15	Estimate costs	2 days	1	0
20	Excavate for foundation	8 days	6	5
21	Build the foundation	12 days	10	6
23	Build the pillars	10 days	7	7
24	Lay the roof	8 days	7	8
25	Build the walls	7 days	6	10
26	Flooring	7 days	6	10
28	Install plumbing fixtures	4 days	3	10
29	Install wires and cables	3 days	2	5

Legend: NT = Initial time estimate, CT = Minimum time (same time unit as NT). Δ$/ΔT = Cost to crash the activity one-time unit ($000s)

Given the plan technical network specification data outlined above, answer the following questions:

A. Clean up the initial network and show results for the following:

   i. Original plan based on network specification

   ii. Identify the critical path for the network

   iii. Finish dates on all tasks

B. Discuss possible duration reductions that could be made without the expenditure of additional resources. What do you believe are the date reductions that can be made to this plan?

C. Assume senior management has instructed you to reduce the plan schedule by five days. Tell them that this might be possible but will cost extra resources. As part of this exercise, crash the network using the data provided and answer the questions below.

   i. What is the required logic to accomplish this task?

   ii. What task(s) will you crash and why?

   Crash 1: ______________________________

   Crash 2: ______________________________

   Crash 3: ______________________________

   Crash 4: ______________________________

   Crash 5: ______________________________

   iii. How much extra will this add to the initial project budget?

D. Using the sample data above, crash this project plan to 63 days and show the cost impact of that. Use MS Project and crash one day at a time. Record the task selected and incremental cost impact in Table 9.4.

E. Using the modified (crashed) plan, answer the following questions:

   i. What is the earliest time you could reach WBS 1.13? ____________

   ii. What is the latest time you could leave WBS 1.9.2 and not affect the CP? ____________

TABLE 9.4 Task and Cost Data

Duration	Task Cut	Total Cost
70		
69		
68		
67		
66		
65		
64		
63		

iii. What is the slack time at node WBS 1.9.1? ___________

iv. Write a formal description of the CP. ___________

1. What does the term "time/cost tradeoff" mean?
2. What is the difference between fast-tracking and crashing? Where are these ideas used in the plan structure?
3. If value engineering were to be applied, when would it be most useful?

REFERENCES

Brooks, F.P. (1974). *The Mythical Man-Month: Essays on Software Engineering.* Addison-Wesley, Reading, MA.

Carroll, T.N., Burton, R.M., Levitt, R.E., Kiviniemi, A. (2004). Fallacies of Fast Track Tactics: Implications for Organization Theory and Project Management, https://www.academia.edu/84039687/Fallacies_of_fast_track_tactics_Implications_for_organization_theory_and_project_management

Institute for Defense Analysis (IDA). (2006). Value Engineering Handbook. https://apps.dtic.mil/dtic/tr/fulltext/u2/a464089.pdf

Kohler, J. (2009). Fast Tracking or Back Tracking. https://www.projectmanagement.com/articles/284073/fast-tracking-or-backtracking-#_=_

Save International. (2007). Value Standard and Body of Knowledge, https://www.value-eng.org/page/Standard_Documents

Uher, T.E. (2003). *Programming and Scheduling Techniques.* Sidney: University of New South Wales.

III

Execution Processes

CHAPTER 10

Communication and Team Management

10.1 COMMUNICATION AND TEAM MANAGEMENT

Most aspects of a project involve communication and engagement. This is especially true for work involving team members and stakeholders. "Communication entails formal and informal communication, in addition to verbal and written communication" (PMI, 2021, p. 73). This chapter will provide an overview of tools to assist the PM in tracking essential aspects of the project to facilitate access to the accurate status of the project and to share information easily.

10.2 ROLES AND RESPONSIBILITIES

Instrumental to the project team's success, each team member must have an in-depth understanding of where they fit into the project tasks. The responsibility assignment matrix (RAM) is a matrix that "is a grid that shows the project resources assigned to each work package" (PMI, 2021, p. 189). An example of a RAM is displayed in Table 10.1. Furthermore, each team member's roles and responsibilities sheet with detailed information about individual responsibilities can also be helpful. Let's start this discussion with the popular RAM, which serves multiple purposes as it is an easy way for the PM to assess which team members may be involved in too many tasks. Also, the Gantt chart provides additional details, such as the amount of time needed to conduct the task, as that information would

DOI: 10.1201/9781003532989-13

TABLE 10.1 Example RAM Chart

Tasks	Richard	Craig	Darrell	Ryan	Lindsay
Develop the database	X	X	X		X
Perform the statistical analysis				X	
Develop the website			X		X
Create the training manual	X	X		X	
Create a training survey		X			X
Conduct the training					

not show on the RAM chart. However, a RAM chart still helps to provide oversight as to staffing levels within the team.

A RAM chart can provide many benefits to a project team. The RAM clarifies different team members' roles and responsibilities. It also provides clarity to the internal team members so that the team instantly knows which other team members are responsible for the same task. It is also helpful to external individuals outside the organization so that they, too, understand which employees are responsible for different tasks. Regarding structure, a RAM chart may be used only for internal team members or could contain all parties involved in a task. This is an individual choice for the PM. Overall, a RAM chart is valuable for all team members to understand the role and responsibility each team member has on the project.

A Responsible, Accountable, Consult and Inform (RACI) "chart is a common way of showing stakeholders who are responsible, accountable, consulted, or informed and are associated with project activities, decisions, and deliverables" (see Table 10.2). The design difference between a RAM and RACI is that the latter breaks out each team member's roles more by displaying who may be responsible and who is ultimately accountable for different roles. It also identifies who on the team may hold particular expertise and must be consulted to ensure the task is carried out correctly and successfully. Furthermore, the RACI chart identifies which team member needs to be informed of tasks that may not fit into the project schedule yet need to be accomplished. Overall, a PM may choose the RACI over the RAM data format because of the additional details that the chart provides.

Another way to clarify team members' responsibilities further is to provide each member with a sheet of their roles and responsibilities. The roles and responsibilities sheet are shared between the PM and a specific team

TABLE 10.2 RACI Matrix

Task #	Tasks	Completed	Date Task Identified	Task Due Date	Darrell	Ryan	Lindsay
1	Develop the Database	X	Month/ Day/Year	Month/ Day/Year	A	R	C
2	Perform the statistical analysis				A	I	C
3	Develop the website	X			A	R	I
4	Create the training manual	X			C	A	R
5	Create a training survey				A	C	I
6	Conduct the training				C	A	R

member as it contains a more specific definition of each work task. This communications artifact helps prevent misunderstandings later as it communicates, more specifically, the tasks assigned to each team member. This helps ensure that the team member is on the same page as the PM regarding what the task involves. Another benefit of this sheet is that it can later evaluate the team member's performance and the progress made within each assigned responsibility. Therefore, such tools are essential in interpersonal goal-setting and performance appraisal actions and are vital to the ongoing team management process. Table 10.3 displays an example of documenting roles and responsibilities for a team member. This format helps to provide clear roles and responsibilities for each team member. The RAM or RACI charts and the individual roles and responsibility sheet provide guidance and direction for each team member and the overall team.

A resource management plan "is a component of the project management plan that describes how project resources are acquired, allocated, monitored, and controlled" (PMI, 2021, p. 186). This plan helps identify the skill sets the project team needs to staff the project appropriately. The content of a resource management plan can vary depending on an organization, but the main categories of this plan are listed in Table 10.4. The introduction section will contain general information on the purpose and scope of the plan. This section will also discuss different processes, procedures, or documentation to manage the staff. The second section is

TABLE 10.3 Examples of Roles and Responsibilities of a Team Member

Suzanne Smith's Responsibilities	Task	Sub-Task
Task 1	Serve as lead database designer responsible for assigning work and monitoring the work of other team members in a database role.	• Identify database tasks. • Assign database tasks to team members. • Monitor each team member's progress on tasks weekly.
Task 2	Serve as lead website developer responsible for assigning work and monitoring the work of other team members with a website role.	• Identify website tasks. • Assign website tasks to team members. • Monitor each team member's progress on tasks weekly. • Update website information. • Design website. • Arrange access for all team members to the employee login of the website and shared drive. • Arrange access for all team members to the employee login of the shared drive.

TABLE 10.4 Resource Management Plan Template

Table of Contents for a Resource Management Plan	Description
Introduction	This section typically contains the purpose of the plan, the scope of the plan, and processes, procedures, or documentation used as part of the management of staff.
Process for Resource Management	This section will contain how the PM will plan for resources, acquire resources, train resources, track team members' work and performance, and transition resources throughout the project life cycle.
Roles and Responsibilities	This section contains the specific roles and responsibilities of different categories of team members (technical staff, engineering staff, etc.) and other project stakeholders.
Glossary/Acronyms	A list of glossary terms specific to the project should be included in the plan and an acronym table.
RACI (or RAM)	A RACI table or RAM table should be included to include all resource-related material in the plan.

centered on the process of managing the staff. This section contains everything from the planning for the resource to what is involved in acquiring the necessary team resources. It also includes how the resources will be trained and how the team members' performance will be monitored.

Finally, it discusses how the resources will be transitioned off the project team at the project close. The third section identifies roles and responsibilities for each resource category, such as engineering and technical staff. This section includes the description of the project stakeholders and the role these stakeholders play in the project, such as providing equipment, facilities, and other team roles. Examples of stakeholder categories may be an advisory team to the project, project sponsor, specific management, or department. The fourth section of the plan is a glossary of all the terms particular to the project and an acronym table. The last section of this plan is to display the RACI and RAM, which were discussed earlier in this section. This plan is an essential definitional document that is useful as input to the managing communications process.

10.3 PROJECT MANAGEMENT INFORMATION SYSTEM

Communication technology must be identified so stakeholders know where project information is stored and available. This may be called the project management information system (PMIS). It comprises electronic project management tools such as software, web interfaces, portals, and collaborative work management tools like Dropbox or Google Documents. Different access levels to the project information can be given to project stakeholders based on their needs. It also consists of electronic communications management, such as email or Teams. Furthermore, it can include social media management, such as private group pages on Facebook or Instagram. Technology is ever-changing, and the ability to share information is growing. Many social media accounts can be linked together so that one meeting reminder posted on one social media platform will result in all the project social media accounts posting the same information so that every team member will see the reminder regardless of which social media accounts they hold.

10.4 TRACKING TOOLS AND TECHNIQUES

Different status tracking tools and techniques exist to help the PM and team stay updated on the ongoing status of various versions of documents, commitments on presentations, or demonstrations promised by team members. The commitment-tracking document helps the PM track the formal commitments made by the team members. Table 10.5 displays an example of a commitment-tracking document filled out with sample information. It is a great organizational tool that tracks mismanaged communications

TABLE 10.5 Commitment Tracking Document

Commitment	Date of Commitment	Team Member Making the Commitment	Person to Whom the Commitment Was Made	Date Commitment Is to Be Completed	Comments
Demonstration of Website Prototype	Month/Day/Year	Project Team Member Brooke Jones	Customer Joseph Thomas	Month/Day/Year	Team Member Brooke Jones will present the demonstration.
Quarterly Status Review to Management	Month/Day/Year	Project Team Member Stan Lewis	Customer Norma Smith	Month/Day/Year	Team Member Stan Lewis and the PM to present.

promises, often by logging them and tracking them to fruition. This is yet another tool for helping to keep track of less well-defined elements related to daily activities associated with properly executing the project.

A communication management plan is "a component of the project, program, or portfolio management plan that describes how, when, and by whom information about the project will be administered and disseminated" (PMI, 2021, p. 237). This tool is helpful in the PM identifying when communication is needed and expected with different project stakeholders. An example of a worksheet tool that can be used to track communication is shown in Table 10.6.

TABLE 10.6 Communication Management Plan

Stakeholder	Type of Communication	Schedule	Responsibility	Communication Format
Project Team	Weekly Status Review	Every Thursday at 10 a.m.	Project Manager	Meeting/Teleconference
Program Manager	Weekly Team Progress Report	Close of Business on Monday	Project Manager	Email
Program Manager	Monthly Team Progress Report	First Monday of every month at 10 a.m.	Project Manager	Meeting
Project Sponsor	Project Review	First Tuesday of every month at 10 a.m.	Project Manager	Meeting
Customer	Customer Reviews	Second Monday of every month at 2 p.m.	Program Manager Project Manager	Meeting with the customer and responsible team members

TABLE 10.7 Stakeholder Register

Name/ Position	Contact Information	Team Role	Degree of Influence	Degree of Support	Expectations	Interests
Madison Carre/ Program Manager	Phone Email Address	Sponsor	High	Medium	Monthly Communication, Status Updates on Customer Satisfaction with Deliverables	Enrichment and Advancement of Project Team Members
Samuel Richards/ Quality Director	Phone Email Address	Quality Assurance	Medium	Medium	Weekly Communication, Immediate Notification of Product Defects	Company Reputation, Quality Improvements

The Stakeholder Register is another crucial "project document that includes information about project stakeholders, including an assessment and classification of project stakeholders" (PMI, 2021, p. 250). An example of a Stakeholder Register is displayed in Table 10.7. It is a chart that contains specific information on each stakeholder, such as contact information, function (i.e., sponsor, customer), expectations, degree of influence on the success of the project, etc. It will be updated throughout the project life cycle as project changes occur.

A stakeholder engagement plan is "a component of the project management plan that identifies the strategies and actions required to promote productive involvement of stakeholders in project or program decision-making and execution" (PMI, 2021, p. 250). This is another input into the management of project communications. An example of a worksheet tool to assist in tracking stakeholder engagement is shown in Table 10.8.

TABLE 10.8 Stakeholder Engagement Plan

Stakeholder Organization, Group, or Individual	Priority	Communication Strategy	Project Phase	Engagement Tools
Risk Manager	Medium	Engage during project kickoff through project close-out	All	Face-to-face Email
Quality Manager	High	Engage during product development	Development Production	Face-to-face Email Demonstrations Product inspection

TABLE 10.9 Issues Log

Issue Log #	Report Date	Issue Description	Reported By	Urgency	Assigned Team Member	Status	Solution	Close Date
1	Month/ Day/ Year	Broken Link on Website	Alec Monroe	High	Pam Mellon	Closed	Link Fixed	Month/ Day/ Year
2	Month/ Day/ Year	Website Navigation Needs Improvement	Sophia Lane	Medium	Victor Arias	New	Perform User Testing	Month/ Day/ Year

The issue log is "a project document where information about issues is recorded and monitored" (PMI, 2021, p. 241). Table 10.9 shows an example of an issue log. Every organization or even project should have its version of this communications thread based on what is most important to track for each unique project. The example shown here has many different columns to track the issue number, report date, description of the issue, and name of the person who reported the issue. It also lists the degree of urgency to resolve the issue, the assigned team member responsible for the resolution of the issue, the status of the issue, the solution to resolve the issue, and the close date that represents when the issue was resolved.

The Lessons Learned Register is "a project document used to record knowledge gained during a project, phase, or iteration so that it can be used to improve future performance for the team and the organization" (PMI, 2021, p. 242). An example is displayed in Table 10.10. Many companies have lessons learned databases where employees can search for keywords to find how the company dealt with a specific issue in the past

TABLE 10.10 Lessons Learned Register

Lesson Name	Category of Lesson	Description	Solution	Action
Improving Team Communication	Communication Management	Challenge: Communication with co-located team members	Increase technology use and weekly communication	Increase team communication by holding two teleconferences a week
Extending the Risk Mitigation Process	Risk Management	Shortcoming: Inadequate risk mitigation of software bugs	Invite more stakeholders to participate in risk mitigation tasks	Submit a Change Request to revisit risk mitigation tasks and increase software testing

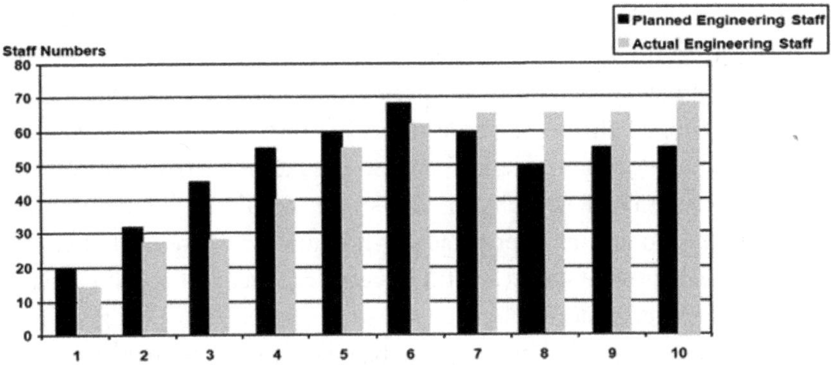

FIGURE 10.1 Staffing profiles.

consisting of lessons learned on any topic, from running effective project team meetings to resolving faulty wiring. These registers contain the root causes of an issue along with corrective actions. The information contained allows employees to learn from others on how to be proactive in preventing an issue from arising or on how to fix an issue that has been presented to the project team.

One of the key management actions is to track the needs of project human resources. Failure to obtain planned resources on schedule is one of the primary controllable factors leading to schedule variation. For this reason, it is vital to track the plan versus the actual status of this item. Figure 10.1 displays the staff for any category needed during the project life cycle. The required staff is categorized as "planned"; in this example, the amount of engineering staff needed is graphically represented. The "actual" category is the staff within that specific discipline available for the project. From this view, it is possible to tell immediately where the resource gaps are throughout the project's life cycle. In some cases, this gap analysis may be instrumental in explaining why the project is falling behind.

10.5 PROJECT CLOSE-OUT

Archiving various project data is an essential aspect of project close-out and part of communications management. Furthermore, an electronic backup is necessary as technology-based storage devices can become damaged, so providing backup files is essential. What types of data are needed to be archived? That may depend on the organization, project type, or other contractual requirements. Think about a construction project where many different homes of the same model may be built over time.

In that case, as much data as possible should be kept as it will reduce the learning curve involved with each additional house being built. However, a one-time project may require less project data to be archived. Generally, reports developed throughout the project, along with documents such as Gantt charts, project charter, Work Breakdown Structure (WBS), budget information, RAM charts, and so on, should all be maintained. Furthermore, a one-time-type project still requires archived data because much of the documentation can still be recycled, even if it is just partially useful for another project.

Another area of project close-out is the recording of status metrics. An organization and even the PM need to maintain metrics regarding the project to help estimate and plan future efforts, which will be discussed later in Chapter 13, Project Performance Tracking. Many different types of metrics can be recorded depending on the project. Retrospectively evaluating a project helps one identify and validate additional metrics. Communicating these metrics can help to document the success of the PM and the project team.

Another aspect of project close-out is to prepare a final report and update any necessary company documentation, such as vendor status and financial summaries. The final project close-out report, as a minimum, should include an executive summary, including information related to scope, cost, schedule, and lessons learned sections. The Lessons Learned Register previously discussed can be consulted for this section of the report. Some organizations have formal online lessons learned databases where employees can search different problem areas, from how to hold an effective meeting to how to effectively and efficiently get airplane maintenance performed between flights. This is also the time to document in the report all project objectives that were successfully achieved and those that were not. The final step of the close-out report is to obtain a formal signature from the customer that project deliverables were received and accepted. As the team finalizes the report, the identification of improvements or enhancements to formal documentation is needed, too. For instance, the team may realize that an organization's IT support policy, process, procedure, or technical documentation format should be modified so that the documentation includes information not previously included. For instance, the project team may recommend using another method to house the project team's meeting minutes. Project close-out is essential for an organization's survival and continuous improvement. Also, the type of formal project documentation, as outlined here, is necessary because all aspects of a project need to be captured and utilized as future intellectual property for internal improvement.

10.6 ORGANIZING PROJECT TEAMS

Now, we move into the team management section of this chapter. Project teams need to have a clear understanding of their role within the team and the overall organization. In many situations, groups can use an input-and-output relationship model to complete project tasks effectively. Figure 10.2 displays how every team member's "output" is "input" for another team member's assigned task. Understanding this concept will help a team, and, specifically, each team member will identify how they fit into the overall scheme of the project and how important each person's role is to the team. This model will help the entire team because each team member will understand the importance of keeping up with their deadlines so that their output gets into the hands of the team member(s) who need it as input before beginning another task. If any team member is late with their output, it could delay the entire project because the next task can't start without the other team member's output. Eventually, all the output generated leads to successfully completing project deliverables.

Project software tools allow for names to be displayed, such as a business analyst job title or a person's actual name, to be listed next to each work task on a Gantt chart. This essentially serves as a more detailed version of the simple input-and-output relationship model. One example of a project software tool where different job titles or names of project team members can be displayed next to project tasks is shown in Figure 10.3. Informing team members of the importance of each team member

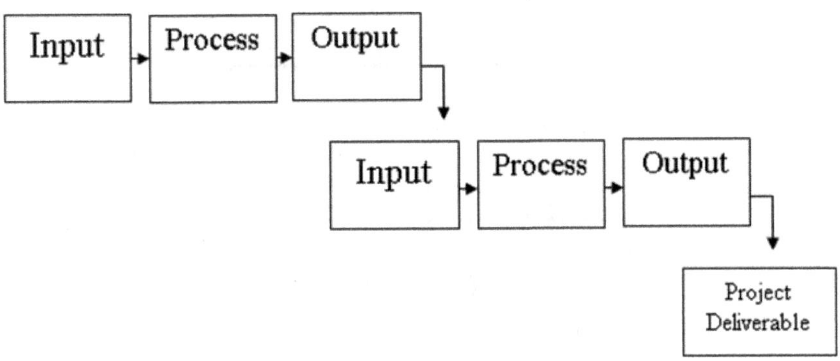

FIGURE 10.2 Team member input and output.

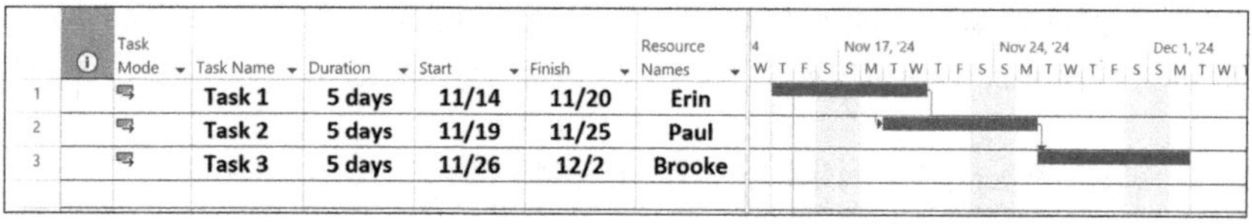

FIGURE 10.3 Microsoft project Gantt chart.

fulfilling their responsibility is vital to the success of the team working together to carry out the project tasks successfully.

A simple organizational chart must be developed for project teams so the PM and the team members know which team member(s) are responsible for different aspects of the overall project work. Figure 10.4 displays a sample organizational chart. An organizational chart can list names under each box so that each team member knows the chain of command relative to their position. This assists in communication flowing to the appropriate team member, resulting in the most efficient way to resolve problems that may occur throughout the project duration. There are three main benefits of having a formal organizational chart. First, it provides all team members with an understanding of how each member fits into the overall project. Second, it offers internal team members information regarding the different levels of authority. Third, it provides external individuals affiliated with the project, such as customers, with an understanding of the roles and authority that other project team members have. Project teams should formally display an authoritative structure. PMs

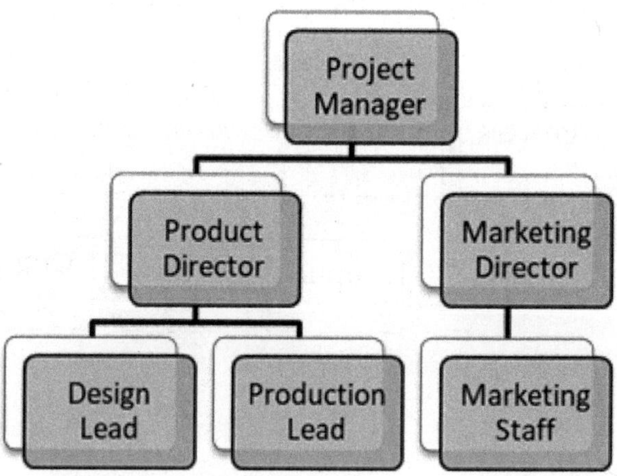

FIGURE 10.4 Project team organizational chart.

only sometimes think of providing a specific project team organization chart because organizations already have an organizational chart. However, the organization chart doesn't contain the details for each project within the organization, necessitating a project team organization chart.

10.7 PROJECT TEAM PERFORMANCE TOOLS

PMs need skills to lead their teams effectively. PMI (2021) describes many valuable skills in this endeavor, such as emotional intelligence, project management experience, conflict management, self-awareness, social skills, self-management, social awareness, and critical thinking. PMs must create a continuous improvement environment to lead a project team successfully. PMs must be good communicators, using open and effective communication. PMs need to help team members develop trust among each other, which can be accomplished through team-building opportunities. Conflict management is another strength a PM needs because there will not just be one or two conflicts but likely daily conflicts between all the stakeholders involved. Table 10.11 is a tool to help a PM resolve conflict. This tool can be used with the stakeholders in conflict and by the PM as the facilitator. Having an encouraging PM that involves the team members as needed in problem-solving and decision-making is vital to the team's success. Another trait that a PM should have is listening actively. It is also vital to have cultural awareness and ensure all team members respect different cultures. Meeting management is another important PM trait. Agendas for meetings should be sent out in advance, and there should always be communication about upcoming meetings.

TABLE 10.11 Conflict Resolution Template

Conflict Resolution Worksheet	
Identify the Conflict	Identify the Stakeholders Involved
Describe Options for Resolution	Describe the Consequence for Each Option
Prioritize Each Option	Select an Option
Create an Implementation Plan for the Option Selected	Assign Dates for Implementation Plan Tasks
Signature of Stakeholder 1	Comments by Stakeholder 1
Signature of Stakeholder 2	Comments by Stakeholder 2
Signature of PM	Comments by PM

There are many tools available to assist a PM. Some of these were discussed earlier in this chapter. There are five distinct phases of team development: forming, storming, norming, performing, and adjourning (PMI, 2021). In forming, teams come together and get to know each other. Team building at this early stage can begin with activities such as brainstorming how to best meet the project objectives. Table 10.12 is a tool for brainstorming tasks and any project considerations that can be asked as a group or a combination of both individual and group activities integrated. Another format to use is to request that each team member list one task per sticky note. There will be several iterations of this as themes emerge and discussions occur among the group. The different themes will then become tasks with subtasks, subsubtasks, etc., to follow. The second stage of team development is storming. During this stage, team members will be jockeying for positions on the teams. This is where team members' personalities become apparent, and their strengths and weaknesses show. Teams learn how to work together and form trust among each other. The third stage of team development is the norming stage. Things start to normalize at this point, and the team functions as a collective body. Cooperation among team members suddenly emerges. The fourth stage is performing because the project work occurs while working together on shared goals. The last stage is adjourning. This is where the team completes work, and team members disperse for other work. A sign of a team with good relationships occurs when team members are sad to depart from their colleagues.

Table 10.13 displays a template for a team performance appraisal form. It is essential to share with the team when performance evaluations take place. Communication is key to success, so be sure to talk with team members at the earliest opportunity when they need to make

TABLE 10.12 Project Brainstorming Template

Brainstorming Template
Brainstorm tasks needed to be conducted to meet the project goals.
Brainstorm input for the quality management plan.
What are the challenges for this project?
What are the project risks?
What are ways to mitigate the risks?
Who are additional stakeholders yet to be identified?
What questions still need to be answered?

TABLE 10.13 Performance Appraisal Template

Project ________________ Team Performance Appraisal				
Team Member's Name:		**Performance Review Date:**		
Position:		**Reviewer's Name:**		
Performance Categories/ Performance	**Exceeds Expectations**	**Meets Expectations**	**Below Expectations**	**Comments**
Job Knowledge				
Productivity				
Quality of Work				
Attitude				
Adaptability				
Interpersonal Relations				
Decision-Making				
Overall Comments				
Team Member's Strengths				
Team Member's Weaknesses				
Recommendations for Development and Improvement				
Team Member's Comments				
Team Member's Signature				
Project Manager's Comments				
Project Manager's Signature				

improvements. However, be sure to point out the strengths you see in all team members.

10.8 PROJECT CLOSE-OUT

As projects near the end of their existence, the PM should take responsibility for reassigning the staff. Each staff member within the organization must be reassigned to a new project once the current one ends. The reality of this process is that new projects may need team members before the current project has ended. This can result in many resource issues for the current project. Sometimes, it is possible to have project team members maintain some percentage of work with the current project while also beginning to take on their new role on their new project. These special activities may consist of a team member needing to begin attending team meetings for the soon-to-be new project team. The more organized a team is, the easier it will be when resources need to reduce their percentage of work time on the current project. If different tasks are well documented, then a PM may be able to negotiate the addition of a new team member even as the project is coming to an end. Often, individuals are between projects, and instead of being idle, they will be assigned to help a team with project close-out activities. Ideally, the same team that began the project will end the project, but this seldom happens.

10.9 CONCLUSION

This chapter addresses the value of various tools that assist the PM with effective communication. It is essential to provide internal and external team members with the class of project artifacts discussed in this chapter. These documents assist project stakeholders with whom to contact for different project needs. Different roles and responsibilities tables were discussed as a means for the PM and the team members to document individual task assignments and which team members work together on specific tasks. The chapter includes a discussion on technology and, specifically, the components of the PMIS. Tracking tools such as the commitment-tracking document and communication management plan were discussed to help a PM and project team track the frequency of team activities such as meetings, work assignments, commitments, and documentation control. A Stakeholder Register and Stakeholder Engagement Plan were discussed, representing a tool for the PM to track crucial demographic information on the project stakeholders and the best way to engage these stakeholders during the project. An Issue Log is another valuable tool in tracking ongoing issues as they emerge throughout the project life cycle. This tool is helpful in assigning a responsible person to follow the resolution of an issue. A Lessons Learned Register was also discussed to help share lessons that could contribute to the efficiency or quality of projects. Lastly, the area of project close-out was discussed, identifying the tasks that the PM and team members need to complete before officially ending a project. With effective communication, a project will avoid many challenges and possibly fail. Effective communication skills tend to make or break a PM's longevity in their field. Also, failure to communicate is often ranked as the number one reason for projects to fail.

This chapter also addressed the dynamics of the project team. Project teams must clearly understand their role and how each team member fits within the team. The organizing project teams section discusses a simple input-and-output relationship model to help team members complete their tasks and assist their team members with their tasks. The Gantt chart and how it can be used effectively in identifying different team members' assigned tasks are discussed. A project team organizational chart is also addressed in that it provides all team members with an understanding of how each member fits into the overall project. In the project team performance tools section, the team development model consisting of forming, storming, norming, performing, and adjourning was discussed (PMI, 2021). A brainstorming template tool was provided. Furthermore, a project team member appraisal form template was also offered. Lastly, project close-out activities were discussed.

QUESTIONS

Student Review Questions

1. Describe and differentiate between the RAM, RACI, and the individual responsibility sheet.
2. What is a Resource Management Plan?
3. Describe a Stakeholder Register and Stakeholder Engagement Plan.
4. Explain the role of each of the different tracking documents:
   - Commitment-tracking document
   - Communication management plan
   - Issues log
   - Stakeholder Register
   - Stakeholder Engagement Plan
   - Staffing profiles
5. Summarize the project close-out activities discussed in this chapter.
6. What is the purpose of an organization chart for a project?
7. Describe the P-CCM and differentiate between the different maturity levels.
8. Discuss what is meant by the five stages of group development.
9. Discuss the purpose of a performance appraisal.
10. Summarize the project close-out activities.

In-Class Questions

1. Discuss other tracking documents outside of those discussed in this chapter that may be useful.

2. Identify templates available online for developing a Resource Management Plan.

    - Identify templates on the Internet for each of the following:
    - Communication Management Plan
    - Issues Log
    - Stakeholder Register
    - Stakeholder Engagement Plan

3. What can a PM do to help their project team go through the five stages of group development?
4. What can a PM do at each P-CCM level to help the project team and organization?
5. Find team-building activities online.
6. Find brainstorming tools online.
7. Identify templates available online to assist in conducting a performance appraisal.

EXERCISES

1. For your student team, develop a RAM or RACI and an individual responsibility sheet for each team member.
2. For your student team, develop a Gantt chart with the names of your team members affiliated with specific tasks.

REFERENCE

Project Management Institute (PMI). 2021. *A Guide to the Project Management Body of Knowledge*, 7th edition. Newtown Square, PA.

CHAPTER 11

Project Financial Analyses

Financial concepts lie at the core of many project decisions and are key variables used to evaluate various project decisions throughout the life cycle. Organizations use multiple forms of the concepts outlined here, representing an important knowledge area for the PM. A primary role of financial measures is to quantify the financial view of project status. In this view, the goal is to assess the relative measure of benefits versus the associated costs.

11.1 INTRODUCTION

There are three classic financial measures demonstrated here. These are payback, net present value, and internal rate of return. Each of these has inherent advantages and disadvantages in its underlying constructs, which should be understood before making decisions based on any of them.

11.2 BENEFIT MEASUREMENT

Step one in the project financial measurement process is to estimate the benefits and costs of a project proposal over its life cycle. This is often done for five years, given the inaccuracy of estimates longer than this. For example, let us assume the project had a cost-benefit estimate, as defined in Table 11.1.

DOI: 10.1201/9781003532989-14

TABLE 11.1 Project Cost-Benefit Estimate

	Project X ($)
0	−605,000
1	305,000
2	300,000
3	15,000
4	15,000
5	15,000
Total	**45,000**

The financial interpretation of the Project X data indicates that the initial project development would cost $605,000 (negative), and the project would be completed a year later, at which point an array of varying positive benefits would occur over the next five years. The project would generate a stream of net positive business over its life cycle, as shown. Summing the total cost and benefit values shows the project would generate $45,000 in net benefits over five years. A fundamental financial question related to this data set is, "Is this a good financial project?" The answer to this depends on how one looks at the data using basic financial measures. This data set will describe the classic financial measures and illustrate how this might yield potentially different answers.

11.3 BASIC FINANCIAL MEASURES

The *Payback (PB)* method is the simplest and most frequently used financial measure. The payback period is calculated by counting the number of years it will take to recover the cash invested in a project. The goal in using this measure would be to favor projects that pay back their investment the quickest.

Net Present Value (NPV) is defined as the summarized present value of the benefit stream. The value is calculated by translating a future stream into the current time at some specified interest rate. The relevant formula for this is $FV/(1+i)^{\wedge n}$, where FV is the future value, i is the monetary value, and n is the number of years in the future. This measure is sensitive to the time value of money and uses that view to compute the net benefit of a projected cash stream in dollars. This measure would favor large benefit streams over smaller ones.

The *Internal Rate of Return (IRR)* represents the percentage rate of return calculated for the benefit stream. It is a quantitative measure of the effectiveness of an investment project. This is analogous to an investment return on the development cost. Organizations often have a percentage *hurdle rate*, which is used to compare across various options. Thus, the hurdle rate is the net benefit rate of return used for financial comparisons. This measure is the most complex financial calculation and was often not used in the past because of the computational difficulty. However, given the introduction of an automated tool, this method has become more common in use. Its advantage is in producing an equal measure for all diverse benefit streams. This measure is the most financially pure of the three options in that the relative return value of the stream is measured regardless of size or payback time.

Recognize that these three measures offer different views of the same benefit stream. Despite the lack of financial purity, the payback method is favored by many organizations. The listing below summarizes the basic calculation mechanics for each of the measures:

1. PB is the time required to pay back the initial investment.
2. A future benefits value can be converted back to the present time by the formula $(1+i)^{\wedge n}$, where i is the interest rate used to calculate the value of money. The term n is the number of years in the future from which the amounts need to be transformed. Project assessments look for the present value of a future benefit stream in this way.
3. IRR uses the same set of mechanics as NPV, except that the value of i is unknown; therefore, it is necessary to repeat the calculation with estimated values of interest until a value of NPV is zero. So, IRR is the value of *I (interest rate)* where the NPV is zero. Spreadsheet utilities such as Excel can perform this calculation.

11.4 TIMING ISSUES RELATED TO ESTIMATES

When using these measures, the question of timing becomes a key issue. Do we assume the cost occurred at the start of the year, in the middle of the year, or spread equally through the year? For the examples shown here, we will follow the Excel default assumptions; however, for critical analysis, the timing question should be reviewed to ensure that the calculated values properly reflect the true nature of the stream. In calculating NPV, Excel assumes that the cash flows occur at the end of the indicated period so that one can see a built-in bias with that assumption.

11.5 SAMPLE CALCULATIONS

The Big G Company is evaluating three potential projects, but only one can be pursued because of resource constraints. Each project is expected to require $605,000 in development costs, but each has a different benefit stream, as shown in Table 11.2.

This set of cash flow data will be used to demonstrate the three financial measures.

Note: Excel contains the built-in function to calculate both NPV and IRR. Demonstrations of these can be found in online videos. Be aware of the timing calculation default in Excel. Usually, these are minor issues, but they are worth reviewing before use.

11.5.1 Manipulating the Data

For this calculation, assume the company wants only to consider projects that return more than 10%. The calculation logic for this and the other measures is summarized below.

Figure 11.1 illustrates the process of moving the future year benefit values back to a current period equivalent using the NPV formula. As an example of the NPV process, the transformation value for year five is (1.10)^5 or 1.61051. So, the present value of $50,000 at year five is transformed to the equivalent current value by the equation $50,000/1.6105, or $31,046. This means that a $50,000 benefit in year five is only worth $31,046 at present because of the time value of money. It is sometimes easier to envision this process in reverse. How much would a savings account of $50,000 be worth in five years? This scenario involves moving a future value into the present time frame, so the equation is $50,000 times 1.61051, or $80,525. The value equation works in both directions.

TABLE 11.2 Portfolio Cost-Benefit Estimates

Year	Project X ($)	Project Y ($)	Project Z ($)
0	−605,000	−605,000	−605,000
1	50,000	325,000	130,000
2	150,000	175,000	130,000
3	300,000	75,000	130,000
4	100,000	50,000	130,000
5	50,000	25,000	130,000
Total	**45,000**	**45,000**	**45,000**

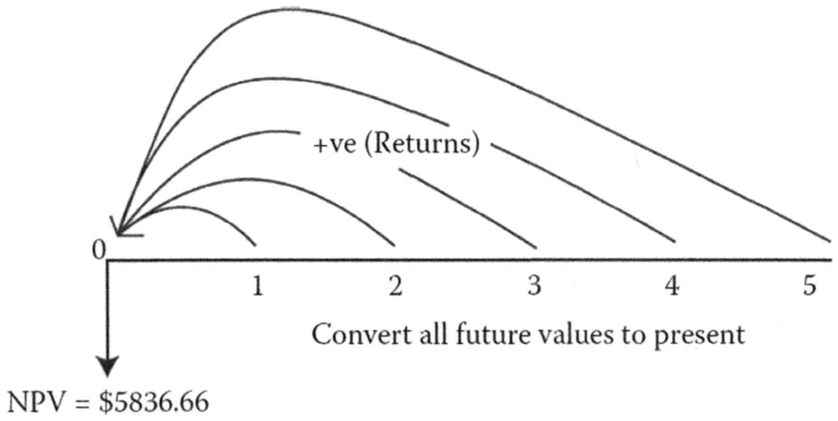

FIGURE 11.1 NPV transformation flow diagram.

The NPV process must be repeated for each year to compute the overall decision NPV. Figure 11.1 schematically shows the time movement process for each year.

11.5.2 Project X, Y, and Z Calculations

Table 11.3 shows the results of the NPV manual calculations for the three projects. This can be replicated using Excel, although not shown here.

The NPV summary calculation shows Project X as the clear winner. It would be worthwhile to manually execute a sample calculation of these values to ensure you understand the formula. Also, compare these values with the Excel output (expect slight differences based on the default timing assumption).

The next step is to use the same cash flow data and calculate the IRR values for each project. This won't be attempted manually, but you can try

TABLE 11.3 Project NPV Calculations

Year	Factors	X NPV ($)	Y NPV ($)	Z NPV ($)
0	1	−605,000	−605,000	−605,000
1	1.1	90,909	309,091	363,636
2	1.21	123,967	144,628	165,289
3	1.331	266,717	75,131	22,539
4	1.4641	204,904	23,905	0
5	1.61051	31,046	0	12,418
Total =	**NPV =**	**112,543**	**−52,244**	**−41,117**

TABLE 11.4 Project Summary Comparison

Project	NPV ($)	IRR (%)	PB
X	112,543	17	3
Y	−52,244	4	2.8
Z	−41,117	5	4.8

a sample to see why no one does this manually. The PB values can be produced by manually inspecting the data. In this case, the question is how long it took to recover the initial investment.

11.5.3 Summary Results

IRR values for the defined benefit stream can be easily calculated using Excel's bulletin utility. Table 11.4 summarizes the financial measures for three projects.

In this review, Project X wins based on the NPV and IRR measures. Also, this project is the only one of the groups that passes the minimum hurdle rate of 10%. Project Y's only positive attribute is quick payback, but the other measures are poor and unacceptable.

The question might be raised about why an organization would accept a payback value over an NPV or IRR calculation. If a project were to have a payback of less than three years and an acceptable IRR, it might be selected over the same IRR but with a longer payback value. The logic behind this is a lack of confidence in the projected benefits. This measure favors short-term benefits over the longer-term ones. On the counter side of this, if all projects are selected favoring payback over IRR, a "market breaker" option that takes longer to get started will lose. If project Y had an acceptable IRR, it would likely be accepted over another candidate with a longer payback.

11.6 MULTI-CRITERIA ANALYSIS

There are many different reasons for a company to select a project other than financial benefits, and for this reason, a more advanced approach can be pursued using weighted multiple criteria. For example, the following list offers some non-financial criteria that might be judged important in a project proposal:

Strategic goal fit

Availability of resources

TABLE 11.5 Multi-Criteria Project Assessment

Criteria	Weight (%)	Project X Criteria Rating 1 = Low, 5 = High	Project X Weighted	Project Y Criteria Rating 1 = Low, 5 = High	Project Y Weighted	Project Z Criteria Rating 1 = Low, 5 = High	Project Z Weighted
Strategic Fit	20	4	0.8	3	0.6	5	1
Availability of Resources	10	3	0.3	4	0.4	4	0.4
Probability of Success	5	4	0.2	3	0.15	5	0.25
Revenue Impact	10	2	0.2	4	0.4	2	0.2
Operation Cost Impact	10	4	0.4	2	0.2	4	0.4
IRR	40	1	0.3	4	0.8	1	0.2
PB	5	3	0.15	4	0.2	2	0.1
	Score		2.65		3.55		2.75

Likelihood of success

Increased revenue potential

Potential to achieve market dominance

Multi-criteria analysis is a technique used to combine additional factors into the overall view. Table 11.5 shows how a broader perspective can be added to the raw financial view. Note that some of the attributes used here are pure numeric, and others are qualitative assessments translated into an integer score.

As can be seen from the weighted factors used for each area, the resulting project score involves more than just the financial metric. From the weighted score, Project Y is still shown as the best option based on the highest weighted score of 3.55. However, be aware that the weights assigned for each category are based on raw judgment, and it is often wise to compare other weightings to see if the answer is sensitive to a particular weighting factor. This type of complex analysis is common in technical selection actions.

11.7 CONCLUSION

This chapter has provided a summary of classic financial-oriented measures commonly used in project analysis. These measures must be understood by the PM, along with the underlying assumption related to the measure. The analysis models described here are considered common financial techniques and should be understood by the modern PM.

REVIEW QUESTIONS

1. In comparing two alternatives, option A had an IRR value of 12% and a PB of 3, while option B had an IRR of 14% and a PB of 5. Which of these should be selected, and why?
2. Is there a situation where PB might be the best financial metric?
3. What evaluation criteria might be added to the matrix?
4. What might be the value in selecting one of the other projects?
5. Do you see any flaws in the manner in which weights are assigned to the various criteria?

CHAPTER 12

Project Performance Metrics

This chapter explores using performance-measuring variables to define the project status. *The mechanics related to project status are heavily based on defining various project variables that represent events.*

12.1 INTRODUCTION

Status variables are metrics, measures, Critical Success Factors (CSFs), or Key Performance Indicators (KPIs). A brief formal definition of each of these follows:

- Metrics are "a description of a project or product attribute and how to measure it" (PMI, 2021, p. 243).
- Measures are the value of a metric.
- CSFs (Key Results Areas or KRAs) are the project's status areas related to successful outcomes.
- KPIs are a quantifiable measure of performance over time for a specific objective.

When all is said and done, these terms are not separated. They each provide some quantification related to a defined performance item about areas such as delivery functionality, resource schedule, or cost.

DOI: 10.1201/9781003532989-15

More specifically, metrics are used to define the attributes of a process or project, and numbers or grades represent them. More important, a measure is a value for a metric. These items are defined to be critical to a project's success and, for that reason, must be relevant and quantifiable. Table 12.1 displays examples of metrics that a project might use for status reporting. Metrics should be linked to the project's defined vision and strategy to form tangible targets and focus work activity on formal CSFs (Richardson & Jackson, 2025). CSFs are related to activities that must occur for the project to produce its deliverables successfully. CSFs include measures indicating adherence to schedules, budgets, product quality, process values, and so on (Kerzner, 2011).

A simple example of a metric is the speed of the car. The measure for this metric is the actual value of the metric (e.g., 55 miles/h or 200 km/h).

TABLE 12.1 Examples of Metrics

Types	Examples
Quality	Number of defects
Customer Satisfaction	Number of positive reviews on social media Number of customer complaints
Resources	Actual project team members' hours and overtime hours
Scope	Number of project change requests
Action Items	Number of actions completed on time or behind schedule
Resource Utilization	Percentage of time project team members spend on work (includes tasks that a company pays for, with examples being proposal preparation and tasks funded by external sources, such as working on a funded project task) *Billable Utilization = percentage of time project team members spend on funded activities
Productivity	Compare the total effort to the budgeted effort Task delays Underperformance
Gross Profit Margin	Gross Profit Margin = (Total Profit) – (Total Costs)/100
Planned Value (PV)	PV = (the hours left scheduled on the project) × (project worker's hourly rate)
Actual Cost (AC)	All project-related expenses
Earned Value (EV)	Resources consumed to complete the planned work
Return on Investment (ROI)	A measure of benefits versus cost
Cost	Planned budget versus actual budget
Schedule	Schedule performance versus plan
Cost Performance	Budget versus actual amount spent

Therefore, a measure contains a defined standard unit of measurement. In some cases, a metric is defined as combining two or more discrete measures, such as using resource hours per function point as the combination metric. Note that a metric "measures" something, and a KPI "identifies" a formal interest in some performance attribute. A KPI is a snapshot of the project's progress toward its targets in some area (Kerzner, 2011). They draw attention to and evaluate a particular situation, such as flagging a particular cost deviation outside predetermined tolerances or control limits.

A common management adage says, "You cannot control what you cannot measure." Another common management adage is "What gets measured gets done." Recognize that it is also difficult to improve something that does not have a defined performance value. In this view, a KPI provides a way to measure current performance against benchmarks to evaluate strengths and weaknesses (Kaskinen, 2007). Once a particular weakness is identified, the project team is better positioned to evaluate corrective actions, such as assigning more resources or improving some operational processes to correct a deviation. KPIs also provide management with ways to measure and manage a company's performance.

12.1.1 Project Benefits Management Plan

Projects provide numerous benefits, but these are often not enumerated formally. The project benefits plan outlines areas where the project will produce improvements. Table 12.2, describes examples of project benefits, and this can serve as a template for the project benefits management plan.

An associated Benefits Management plan identifies specific measurements that will be used to *quantify benefits realized from the project*. The components of this plan are target benefits that consist of value gained through the implementation of a project, such as financial value. This plan will also have a strategic alignment section describing how the project aligns with an organization's business strategy. The timeline for realizing these benefits is also discussed in the plan. A benefits timeline is defined as short-term, long-term, or more continuous ongoing benefits. The benefits owner section describes who is accountable for tracking and recording the status of a specific benefit. There is also a metrics section to discuss the direct and indirect measures of project benefits. Two other plan sections relate to risk areas that can affect the receipt of benefits and a list of key design assumptions that can negatively impact the project outcome if proven wrong.

TABLE 12.2 Examples of Project Benefits

Benefit	Examples
Social	Increase public safety Increase citizen engagement
Environmental	Increase traffic flow Reduce pollution Increase recycling efforts
Economic	Increase business growth Increase jobs
End User	Improve safety Improve efficiency Improve utility
Health	Increase preventative and wellness programs Reduce the length of stay for patients in hospitals
Energy	Enhance the quality of life for communities Lower greenhouse gases
Revenue	Increase profits from a new product Increase profits from a new process
Efficiency	Improve systems Improve infrastructure
Reputation	Increase repeat business Increase new customers
Organizational Culture	Increase employee satisfaction Reduce turnover

This section should detail all expected values produced from the implemented project. This includes all tangible and intangible benefits. Tangible benefits, such as cost savings, can be easily measured, while intangible benefits are more difficult to measure, such as the positive impact on a business due to employees having increased job satisfaction (see Table 12.3).

12.1.2 KPIs

KPIs can relate to broad areas of project performance and, therefore, need to be interpreted differently. This makes it difficult for companies to develop meaningful and valuable KPIs (Enoma & Allen, 2007). However, a KPI is an indicator signaling the status of that defined area. Based on this design characteristic, a collection of KPIs quantifies the project's performance (Chan & Chan, 2004). Not only is it challenging to develop meaningful KPIs, but it is also often a mistake to define too many KPIs, resulting in the generation of too much data beyond what is needed to

TABLE 12.3 Project Benefits Management Plan Template

Target	Description
Strategic Alignment	This section should contain how each benefit to be realized by the project aligns with the company's mission, goal, and objectives. It should also include how any process changes and initiatives to produce the benefit align with the company's strategic direction.
Realizing Benefits	This section outlines the timeframe of when the project will produce each benefit. This can be expressed in calendar dates or listed as short-term, long-term, or ongoing for each project benefit.
Benefits Owner	This section should detail the person(s) accountable for tracking, recording, and reporting each benefit the project provides during the life cycle of the project. This section should also identify a sign-off process for the measures.
Metrics	This section will list the measures used to show how each benefit has been realized. The measures include both direct and indirect measures. A direct measure includes easy-to-gather assessment information, such as the number of forms processed. This could then be compared to the number of forms processed before implementing the new system. An indirect measure is more difficult to gather, with an example being the perceived value of the new system. This section should also identify tools, like satisfaction surveys, to collect the metric data.
Assumptions	This section lists all assumptions, which are anything believed to be true for the project. Assumptions can be about anything, such as the project budget or resources. This can include project limitations such as the budget or scope of work.
Risks	This section should identify all risks to the project benefits being realized. Risks are discussed in more detail in Chapter 16. However, an example could be the use of a specified new technology.

collect and interpret (Carlucci, 2010). Excessive volume of this does not improve understanding of the project.

KPIs help to track project results that align with organizational goals (Shahin & Mahbod, 2007). For instance, a company with the highest market share in its industry should have a KPI that measures some related attributes of that goal. All organizations should have the goal of continuous improvement, a standard KPI (Kaskinen, 2007). A university might want to measure its education program alignment by tracking the employment rate after graduation. Regardless of the KPI target, it assesses progress as measured. Whether the term is metric, KPI, or CSF, these status variables focus the project on established performance values.

The remainder of this chapter will describe more approaches for developing well-defined KPIs that will support a better understanding of the project's performance.

12.1.3 Developing KPIs and KPQs

Before developing KPIs, an organization needs to identify key performance questions (KPQs), such as the project's end goal (Nixon et al., 2012). Identification of KPQs provides a base for developing relevant and meaningful KPIs. Too often, KPIs are developed without first understanding what areas need to be tracked. Once there is an understanding of the higher-level goals, an appropriate metric can be specified. Another important characteristic of a KPI is its role in monitoring antiperformance, determining present status, and assisting in the control process. However, KPIs should not be considered universal generic variables used for all projects. Since projects have different characteristics and goals, they must apply to the specific project or organization to be useful (Harris & Mongiello, 2001).

There are four common categories of KPIs:

- Costs and expenditures
- Physical assets—facility, hardware
- Functional performance of the project deliverable
- Customer feedback

Other KPIs identified by Practical Software and Systems Measurement (2012) relate to schedule, work progress, resources, cost, product characteristics, process performance, technology effectiveness, and customer satisfaction. Buried under these high-level factors are KPIs related to basic status, product testing and verification, personnel, requirements analysis, process improvement, defect prevention, and various KPIs focused on employees and customers. Other option areas include risk status, reserves, baseline comparisons, and buffers. Also, completion forecasts and Earned Value (EV) parameters can be used. More specifics related to EV parameters will be discussed in Chapter 14.

KPIs should be easy to calculate and compare (actual versus estimated or comparable from month to month), clearly defined, and focus on their purpose (Hursman, 2010). One of the tests of a good KPI is that it is specific, measurable, attainable, realistic (relevant), and time-sensitive (Hursman, 2010; Shahin & Mahbod, 2007). The first letter of each of these characteristics makes up the acronym SMART. The SMART method for evaluating a particular KPI is a helpful guideline. Indicators must be *specific* and clear so that no two individuals would interpret the indicator differently. Indicators also need to be *measurable* so that it is easy to gauge the trend performance of increasing, decreasing, or no change. The *attainable* attribute serves to make sure that the indicator is something that can be obtained with reasonable effort. The indicator must also be realistic, which is slightly different than attainable, as *realistic* means that the cost of collection is within constraints (Shahin & Mahbod, 2007). Without resource support, some indicators are not realistic. The realistic criteria can also be referred to as relevant, as it must be relevant in measuring something of value but also relevant in measuring performance where the KPI applies (Hursman, 2010). In other words, does the metric relate to a valid area of concern? All indicators should be *time-sensitive* in terms of the period in which the indicator is reviewed and measured (Shahin & Mahbod, 2007). KPIs should be developed by defining the goals in terms of what the KPQs indicate: time, cost, functionality, etc. KPIs should also focus more on tracking output versus input. For example, milestone-based targets for a project or organization are valid for KPIs. Finally, the number of KPIs tracked is an important constraint discussed in the next section.

12.1.4 Selecting KPIs

When selecting KPIs, the focus must be on a small set of metrics (Richardson & Jackson, 2025). Producing more status measures does not necessarily help one understand the current status. Albert Einstein said, "Not everything that can be counted counts." Therefore, the focus should be on those factors that are key to the project's success. The frequent tendency is to collect too much KPI data, which becomes unmanageable to analyze. Just because a company or project measures many KPIs does not mean that these provide valuable information to justify their collection expense. Reh recommends a limited number of standard KPIs because too many become overwhelming to manage and costly to produce. Each project can support this by defining three to five specific KPIs that support the organization's overall KPIs. Generally, project KPIs should link to departmental or organizational goals. One should question the project selection process if that is hard to define.

Let us take the example of the organizational KPI of "increase sales" and translate it downward to lower-level units into one or more KPIs that contribute to that higher-level goal. For example, a customer relationship management (CRM) development project could contribute to the "increase sales" KPI through its improved linkage between the sales force and customer. In this case, a CRM development project metric would also fit one or more departmental goals that might, in turn, be translated across multiple business units, which are indicated below:

- Product department KPI—"new product time to market."
- Sales department KPI—"number of new customers."
- Marketing department KPI—"number of telemarketing calls completed."
- IT department KPI—"online website up-time."

Organizations are often tempted to overly focus their KPIs around cost and schedule because those are common interests to management. However, these do not address the output side of the project very well, such as the quality or performance of the deliverable. Also, schedule and cost variances are the result of other issues. Specific causal factors are found elsewhere. It is also important to have KPIs for different segments of the project life cycle so that all the focus again is not in one area. If the KPIs are unbalanced, it could unintentionally affect and drive the project toward undesirable behavior.

12.1.5 Analytical Hierarchy Process

Shahin and Mahbod (2007) established a process that results in KPIs most relevant to an organization's goals. The SMART criteria discussed above should be used to measure each KPI using their analytical hierarchy process (AHP) to identify which KPIs are best. AHP is a way to model complex problems in a simple way that assists decision-makers in selecting different status measures ranging from a selection of projects for an organization to even the selection of KPIs. AHP allows weights to be assigned to different criteria, such as the SMART criteria, as each area may be more or less important than other options. In this instance, AHP, combined with the SMART criteria, helps to establish priorities among options and identify appropriate KPIs that measure the progress of achieving those goals most important to the organization. Successful KPI programs often have achieved buy-in from the stakeholders charged with using the KPIs, and the same type of buy-in should also exist from higher-level management that receives these measures (Kaskinen, 2007).

12.1.6 Individual Performance Measures

KPIs are also linked to individual performance attributes. Therefore, these are often linked to compensation related to salary or bonuses (Hursman, 2010). Employee motivation can occur through bonuses or alternate rewards for those progressing toward specific KPIs. Continuous evaluation of defined KPIs is needed to ensure they remain appropriate for continued use. This process will also provide meaningful feedback to employees and management on the progress of the KPI target and ideas for continuous improvement. It is helpful in the development of KPIs, and the measurement is highlighted if there is knowledge of the business processes and environment where the KPI is affiliated (del-Rey-Chamorro et al., 2003). Therefore, those closest to the measured area should be involved in developing the KPI and measurement. Continuous improvement will not occur if a KPI measure does not impact improvements in processes, resources, etc. If there is no continuous improvement, there is a suspicion that this metric is not worth being collected and may need to be eliminated from the KPI program.

12.1.7 Data Collection

The data collection step should be as simple and automated as possible to minimize cost. Also, realize that the data collection process may not be perfect or even arithmetically accurate according to accounting standards. However, any data that helps identify trends promptly is better than waiting for accurate data to exist too late to support management needs. In many cases, "directional" data will still support the needed decision process. The cost of data collection must be reasonable, too, or it will not be supported in the long term. The data also needs to be consistent and directionally accurate. Please realize that the term *accurate* must be measured regarding its usage and audience. For long-term KPI success, there should also be designated times and schedules for the data collection. Without reliable timelines, the measure will decline in usage and value. The data should also not include excessive qualitative measures, which can create invalid results. One of the more common examples of this is asking

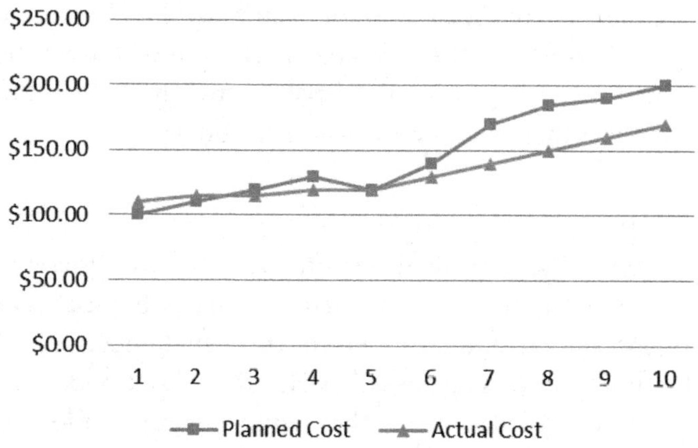

FIGURE 12.1 Planned versus actual cost.

a team member to define the percentage completed for a work unit. That result may be worthless unless it is based on some objective criteria.

12.1.8 Graphics

Collecting a KPI measure over time produces output, as illustrated in Figure 12.1.

This example displays a timeline view of the plan versus actual costs over ten weeks. Graphics of this type provide a clear and understandable way to present the KPI in an easy-to-understand format. Displaying KPI information visually is a powerful tool as it provides a quick way to see trends and status in an easy-to-understand format.

12.2 CONCLUSION

The presentation of metrics will not replace management, but it can be the source that helps identify gap errors in the current state, as the examples illustrate. This process is an important component of the successful delivery of planned outcomes.

The absence of a valid array of project metrics leaves the project team and its stakeholders in the dark, guessing the source and magnitude of deviations from the baseline plan.

This chapter has described the differences between metrics, CSFs, measures, and KPIs. When effectively used, these status-defining tools provide data about potential problem areas that help focus management's attention on the correct targets. They help the project team perform their role more effectively and present data to company management, outlining how the project aligns with the organizational goals defined for that effort. Therefore, these tools help the project's overall health and ensure it meets company goals. Status reports to management should include the tools used and their current and approved baseline values to communicate an accurate performance picture. In essence, the KPI collection measures achievement in alignment with a company's strategic target (Chen et al., 2009). In addition to KPIs, a related comprehensive performance management system will support other management issues such as cost, productivity, quality, employees, supplier-specific issues, and strategic alignment (Kaskinen, 2007). While metrics will provide performance information, they do not provide a solution to resolve a deviation from the approved baseline; therefore, these need to help define the necessary corrective action to resolve identified performance gaps. The next chapter will show more illustrations of the types of status variables used in projects.

QUESTIONS

Student Review Questions

1. What is the difference between metrics, measures, CSFs, and KPIs?
2. What are the components of the Project Benefits Management Plan?
3. What is the difference between a KPQ and a KPI?
4. How does a company develop a KPI?
5. How can a company keep employees interested in achieving a KPI target value?

In-Class Questions

1. Perform an internet search to identify a metric and the affiliated measure used by a company.
2. Perform an internet search to identify five project benefits.
3. Perform an internet search to identify a template other than the one discussed in this chapter for a Project Benefits Management Plan.
4. Identify two examples of a KPI.
5. Table 12.4 shows the number of estimated versus actual defects in producing a new product. Using this data, create a status trend chart similar to Figure 12.1. What does the chart indicate?

TABLE 12.4 Defect Data

Defect Data										
Week number	1	2	3	4	5	6	7	8	9	10
Estimated defects	34	24	20	15	12	10	8	6	4	2
Actual defects	50	42	30	21	18	17	14	10	5	1

EXERCISES

1. Identify three key KPIs to track project performance when designing a new bike.
2. Define a metric to trace scope change.

REFERENCES

Carlucci, D. (2010). Evaluating and Selecting Key Performance Indicators: An ANP-based Model. *Measuring Business Excellence* 14(2), 66–76.

Chan, A.P.C., & A.P.L. Chan. (2004). Key Performance Indicators for Measuring Construction Success. *Benchmarking: An International Journal* 11(2), 203–221.

Chen, S.-H., H.-H. Wang, & K.-J. Yang. (2009). Establishment and Application of Performance Measure Indicators for Universities. *The TQM Magazine* 21(3), 220–235.

Del-Rey-Chamorro, F.M., R. Roy, B. van Wegen, & A. Steele (2003). A Framework to Create Key Performance Indicators for Knowledge Management Solutions. *Journal of Knowledge Management* 7(2), 46–62.

Enoma, A., & S. Allen (2007). Developing Key Performance Indicators for Airport Safety and Security. *Facilities* 25(7), 296–315.

Harris, P.J., & M. Mongiello. (2001). Key Performance Indicators in European Hotel Properties: General Managers' Choices and Company Profiles. *International Journal of Productivity and Performance Management* 61(2), 204–212.

Hursman, A. (2010). Measure What Matters; Seven Strategies for Selecting Relevant Key Performance Indicators. *Information Management* 20(4), 24–26.

Kaskinen, J. (2007). Creating a Best-in-class KPI Program. *Strategic Finance* 89(4), 29–33.

Kerzner, H. (2011). *Project Management Metrics, KPIs, and Dashboards*, Hoboken, NJ: John Wiley & Sons.

Nixon, P., M. Harrington, & D. Parker. (2012). Leadership Performance is Significant to Project Success or Failure: A Critical Analysis. *International Journal of Productivity and Performance Management* 61(2), 204–212.

Practical Software and Systems Measurement. (2012). http://www.psmsc.com

Project Management Institute (PMI). (2021). *A Guide to the Project Management Body of Knowledge*, 7th edition, Newtown Square, PA; Project Management Institute.

Richardson, G.L., & B. Jackson (2025). *Project Management Theory and Practice*, 4th edition, Boca Raton, FL: CRC Press.

Shahin, A., & M. A. Mahbod. (2007). Prioritization of Key Performance Indicators an Integration of Analytical Hierarchy Process and Goal Setting. *International Journal of Productivity and Performance Management* 56(3), 226–240.

CHAPTER 13

Project Performance Tracking

A project's status refers to its progress toward an end goal. This chapter describes a model-based status-tracking framework that monitors common project problems and general progress tracking. Basic examples for each of these areas will be offered. The intent here is to build a foundation on which a new or existing PM can develop an effective strategy for managing project execution, tracking, and control.

13.1 INTRODUCTION

Each industry and organization has individual nuances regarding techniques to control project execution. No one source or approach can claim that it has the silver bullet or magic list of metrics and Key Performance Indicators (KPIs) that will fit every tracking and control requirement. The concepts and examples described here are intended to provide reasonable insight into general project control functions. In addition, examples of control parameters are discussed, along with formal methods for dealing with each.

At the macro-level, project control activities related to status tracking can be viewed through the following management processes:

1. Design of the process
2. Collection of raw data

DOI: 10.1201/9781003532989-16

3. Compilation into a coherent status view
4. Management corrective actions
5. Data storage
6. Information distribution

The processes related to evaluating project status represent much of the value and role of project planning. The status assessment process defines approved targets for the project and uses that knowledge to trigger management actions focused on moving the project along the approved track. In addition to this active management role, customized communication of status to various stakeholder groups is another important associated role. Left unchecked, the project outcome would be even worse than noted in industry surveys, philosophically, status-tracking attempts to make sense of project complexity and seek root causes for variation.

13.2 PROCESS DESIGN

Selecting an appropriate tracking method for a project depends on several factors, including the following:

Size of the project: If the project is tiny, the PM may track task progress manually. Alternatively, if the project has more than 50 tasks, a more automated tracking tool would be desirable.

Communication tool availability: Email and other organizational communication tools are often used to distribute progress tracking status. Each tool provides a level of capability in delivering methods to collect and disseminate task status. The features offered by each tool add incremental value to the overall monitoring and control process.

Tracking granularity: The control needs of the project determine the level of detail needed. For example, if the project plan has scarce resources formally assigned to tasks, a tight budget, or a critical deadline, the required control approach may dictate detailed tracking of the work and costs associated with a task. However, if this tight level of detail is not needed, the PM may choose to track status at higher levels in the Work Breakdown Structure (WBS) (such as summary tasks).

Realize that a higher tracking level will omit status details at levels below this, which represents the fundamental management decision question.

No fixed control method will suit every situation. This means that a specific analysis of control goals is required to provide the proper focus and visibility. Each metric chosen for tracking supplies a narrow view of the status picture, so a suite of these needs to be used to provide an overall tracking scheme. Another metrics selection driver involves the project profile characteristics and project control culture.

13.3 STATUS-TRACKING MECHANICS

There are four very visible variations in project performance versus its plan:

- Schedule and cost variations for work tasks
- New work is being added through approved change requests
- Unplanned events occurring in the form of risk events
- HR-related staffing status

Beyond these visible events, a broader view of the status-tracking goal is to measure and evaluate the project in the general areas of delivery efficiency, output quality, work execution efficiency, and team efficiency. Each of these groups will be compared to the baseline project plan, and status evaluation will be measured using both quantitative and qualitative techniques. The essence of control involves measurement, comparison, and corrective action. A key management component is tracking ongoing performance against the project baseline.

Chapter 12 described the selection of specific metrics to track status. The focus of this chapter is the use of these parameters in the control process. Specifically, the measurement of defined project metrics is the base mechanic for monitoring and control. As the execution process unfolds, project tasks will typically deviate from planned values. The four basic control steps are defining actual task status during execution, measuring variances against baseline targets, reporting performance results, and taking appropriate management action to bring the project back into alignment with the plan.

13.3.1 Tracking Mechanics

From a project management standpoint, the monitoring and control process requires the following six essential elements:

1. An approved project plan.
2. A WBS that reflects the level of control appropriate for a particular project. This includes data related to Work Packages (WPs) and defined Control Accounts (CAs).
3. A defined method for recording work performed using some form of work accomplished for each WP or CA will be monitored according to the defined measurement technique. This process requires measurement of deviations in schedule, work effort, cost expended, resource status, risk events, and/or HR issues. In many cases, the presentation of these items in status mode would be as a planned (baseline) value versus actual number format, often in graphical views.
4. Estimating work remaining on each active WP provides a cross estimate for the mechanical calculation of work left to be performed.
5. One management strategy to deal with variances is to assign follow-up delegation through the Issues log. This action would help ensure continuity and timely action. These delegations should outline what needs to be resolved to improve future performance and any guidance on the solution (i.e., cost, timing, technical). A prioritized list of these issues becomes a formal work and review list for the project team, representing a frequent gap in the status process.
6. A project communications plan must be established during the planning stage to formalize how status data will be distributed. Mature organizations will have this structure in place, but in some organizations, this will have to be cobbled together by the project team. This planning format should outline formats for stakeholder status reporting, along with the timing, media, and mechanism to deliver that status to specific stakeholders.

Figure 13.1 describes a schematic model overview of the basic project monitoring process.

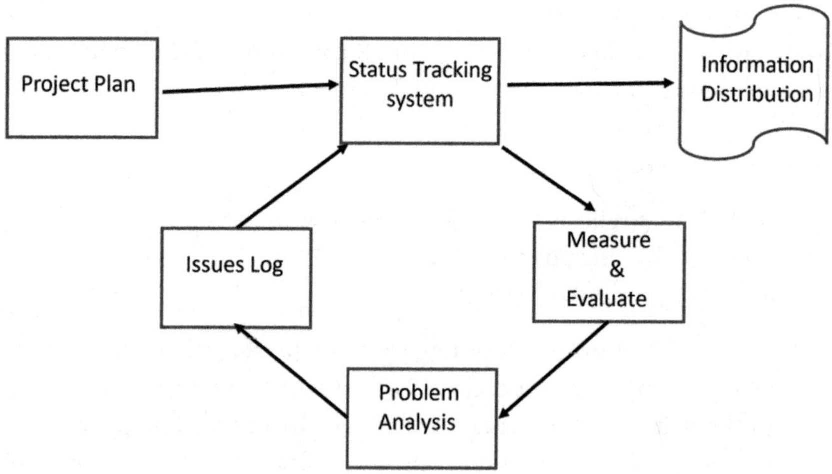

FIGURE 13.1 Project monitoring and control cycle.

This process is driven by a set of defined control parameters that are the focus of the control process. The Issue Log box represents the internal follow-up based on the defined variance corrective actions outlined in the previous step. The Issue Log represents an essential internal control process to keep corrective actions visible and monitored through completion.

The "information distribution" process reflects the physical delivery of control metrics, which can be hosted using various technologies. The historical format of status distribution was a printed standard template manually delivered to the recipient; however, the more modern approach for communicating status is using various Internet-based technologies, including dashboards, webinars, and personal video conferencing.

13.4 PROCESS DESIGN

KPIs represent important longer-term performance targets and are often defined based on a competitor's product or process. These can be set in various ways but should be formally recognized and reported in the status-tracking domain. In some cases, the target is important enough to be identified in the initial Charter, or it might be derived as part of the planning process. Various performance metrics can be set from past practices to compare industry performance and establish a stretch goal for a new product. Stretch goals can be set to motivate performance improvements and, in that vein, are not status-tracking so much as stretch targets to pursue. Comparing internal metrics with industry equivalents will add motivation to improve and validate that the selected goal levels are attainable. For example, a 100% customer satisfaction goal sounds admirable, but when the industry standard is 80%, factors make perfect performance scores unrealistic. This suggests a more realistic target because setting unrealistic and unachievable levels defeats the motivation goal.

During execution, work is completed in various WBS segments. The results of this activity can be translated into a status measure (i.e., schedule date, cost, percentage complete). Actual work unit data is collected and summarized to produce an overview in the following typical areas: scope, schedule, budget, quality, risk, procurement, and team resources. Comparing these values to the approved plan is an everyday activity related to status tracking. As the overall work view is compiled into a broad project-level performance picture, variations from the approved plan trigger management actions (review the flowchart in Figure 13.1). Given the Performance Measurement Baseline (PMB) described in Chapter 12, this information can be translated into overall status. Using this comparative data, the listing below summarizes various formal project control-oriented activities that collectively deal with project status data and analytics. These collectively are the PM's information sources that will be used to support any needed corrective action. The significant areas of concern are summarized as follows:

1. *Scope control* involves factors influencing project scope changes and corrective actions associated with that activity. The change control system is an integral part of this process.
2. *Schedule control*: This involves using various tools and techniques to evaluate and influence the status of the project schedule and related variances. Work performance data is used to help assess project status.
3. *Variance and trend analysis*: This process translates the raw data into meaningful decision format. The degree of the variance and the trends are essential metrics for the PM. These also lead to corrective actions and completion estimates for various project parameters.
4. *Cost control*: This process involves using various tools and techniques to evaluate and influence the status of the project baseline budget and related variances. It is also part of the Integrated Change

Control (ICC) system. The goal of this activity is to keep the project within budget limits.

5. *Quality control*: This involves monitoring project results to determine whether they comply with relevant quality standards and identifying ways to eliminate causes of unsatisfactory results. Another significant supporting activity is to influence process quality improvements.
6. *Project team performance*: This involves several quantitative and qualitative processes to track team and individual performance and provide feedback.
7. *Status meetings*: Status meetings provide an unstructured communication forum for team members to share their experiences and inform other team members of their plans.
8. *Risk monitoring and control*: This involves monitoring the risk register and other sources for the emergence of both identified and unidentified risks. A management activity must handle each category correctly once previously identified or new risk items emerge.
9. *Third-party deliverable status*: Monitoring third-party activities is done through contract administration. Key metrics here are deliverables, funds flow, and status reports.
10. *Risk audits*: This examines and documents the effectiveness of risk responses in dealing with identified risks and their root causes, as well as the effectiveness of the overall risk management process.
11. *Reserve analysis*: This compares the project's contingency and management reserves to ascertain whether the remaining reserves are enough.
12. *Technical performance measurement (TPM)*: This represents a set of techniques to identify deficiencies in meeting system requirements and provides early warning of technical problems and related technical risks. This process attempts to compare the ongoing project with specified performance goals to estimate the degree of success in achieving the project's scope.
13. *Performance data*: This involves collecting data from various sources regarding project performance and comparing these values to

established baselines, which will be reported to appropriate stakeholders.

14. *Managing stakeholder expectations*: This activity involves appropriate stakeholders in the project.

The status data and analytic-oriented processes outlined above are internally focused on project results versus the baseline plan. In addition to these sources, insight into the project status is obtained from a less formal set of external sources. These relate to the project sponsor, senior management, future users, and possibly outside regulatory-oriented sources. Each of these perspectives can have a bearing on the project's future and needs to be assessed by the PM, even though that data may be challenging to obtain and less formal in format.

As one can see from the list above, one of the metrics design issues involves where to focus data collection to support the chosen performance areas. Some metric choices are predetermined and required by the local environment. Almost universally, this includes cost and schedule data. It is common to need interim performance evaluation of the output target, especially for a product goal. After that, scope change and human resources data may be required for display. Stakeholder and organizational communication requirements are then added to the data collection list. Each of these data-driven perspectives can have a bearing on data collection performance. The project's future needs to be assessed by the PM, even though that data may be challenging to obtain and less formal in format.

13.4.1 Management Actions

Management actions are triggered by measured variances in the plan or unusual events outside of the data collection process. As described earlier, the project schedule consists of detailed plans outlining the estimated outcomes for each WP (e.g., schedule, cost, functionality). This aggregation of scheduled WP planning data represents the bulk of status-tracking targets for project control and reporting.

13.5 CORRECTIVE ACTION TARGETS

The primary focus for status tracking is to identify areas of the project that require management action. The control aspect of status tracking is exercised through formal and informal actions undertaken by the PM, project team, or senior management. Also, conducting reviews and monitoring reports is a companion to the control process. Several major process areas represent the major control levers. This section will summarize how actions in these areas can affect the project direction.

13.5.1 Project Plan

The project plan establishes the target goal and sets the project schedule, work requirements, and resources needed to complete the project. From this definition, the plan is then used to measure the ongoing status of individual WPs. Also associated with this are procedures for dealing with status related to quality, risk, communication, and change management. Because of this broad and diverse view of status, the active project plan must be flexible and open to modification as new insights emerge from ongoing status. Effective status tracking and timely corrective action are keys to future delivery success. The more that planning is based on sound analysis, the better the chances of ultimate success. The entire life cycle plan considers everything in an organization and its long-term achievement of a desired objective. In this view, planning may consist of an orchestrated collection of sub-plans for specific short-term projects.

13.5.2 Knowledge Area Controls

This section outlines control steps related to controlling various project process areas. These process areas represent the fundamental management groups that drive the defined output. A summary of tracking issues for each is provided below.

13.5.3 Integrated Change Control

ICC and Change Management (CM) are siblings in the control process. One way to separate these two processes is to view CM as the static control mechanism and ICC as the workflow modification and decision-making component. The ICC process allows the PM, sponsor, and stakeholders to be aware and involved with workflow changes through the actions of a diverse approval board that includes both internal and external decision-makers. It is common to find scope changes in the 2% monthly growth range. With this rate of change, it is easy to see how the overall project control process can get out of hand. Change control is the process designed to manage change requests and ensure that any changes approved are appropriate to the project's goals. Approval for these changes should occur externally to the project team, and a project board representing management usually handles the decision-making part of the approval process. Without this type of control, a project can become a runaway effort. In addition, this activity adds the required higher-level management oversight to the project. The project board keeps the project within the boundaries established by the Charter or approved baseline. Within this boundary, they attempt to balance user needs and maintain a balance of minimizing disruption of an orderly execution.

13.5.4 Scope Control

Scope changes can be stimulated by either external requests or activities within the project team. In either case, the control process is designed to manage approval of any changes and then manage them as they progress to ensure that they are appropriately integrated into the overall project deliverables. Executing this process requires a formal change procedure (i.e., ICC) to document the request and map the process through which such requests flow. This includes project board approval, tasking the team for execution, checking results to ensure proper results, and monitoring the activity through its life cycle. The term for changes in a project is "scope creep." Indeed, one of the significant goals of scope control is to minimize this activity based on its potential adverse impact on project success.

13.5.5 Schedule Control

The PM manages schedule variations at the project level and should be proactive, as with all control strategies. When variations from the planned schedule are identified, they must be evaluated as to the root cause; then, appropriate action must be taken to solve the issue with the least negative impact on the project.

13.5.6 Cost Control

This process involves monitoring the status of project direct costs that affect the defined baseline value. An additional aspect of this process is to ensure that any changes are beneficial, determine that the project budget estimates have changed, and manage the actual changes when they occur. The following represents key activities related to the cost control process:

- Manage the overall project cost budget within the approved level and tolerance.
- Manage the corrective steps necessary to deal with measured variances when and as they occur.

- Monitor various cost categories to ensure that they do not exceed approved levels by total or by other control categories.
- Monitor cost performance to isolate and understand variances from the approved cost baseline.
- Ensure that all change requests are acted on promptly.
- Monitor work performance against expenditures.
- Monitor and deal with unapproved cost allocations.
- Provide format cost status information as defined in the communications management specifications.

The budget management process should be considered a strategic activity, considering its custodial view of issues. This includes not only the way funds are expended in the project but also dealing with the sources of such funds that are planned to be supplied by various external sources.

13.5.7 Quality Control

Emphasis on quality management is a process performed throughout all the project life stages. From a project management model standpoint, a quality management plan will be defined during the planning phase to measure the attributes of work performed at each stage and provide control specifications and guidelines for team members. The quality process deals with productivity, which, in turn, can impact the project budget and schedule.

13.5.8 Configuration Management

CM is a key process to manage and control document and product changes throughout the project's life. Various work activities result in continually modifying most product and management documents, a normal part of every project. Many consider failure to monitor this activity and provide appropriate management oversight as the number one operational control requirement.

The CM process ensures that the process and product artifacts are appropriately managed. For example, in a durable manufacturing project, this process would ensure that a drawing for a part was the latest version. A common problem source in product-oriented projects is having two individuals work on a document simultaneously, but only changes made by the last person saving the document are reflected. In the project management domain, the control process would ensure that all management documentation is the latest version. Most CM solutions require a formal controlled repository for managing the flow of these project artifacts. Some organizations call this activity *Version Control.* In the contemporary environment, automated document management software is often used to manage the flow of documents. Before this, manual processes struggled to ensure that changes to descriptive documentation created in the project were tracked effectively.

13.5.9 Issue Management

Issues represent small, unplanned events that disrupt daily activities and must be quickly resolved to keep the project moving forward. Project issues represent anything that can go wrong. PMs cannot predict what or when a specific issue will occur, but all agree that such events will occur during the life cycle. In many ways, these are like small risk events that can be resolved reasonably quickly. Issues arise from many sources, such as natural events, technological glitches, supporting resources, and economic fluctuations. Within this unplanned domain, an issue management process provides a mechanism for organizing, maintaining, and tracking the resolution of these ad hoc items that might still need to be resolved promptly or in the normal course of activity.

Deviations can affect the project's ongoing performance and need management attention. In the case of status-tracking identified items, these represent variances that have been identified and need follow-up action in hopes of improving that area of the project. Action items in the issues log are assigned to a person for continued action. This item will be monitored by the PM until resolved. Some theorists categorize an issue as a triggered risk. The authors choose to keep this list separate, with the logic being that an event can be resolved by some short-term management technical action rather than by adding a formal work unit to the project plan. Many such events fit this category. For instance, if the copy machine is broken, it needs to be fixed, and if it can be fixed quickly, that is a minor, easily managed event. It's not something in the triggered risk category. This is one of those project vocabulary and process items that some would see fit to argue about. The Issue Log and quick follow-up are the process options chosen here.

The issue management process entails establishing appropriate identification and control mechanisms and defined processes to aid the project team in identifying, addressing, and prioritizing ongoing problems and issues. For this process to be effective, documentation and communication is key. A detailed understanding of the issue that needs resolution triggers the process. From this base point, the PM ensures that the item has appropriate documentation and priority for resolution and that a named person is responsible for tracking that resolution. Without a system of this type, various daily issues get lost in the maze of work and can lie unresolved until they become major roadblocks. Managing this class of activity is a critical success factor for the PM.

13.5.10 Control Summary

Different schools of thought regarding a project's proper suite of control metrics and processes exist. However, the control examples outlined above should be considered a core requirement for keeping the project on task. Do recognize that every approach to standardization brings a potential negative if understood or managed. Some feel that the style of control outlined by the traditional model described here is inefficient and causes more harm than good. For example, Koskeia and Howell (2000) state that control, as described in the PMBOK® Guide, causes problems (PMI, 2021). One way of translating this negative statement is to say that any preset management structure is wrong and needs customization. That certainly is true for the project environment.

13.6 DATA REPOSITORY

Easy access to technical and status information regarding project work products has broad communication value to the various stakeholder groups. It also helps to support basic work activities, collaboration, decision-making, and information sharing. Organizations that accomplish appropriate levels of information sharing across project boundaries report significant improvements in customer relationship and scope verification. In addition to this, the process of learning lessons is greatly enhanced by having ready access to project archive data.

An appropriate enterprise technical repository strategy consists of six essential parts:

1. Application architecture (including project work activities)
2. Operational metrics for work units

3. Project Portfolio (Charter, Business Case, strategic alignment, resource overview, approved plan, etc.)
4. Status documentation (formal reports)

Each repository segment has value in executing and managing the overall workflow. Also, other operation data repositories are dedicated to topics such as scope change and risk registers from which summary information could be exported. Conceptually, the underlying data represented by the repositories outlined is logically interrelated across the four repository data domains. For a complex environment represented by this collective set of data items, it is recognized that integrating these various data sources is a significant challenge but worth pursuing.

Existing database technology allows for the storage of extensive data and flexible retrieval. Software attached to this physical data store adds the level of presentation desired. The support architecture must be improved if an environment is less mature than the one outlined above. The concept of a learning organization needs to be recognized in such situations. Project data should not be physically owned and managed by a single individual. This is an organizational resource that deserves proper control.

13.6.1 Repository Schema

Table 13.1 offers a more detailed explanation of the project data library that will be used to support the status-tracking process.

TABLE 13.1 Status Data Repository Schema

Project Plan Activity and Work
1. Initiation—SOW, Charter, Business Case
2. Requirements documentation—WBS and WBS Dictionary, scope documentation
3. Design documentation
4. Integrated change control
5. Quality management documentation (test strategies, unit test cases, system test plans, test scripts, test reports, etc.)
6. Operational metrics (link to the metrics data warehouse)
7. Implementation plans and issues
8. Stage reviews
9. Project closing (lessons learned, migrating working documents to the application repository, final sign-off, etc.)
10. Issues Log (tactical items requiring attention)

(*Continued*)

TABLE 13.1 (Continued) Status Data Repository Schema

Business Overview of the Project	
1.	Design documentation
2.	Testing and acceptance program
3.	User documentation (user guides, bulletins, etc.)
4.	User and technical documentation

13.7 INFORMATION DISTRIBUTION

Distributing project status to appropriate stakeholders is integral to the monitoring and control phase. This activity effectively relays status metrics to proper team members and stakeholders. Formal status reporting should be a regular event throughout the project's life cycle. This process should contain appropriate information summarizing activity status, accomplishments, upcoming activities, nontechnical and technical project issues, defined actions to mitigate critical problems, resource usage, and risk status. PMs not only have an obligation to manage a project effectively, but they must also be an honest broker of status information.

13.7.1 Distribution Strategy

Few technological changes in the project management environment have been more visible than those related to communicating status information. Paper, standard forms, and hand-carrying might have been the norm in the 1950s, but by the 2000s, electronic spreadsheets and email were the preferred methods. Since then, the process has continued to migrate into Internet-based technologies with global connectivity and greatly improved communication software. Enhancing capability means that everyone has a connection to everything. Display hardware has gone from desktop PCs to cell phones, watches, and small appliances. Data is now presented in various graphical and color formats. A power user can extract status reports and "drill down" in the data to obtain details of multiple items. An information retrieval environment such as this has many advantages, the main one being that it is the one place to look for the truth. Also, this is a "pull" oriented method, with the user deciding more about what they want to see. Admittedly, there will still be others who must be handled more carefully with receiving access. We are reminded of a historical experience with a CEO who had his administrative person print out status reports, then handwrite his comments on the printed paper and hand them back to the administrative person for email to the project team. This is an example of how theory and reality sometimes cross paths. The key design message for data storage is to develop a user-friendly, mature approach to data capture, storage, and distribution. This is the key to effective communication.

13.7.2 Status Distribution Strategy

The emergence of a wired Internet world has changed the approach to personal communications. The popularity of employees working from home is evidence of robust communications technology. The new status delivery techniques reduce the need for a face-to-face meeting. However, some elements need to be included in this automated communication model. The biggest challenge for the system described is interpreting the data. Getting an explanation or background explanation becomes more complicated than two individuals discussing the same situation. On a positive front, remote communication technology, video conferencing, is maturing and has become quite popular since COVID-19. This technology still has space to improve, but PC video conferencing and cell phone "Face Time" add a richer approach to interpersonal communications than text or audio. The inventors of the cell phone did not envision the breadth of applications now resident on that device.

Much of the current communication model has been described as a "non-meeting," meaning that the availability to fetch data on command eliminates much of the need for general person-to-person communications. Communications that used to require heavy support via physical group meetings are now moving into the stored data files and video conferencing collaboration format. On the positive side, there are significant resource savings from not having to attend such sessions physically. The younger generation is familiar with this remote format, so the precedent for the communication model has been established.

The use of color graphics is particularly notable in status distribution. Figure 13.2 illustrates a graphics presentation showing Earned Value (EV) parameters that might be delivered in the status report.

At the project team level, there is still the need to hold traditional problem-solving sessions, but the members can be in different physical locations for this to be effective. In this mode, the stored data becomes the background source to retrieve details for discussion. Group meetings are one of the biggest drains on team productivity when improperly handled. Several rules of protocol and behavior need to be added to this historic form to keep it as an effective method of information sharing. The long-term message is clear: more of the future communication role will be

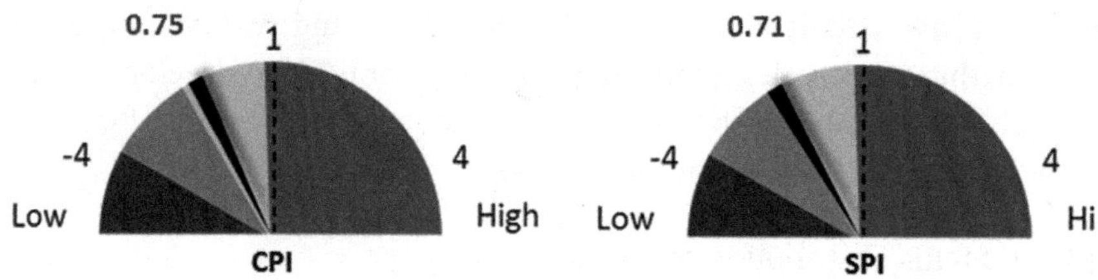

FIGURE 13.2 Sample graphical status view.

handled by evolving technologies, including artificial intelligence's ever-growing function and popularity.

13.8 REALITY-BASED STATUS TRACKING

Up to this point in the chapter, the process of tracking status has stayed close to standard model theory. It seems timely now to deal with more advanced reality points regarding traditional status tracking and describe why the traditional model does not handle this process accurately. The conventional model defines a variance as an actual value compared to a corresponding value in the approved baseline plan.

The first step in this reality review is to use the baseline plan structure. This plan is usually considered the final planning version that approves the scope definition with related cost and schedule values. At least three questionable elements make this an erroneous status measurement approach. First, regular scope change is often not estimated or represented in the baseline view, but these future changes are part of what the project evolves into. However, they need to be explicitly known at the initial point. Second, future risk events are often not estimated or presented in the plan; however, they are also part of what the project will become. Recognizing these future events is required to make the baseline plan represent a status benchmark. A third deficiency exists when task events are almost universally defined by static time and cost estimates. It is all right to compare actual values to these fixed values, but the comparison has no logical status meaning, as will be illustrated shortly. If the baseline plan does not show these three future components, it has little value as a comparison base for the status-tracking process, yet it is the traditional method. It compares to the initially intended project but has little value in a valid status picture. What does one do with such a variance?

One of the first cultural management issues is recognizing that a task is not a fixed event, and variations in that task do not signal anything other than the time difference. It does not say the original estimate was performed poorly or must be corrected. For example, assume that the actual duration time for Task 123 is 15% higher than the plan. Why might a task take longer than planned? Bad estimate, worker skill, lousy weather, or some other reason. There is some potential learning value in comparing WP performance at the estimating level, but more analysis is required to derive any value from such an evaluation. Variations in cost at the WP level are also interesting variances to pursue, but the root cause of such variances will also require additional research. This class of variance will require significant review time to uncover underlying factors. The description above is that historical status measures may have questionable short-term value and are likely minimal if there is no formal review and examination of those variances. This is the learning organization concept where past issues are identified and corrected. Such a technique is less important in the short term but vital in the strategic view of improvement.

A third area where traditional status tracking is wrongly used is the formal presentation of planned versus actual values. One popular version of this is to show the planned cost of the project at some point and compare that to the actual expenditures. The implication here is that spending less than planned is a positive performance, yet there needs to be an associated indication to show what was accomplished. Doing nothing will consume very little, as an extreme example. In general, plan versus actual comparisons need to be viewed carefully. The EV model is not looked at with favor by most project participants, and even when used, the method needs to be better understood. This is an area where computer software can produce a status value that is produced quickly, yet the interpretation of that metric is not what it is thought to be. Examining Chapter 14's EV parameter formulas will make this point more straightforward.

As a final point on the traditional approach to status tracking, there is too much emphasis on plan versus actual measures that do not mean what they are reported to mean. Also, the traditional tracking approach is narrow in its perspective, focusing too much on cost and schedule. Realize that these two metrics are output-oriented, and the associated data used for status reporting is often five or more weeks in arrears. The previous chapter described industry-standard models that showed status linked to organizational visions with KPI-type goals. Status tracking needs to move in forward-looking mode as opposed to the historical one, which produces data that does not help much with corrective action. Using fixed task times lies at the heart of poor status tracking, but a broader perspective is also needed. An expansion is required into the tracking metrics suite that deals with risk learning organization (improvement) and external goals that the project should support.

13.9 CONCLUSION

Successful execution control of any project is complex because of the wide variety of causal deviation factors. Designing an appropriate tracking suite of status metrics is key to success when it helps management promptly identify causal issues. The methods discussed here represent a cross-section of core monitor and control concepts used in the project management arena. Beyond the raw mechanics of this activity, it is essential to recognize that the collection of planned versus actual status metrics neither hinders nor enhances project success. These passively describe what is or has been occurring. It is up to the management process to analyze this data and use it properly to influence an improved outcome. Also, this discussion has highlighted that the design of the tactical control process must be reviewed for each project to decide which combination of methods and degree of granularity will work best in that set of project characteristics.

The concept of management control flexibility is an essential ingredient to effective control and one that many organizations only sometimes adhere to. In these situations, the goal of control standardization is so decisive that management flexibility is removed. For average-skilled PMs and less technical projects, this may be a suitable strategy, but for complex projects being managed by a highly experienced manager, this overly structured status reporting approach can result in wasted resources in data collection and a less effective outcome. It is essential to avoid tying the hands of the PM, where he or she is left with little freedom remaining in tracking and resolving the deviant issues.

REFERENCES

Koskeia, L. & Howell, G., (2000). Reforming Project Management: The Role of Planning, Executing, and Control, https://iglcstorage.blob.core.windows.net/papers/attachment-748e6687-f05e-46ba-b06b-a50f71dd7c6d.pdf

Project Management Institute (PMI). (2021). *A Guide to the Project Management Body of Knowledge*, 7th edition, Newtown Square, PA Project Management Institute.

CHAPTER 14

Earned Value Management

Project N Motion Videos for This Chapter

4EV1	Earned Value A	Intro—Part A Mechanics
14EV2	Earned Value B	Intro—Part B Mechanics
14EV3	Earned Value	EV Conceptual Overview
14EV4	Earned Value	EV in MSP Mechanics

Earned Value Management (EVM) is a contemporary project management cost and schedule analysis technique that is considered the premier performance measurement tool but has struggled to gain broad industry acceptance over the past few years. Originally introduced into the project management sphere in the 1950s, this topic lay essentially dormant for over 40 years before the commercial world began to embrace its usage. There is now broad evidence to indicate that Earned Value (EV) provides one of the most effective and meaningful project status analysis tools available today to measure and report cost, schedule, and performance. Stratton reports that "… use of EVM increased in 2011 by 13% over 2010" (Stratton 2012). Also, this growth is occurring well beyond the boundaries of the United States. EV has the unique ability to combine cost, time, and scope completion measurements within a single integrated methodology.

DOI: 10.1201/9781003532989-17

14.1 INTRODUCTION

During the initial definitional stage in the late 1960s, the Department of Defense defined 35 supporting management capabilities (criteria) to be used by their contractors to support the production of EV parameters. Given the contractor level of management process maturity at this point, these requirements were found to be unrealistic; however, over the next 40 years internal management process maturity evolved, and, eventually, the value of the EV status parameters began to be accepted as a contemporary status creation process. In 1987, the Project Management Institute (PMI) added a reference to EV in their PMBOK® Guide and further expanded that discussion in subsequent editions. Also, in 1998 ANSI/EIA-748 standard was released, which further stimulated a broader recognition of the technique in the non-governmental project environment (Fleming and Koppelman 2006). Efforts to simplify and implement EVs continued to gain further momentum into the early 2000s. There is now a recognized body of research data that validates EV as an effective tool to support the status evaluation of projects (Christensen 1998, 13). In the hands of a trained technician, EV represents the most robust performance management tool available for evaluating project cost and schedule status and completion forecasting. For this fundamental reason, it is an essential tool for the modern project manager to understand and use.

During the middle evolutionary stage (1970–1990), EV was often viewed as requiring excessive overhead compared to its value and hence was rejected. This attitude still exists in some project management circles, but the author's opinion is that avoidance of using EV simply means that the project control process is lacking. EV is now recognized internationally in governmental projects, construction, information technology, and more. One important support driver for this acceptance level is the modern suite of computerized support tools that helps to minimize the computational overhead related to this process. However, as in many cases, using a computer to generate results can lead to misuse as much as it helps perpetuate the process. Invalid values quickly computed remain invalid values!

14.2 BASIC EV PRINCIPLES

Based on its underlying modeling constructs, EV principles deal with many core project management areas, including formal scope definition, resource analysis, schedules, budgets, resource accounting, status analysis, reporting, and change control. EV's specific mechanics favor the use of a Work Breakdown Structure (WBS), formal performance measurement baseline, and defined set of work unit performance measurement rules. Before delving further into the computational methodology, it is important to review the role of one framework support concept critical to EVM implementation—that is the WBS.

As previously described in Chapter 5, the WBS is a fundamental technique for defining and organizing the total project scope into a hierarchical tree structure. Through this structure, the WBS defines a set of approved project deliverables and related subordinate work units that collectively represent 100% of the approved project scope. Arrayed through this structure is a collection of boxes into which planned work and related resource allocations are defined. The EV model then uses this structure, mapped to an approved project schedule, to evaluate the status of project time and cost. Once the WBS has been approved, it is baselined, or frozen, to represent a control framework for the project. These baselined work units are then time-phased into an approved schedule and budget. From this view, actual work completion and resource consumption are compared to the approved plan to assess how the project is performing. The major difference between EV and the more traditional simple plan versus actual resource consumption status method relates to how EV also evaluates work accomplished instead of just resources consumed.

Figure 14.1 shows a status trend curve of the project's planned versus actual resource consumption. This display shows that the project is using less than the planned level of resources; therefore, one could conclude that all is going well, right? Well, maybe not! Too often status curve such as this

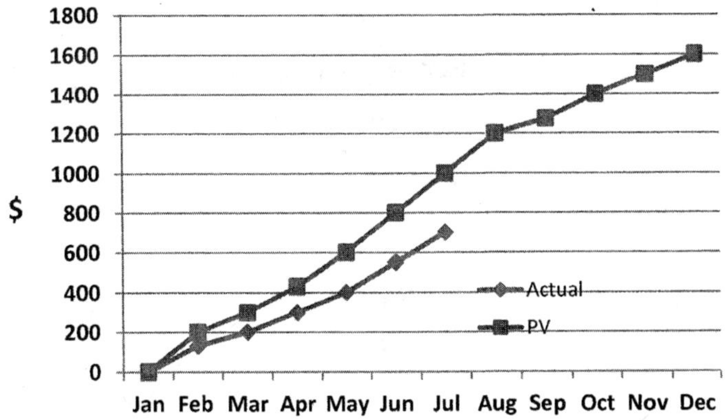

FIGURE 14.1 Traditional status report.

is used to suggest that a project is doing well. However, the missing ingredient in this scheme is "What has been accomplished during this period?" If half of the budget has been consumed but only 25% of the required work is completed, the project is not doing well. EV offers parameters to properly evaluate this situation in quantitative terms.

A third dimension included in the EV status measurement problem is the concept of *earning*. Earning is measured by the completion of defined work *before the status point*. In the examples used here, work is defined by boxes in the WBS. So, the project status measurement simply becomes a two-way comparison view between the baselined plan and earned box measurements. At a defined status time point, the following three sets of performance status data would be collected:

- Actual Cost (AC)—The actual cost incurred for the planned work packages up to the status point.
- Earned Value (EV)—A measure of completed work units and partially completed portions of work units up to the status point. For example, a 50% complete work package would receive 50% of the planned value as earned.
- Planned Value (PV)—The sum of planned baseline cost for all or partial work units scheduled to be accomplished up to the status point.

A physical model view of this is shown in Figure 14.2 with three "cans" of data. This data view is valid for both the entire project and down to

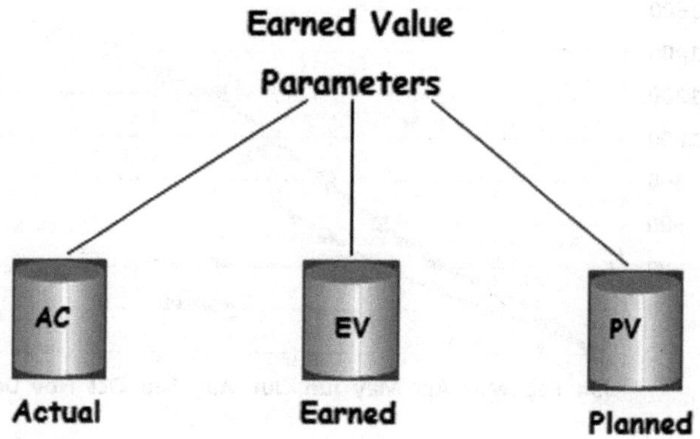

FIGURE 14.2 EV parameters.

the lowest point at which actual data is collected. Recall the term Control Account Package (CAP), which was defined in Chapter 8. Here we see that term come into play. The EV evaluation scheme requires that plan and status data be collected into these three groupings. For future use, it is important to note the alphabetical sequence (A, E, and P) of the three data categories shown, as this will be important to help remember the cost and schedule relationships.

With the addition of a third evaluation variable, it is now possible to produce three separate measures of project status. A typical project EV trend curve is shown in Figure 14.3.

Using these three relationships, EV status parameters can be calculated to help the project manager identify key problem areas and help focus on project work package areas that need corrective action. These three trend views offer insight into how the project is doing, but for operational analysis purposes, this raw data needs to be converted to more quantitative terms. Both schedule and cost status measures can be derived from this data.

Beyond revealing specifics about historical performance, these same data relationships can be used with reasonable assumptions to forecast the outcome of the project. One typical forecasting assumption is to assume the project will continue to exhibit the same trends in the future that it has up to the status point. Of course, every project manager hopes that the

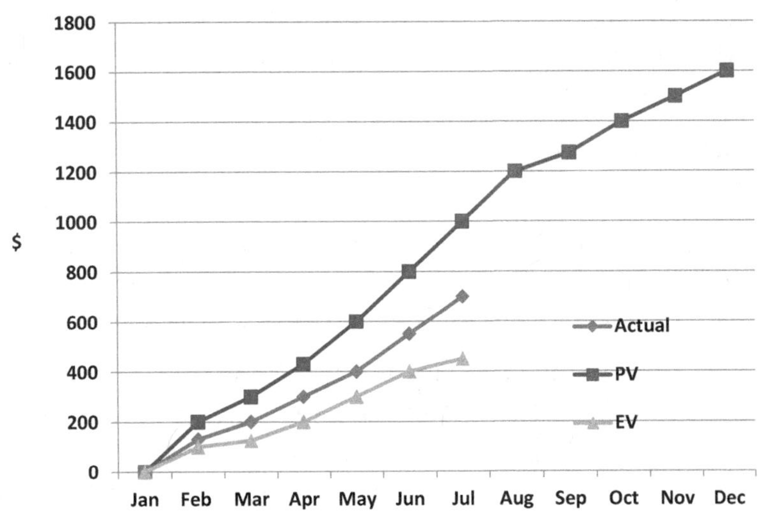

FIGURE 14.3 EV graphical status view.

future will be less bumpy, but there is research evidence that this is not the norm. Trained manipulation of these data relationships offers very sophisticated insight into project performance now and into the future. Learning to use this type of analysis for these purposes makes the EV process an important monitoring and control technique for the project manager. Multiple examples of this will be shown in this chapter to illustrate how EV parameters can be produced at various levels of detail and plan formats.

14.3 EARNED VALUE FRAMEWORK

When viewed in its raw theory state, the framework and mechanics required to support earned value project tracking may seem complex. However, when broken down into fundamental project process elements that are suitable to support most projects, it is much more understandable and reasonable to accomplish [14EV1]. Fleming and Koppelman (2006) studied the fundamental control requirements contained in the Earned Value ANSI/EIA-748 specification standard and, from this review, distilled ten basic management steps necessary to implement a reasonable set of operational procedures that will satisfy control requirements for most moderate to small projects in any industry. These ten basic management process steps are summarized below:

1. Define the scope of the project. One of the most useful framework tools available to the project manager is the WBS and its Dictionary, which fit the fundamental scope definitional requirements for EV quite well.

2. Define the project work units that collectively constitute the defined scope.

3. Estimate the required resources for each work unit. *Individual WBS work units will not contain contingencies or management reserves in the EV analysis portion* as these two reserves are held separately.

4. Plan and schedule defined work. The EV methodology requires a resource management system that is capable of tracking authorized resources (budgets) through the approved schedule. The comparison plan must reflect a time-phased budget mapped to the authorized schedule. From this time-phased formal baseline, work completion

status is measured as work is accomplished. Evaluations are made at defined status points.

5. Define a project Performance Measurement Baseline (PMB). EV requires the formal definition of an integrated project baseline containing defined work units, baseline schedule, and related time-phased budget. Calculation of EV status parameters focuses on comparative measurement between baseline and actual values for each of the selected WBS elements referred to as CAPs. A CAP can be best described as an arbitrary point in the WBS where actual cost and status data are collected (refer to Chapter 8). In some projects, the total baseline cost may include such things as indirect costs and even profits or fees to match the total authorized project commitment. The cost baseline must include whatever executive management has authorized for the project but realize that the *EV analysis best applies to the direct work portion and is not so useful in evaluating the status of such items as the level of effort tasks, contractual, and material.*

6. Record all direct costs consistent with the authorized baseline work units. This criterion requires project managers to have access to current resource expenditures at the level of detail required. Direct costs must be tracked to a CAP. To employ EV metrics on a project, the actual costs must be aligned with the baseline budget. For instance, work accomplished must be relatable to actual resource consumption by the work unit to determine the related EV parameters. This is the single most important EV performance process evaluation step.

7. Convert the work unit results into EV status parameters. The status measurement challenge here is to identify a viable method to quantify work unit completion status. There are various methods used to measure work unit accomplishment, and the most respected ones use some type of discrete physical completion measurement. However, this is not possible in all situations.

8. Continuously monitor EV parameters against the authorized baseline values throughout the project. The primary management focus should be on work unit variances, particularly those that fall above defined control tolerance levels. One of the most important aspects

of employing EV is its ability to monitor the ongoing cost and schedule status of the project at CAP levels of detail.

9. Forecast the final project status. EV is the most sophisticated project management tool currently available to forecast cost and schedule status based on current work performed.
10. Manage authorized scope. The project PMB is defined by the approved plan and is used as a control benchmark value. A formal change control process is required to manage the project baseline. Recognize that approved scope changes often represent new work units added to the WBS, and this can complicate the EV computation mechanics. However, once added to the baseline, the process remains essentially the same.

By utilizing these operational process steps, the project manager is provided with a reasonable mechanism to satisfy most basic schedule and budget evaluation requirements. If the scope and complexity of the project dictate, these abbreviated steps can be expanded to satisfy the more rigorous 32 criteria contained in the ANSI/EIA-748 Standard (NDIA 2014). However, the more simplified structure outlined above should be satisfactory for most small- to medium-sized projects.

Before delving further into the computational methodology, it is important to review one underlying concept critical to EVM implementation—that is, the WBS and its associated Dictionary. As described previously, a WBS is used here as the fundamental technique for defining and organizing the total project scope into a hierarchical tree structure of required work units. The WBS defines a set of project deliverables and related work that collectively represent 100% of the project scope. At each subsequent level, the children of a parent node represent 100% of the scope of their parent's work requirement. The lowest level of decomposition for each parent node is called a Work Package (WP), and these represent the lowest level of control in the structure. Figure 14.4 shows a skeleton representation of the WBS. In this type of structure, each of the WPs defines work estimated, resources allocated, duration schedule, and is linked to the next WBS element.

In this fashion, the project schedule and budget are integrated and defined. Once this overall scope is approved, the related data will become the baseline against which the project performance will be measured.

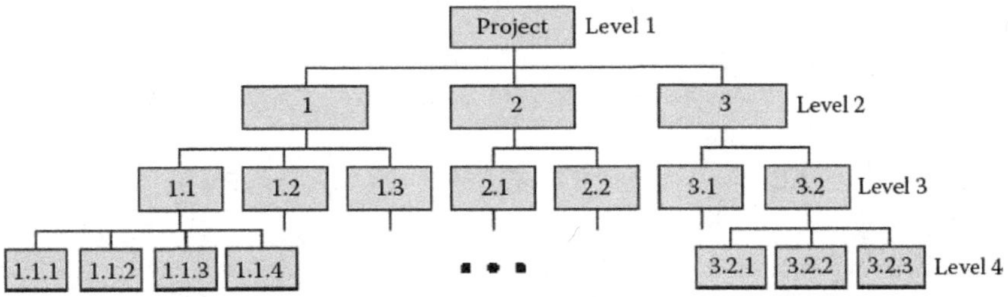

FIGURE 14.4 Work breakdown structure.

In EVM terminology, the WBS boxes represent the PV for that work unit, and this value will be used for actual performance comparative purposes.

In the baseline schedule, WBS work units become discrete measures for assessing periodic progress. As the project progresses, work unit planned performance through time is measured against actual work accomplished to yield a metric called EV. AC for the work unit is then compared to provide both cost and schedule measures for the project at that status point.

14.4 EV CORE STATUS FORMULAS

Before simply listing a set of formulas to show cost and schedule status, it seems worthwhile to show a simple memory schematic that helps produce the core formulas and avoid the need to memorize them. This memory device will structure the key variables in such a way as to produce cost and schedule index and variance formulas.

14.4.1 Magic Box Formula Generator

The first step in this memory structure is to remember "AEP" in alphabetical order. Remember, these are the first letters of the three required measurement parameters (see Figure 14.2)—that is, actual cost, earned value, and planned value. With this as a starting point, let's introduce the "EV Magic Box" in Figure 14.5.

This figure represents a visual memory schematic that will produce the core status formulas from the box design. First, note that the top columns of the box contain AEP data in alphabetical order, which are the three EV status variables AC, EV, and PV. To finish populating this view, we add two status variables for cost (C) and schedule (S) in alphabetical order, going from the left side to the right side, respectively (see magic box formula figure). From this five-symbol structure, the basic EV formulas are essentially defined. To generate the formulas, start in the EV column and

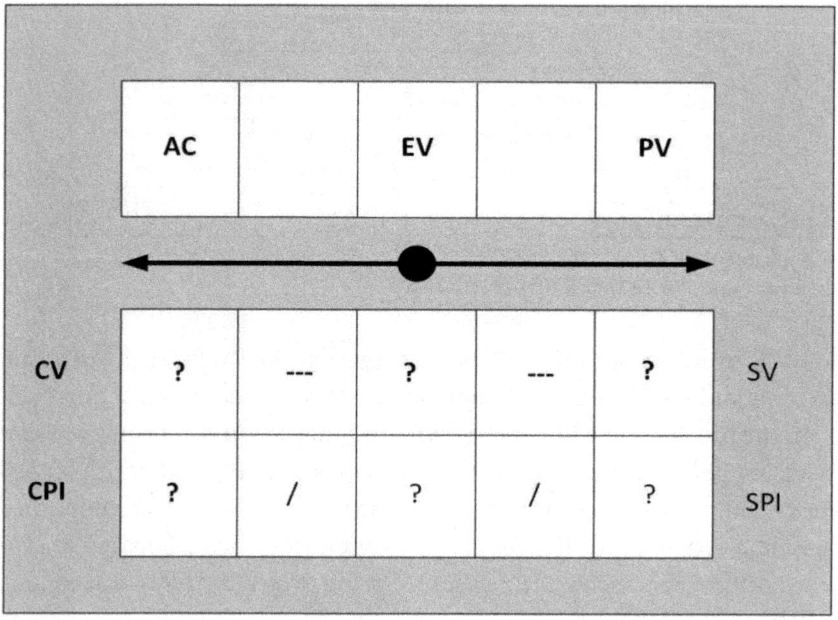

FIGURE 14.5 Magic EV formula box.

work outward to the left and right. Using this logic, the following four core status formulas can be produced:

CV = EV – AC

CPI = EV/AC

SV = EV – PV

SPI = EV/PV

Most would agree that these relationships are not easy to memorize; however, the associated memory logic of the box makes this much more doable. So, if you can arrange the raw data into the logical magic box structure, the formulas are easily derived, and the EV parameter values seem more logical. The process of producing the cost and schedule formulas can be summarized as follows:

> Each of the core formulas is generated by moving from the center EV cell outwards left and right towards a target parameter (cost or schedule) as indicated by the box arrows. Moving to the left would generate cost parameters while moving to the right accomplishes

the same for the schedule. Generation of variance parameters CV and SV require a negative operator sign between the two data items, while the index parameters CPI and SPI require a divisor.

Follow this explanation and note how the formulas "fall out" of the process. The reader should practice this data organization box structure as we further illustrate the EV mechanics. This same structure and mechanics can be applied to any project situation in which the three variables can be defined. As a final reminder, note that all EV variance formulae have EV as the first variable, while the index formulae show it in the numerator. Examine the index and variance formulae shown and compare them to the memory table and the associated memory tricks described. Experience suggests that having a method such as this to organize the raw data will save having to memorize formulae and make the overall analysis more logical to follow. Once you have mastered these basic mechanics, the next challenge is to learn how to interpret the quantitative results. This is the challenge for the following sections.

14.4.2 Gantt Bar Example

This initial example will illustrate the basic EV quantitative mechanics using a simple project Gantt chart plan shown in Figure 14.6. It will be instructive to compare the process described here with the previous ten-step minimal process outlined earlier. One would certainly categorize this example on the low end of complexity, but it is useful for illustrating how the data are collected and used. [14EV2]

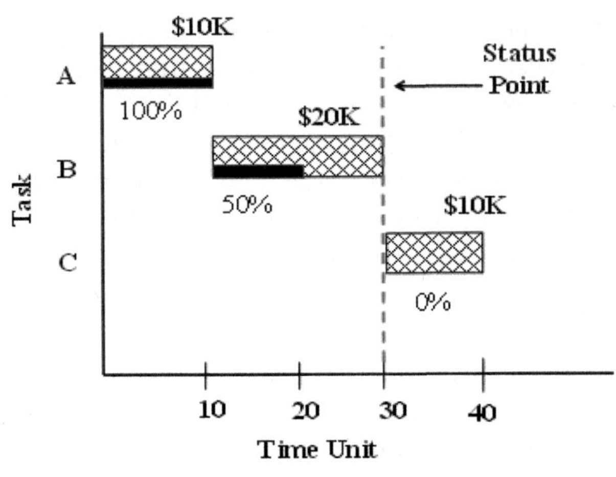

FIGURE 14.6 Gantt bar example.

The hatched Gantt bars represent planned work, and the solid black embedded bars represent actual accomplishment for their respective task. Each bar is labeled with the planned cost (PV) and actual percentage completion. A dashed vertical line indicates the status review point (period 30).

First, note that EV performance is measured using only planned baseline task values before the status point, so in this calculation, only Tasks A and B will have status values. The project status data for the two active project tasks is as follows:

Task A—PV = \$10K and 100% complete

Task B—PV = \$20K and 50% complete

Task C—Not active, therefore not included

The basic status question for this data is "How is this project doing?" In this case, we can see that the actual work accomplishment is less than planned, but we do not yet see the actual costs provided. That data will be added soon. All that can be concluded with the data shown is that the project is running behind schedule. Let's now evolve this example to show how EV mechanics can provide a quantitative set of measures for both cost and schedule, plus the ability to forecast completion under some defined assumptions. The next step will generate that set of parameters.

14.4.3 EV Solution for the Gantt Problem

The next step in the evaluation is to move the required data into the magic box structure. For this, we simply fill in the cell values under the named columns. Review the Gantt chart model shown in Figure 14.5 to see if you can identify actual, planned, and earned values from that visual picture—this calculation is often confusing for a new EV user. The value for PV can be derived directly from the Gantt bar status. Note that for the status point, tasks A and B were planned to be completed, so the project planned value would be \$10 + \$20 or \$30. Earned value is a little trickier. The rule says that you can only earn what you plan. Assuming the %Complete values are correct, the earned value would be \$10 for task A and \$10 for B (half of the plan value). So, the project EV would be 20. We now have defined two of the three required parameters.

The last variable needed is the actual cost. The question in your mind at this point might be "Where would one find actual costs for the project? Reviewing the discussion from Chapter 8, recall that the WBS had special tags for key points in the project where actual resource costs would be collected. These points were called CAPs, and the enterprise accounting system would be linked to these points to deliver that data. Let's assume for this example that the accounting system returned the total actual cost value for work units A and B as $35. This completes the required data cell values for the magic box, and the EV parameters can be produced from this point. In this example, we are only illustrating the mechanics of the total project, but in a real-world example, it would be more beneficial to have EV parameters calculated for each task as well. Keep that thought in mind as we move forward [14EV4].

The resulting data is now shown in the EV schematic boxes in Figure 14.7.

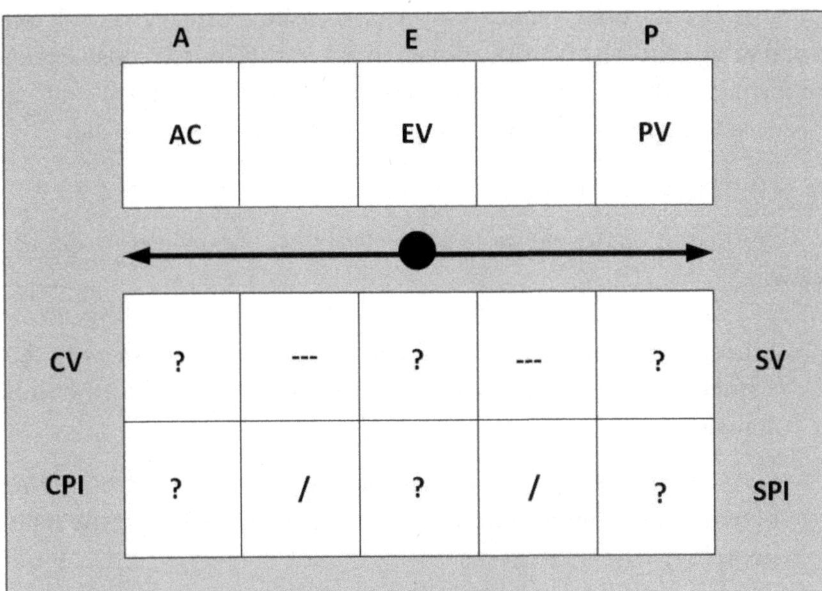

EV Status Formulas

CV = EV – AC

SV = EV – PV

CPI = EV/AC

SPI = EV/PV

FIGURE 14.7 Earned value parameters.

Using the memory box logic explained above, the calculated core parameter EV values are as follows:

CV = 20 – 35 = –15

CPI = 20/35 = 0.57

SV = 20 – 30 = –10

SPI = 20/30 = 0.67

As this example illustrates, the generation of EV core parameters is not that difficult to produce once the three driver data items are identified. A more important goal is to understand what these results mean and use that information to help take meaningful corrective action.

Remember, the design concept of EV is to create a formalized work unit structure against which work measurement and resource consumption are matched. The parameter values above accomplish that goal, but we now need to understand what the values mean. That is our next step.

14.5 INTERPRETING EV CORE PARAMETERS

Working definitions for the basic EV performance parameters are defined below:

- Schedule Variance (SV)—The difference in dollar value of work that should have been completed in each period compared to the work completed.
- Cost Variance (CV)—The dollar value by which the project is either overrunning or underrunning its baselined cost plan as of the status point [14EV3].
- Cost Performance Index (CPI)—The ratio of budgeted cost of work performed (EV) to actual cost (AC). A CPI greater than 1.0 indicates work is being accomplished for less cost than planned or budgeted, while a CPI less than 1.0 indicates the project is running above the planned cost.
- Schedule Performance Index (SPI)—The ratio of work accomplished (EV) versus work planned (PV), for a specific period. SPI indicates the time rate at which the project is progressing. This index value has the same numerical interpretation as CPI, except it now represents schedule performance.

In summary, the basic quantitative interpretive meaning for the core EV status parameters is as follows:

- Negative variance values are bad (below plan efficiency)
- Positive variance values are good (better than plan efficiency)
- Index values > 1.0 are good (better than plan performance)
- Index values < 1.0 are bad (below plan performance)

The variance parameters are reasonably straightforward in their interpretation; however, the index parameters offer more status insight and therefore deserve more comment. CPI for the example program was computed to be 0.57. That indicates the project is only running at 57% cost efficiency compared to the plan. The inverse interpretation says that the project will overrun its budget by 43% if this trend continues at the same rate. Both interpretations are valuable status indicators. For large projects, Christensen found that CPI values tended to stabilize by the 20th percentile point into the project life cycle to within ±10% accuracy (Christensen 1998, 10). There is some more recent research that questions the validity of this statement, but if true for specific project types, the CPI is a valuable predictive completion parameter once the project reaches this point. In other words, the underlying assumption of this statement is that a project develops an environmental, cultural, and technical profile that is not likely to suddenly change after the stability point.

SPI for the example problem was 0.67. This status interpretation is like that of the CPI, with one slight difference. In this case, the project is running at 67% efficiency compared to the plan, and if it continues at this rate the project will overrun its schedule by 33%. However, there is one important warning related to this parameter. Recall that the SPI formula is EV/PV. Earned values cannot exceed planned values. So, as the project moves around to the 70th percentile point in the life cycle, this formula begins to become misleading in that it will begin to migrate toward 1.0 regardless of actual project performance. As a result of this idiosyncrasy, calculations for SPI after the project midpoint will indicate improving results when in fact that is not necessarily the case. For this reason, care must be exercised in interpreting SPI values after the midpoint. We will review an alternative approach for schedule evaluation later in this chapter. For now, the key is to understand the basic interpretation of the values.

14.6 PERFORMANCE EVALUATION AND FORECASTING PARAMETERS

As we move into the forecasting aspect of EV analysis, a new set of three and four-letter acronyms emerges. Working definitions for these are as follows:

- Budget at Completion (BAC)—Baseline budget for the project. Components such as reserves, overhead, profit, significant material costs, etc., are included as part of the total project cost budget, but they should be removed from the EV status analysis since they are not "productive" in the sense of the WP delivery performance.
- Estimate at Completion (EAC)—The completion forecast for the project is based on the computed value of CPI. Similar time forecasts can be produced using valid SPI parameters.
- Estimate to Complete (ETC)—A forecast cost estimate for the remaining period of the project from the status point. It is calculated as the difference between EAC and AC. Senior management and financial groups are the ones most often concerned about this number.
- Variance at Completion (VAC)—Represents the difference between Budget at Completion (BAC) and Estimate at Completion (EAC). VAC represents the forecast variance compared to the baseline.
- Performance Measurement Baseline (PMB)—The sum of all Planned Value (PV) work packages/units aggregated by period for the total duration of the program. The PMB forms the time-phased budget plan against which project performance is measured. This curve is often called the S-shaped cost curve and may include a management reserve buffer. Be careful in interpreting the role of the reserve funds in project evaluation.
- Let's now go back to the initial example Gantt chart (Figure 14.7) and illustrate the EV forecasting metrics. EV input data and related status parameter results are summarized below from the Gantt chart example:

  BAC = 10 + 20 + 10 = 40 (review Gantt chart for these values)

  PMB = BAC = 40 (same value since no reserve is defined)

PV = 30

EV = 20

AC = 35 (from accounting data)

CPI = 0.57

SPI = 0.67

Derived forecast parameters:

EAC = BAC/CPI = 40/0.57 = 70 (projected completion forecast—basic option)

VAC = EAC – BAC = 70 – 40 = 30 (projected overrun)

ETC = EAC – AC = 70 – 35 = 35 (amount forecast to finish project)

EAC (time) = 40/0.67 = 59.7 (forecast schedule duration)

**Note that the EAC (time) calculation violates the rule on life cycle percentile completion but is shown here to demonstrate the mechanics.

Collectively, this set of parameters outlines the basic current and forecast time and cost status of the project and is like the parameters that would be generated by MS Project. At this point, it is time to raise a warning flag. All the forecasting results to this point include assumptions that may not be true. At this point, we complete the overview that most technical sources would define as EV; however, that opens the Achilles' heel of this concept. A simple and important EV concept statement is offered as follows:

> Don't view EV as an arithmetic exercise, but rather an analytical one. This means that forecasts must be made with an understanding of the behavior characteristics of the various WP types. No single formula can accomplish a goal.

There is a lot more remaining to the EV model in terms of proper usage techniques, but this overview will offer a good starting point for the raw manual mechanics.

14.7 EV DUAL NOTATION

The original 1960s EV notation was based on a four-letter acronym set, and this got heavily ingrained into the DoD documentation that drove the concept through its early period from the 1960s into the 1990s. The commercial project world shortened this notation with the result that traditional literature may show either the class or the more contemporary parameters. This dual documentation base creates some user confusion that is not likely to be resolved in the near term, given the associated cost of translating all of the older training and technical documentation into the newer simplified forms. As a result of this historical dichotomy, it is necessary to translate here both notation sets. The newer notation has been used in this chapter up to this point, but we now need to cross-reference both formats so that one can read both new and old technical literature. Also, it is important to recognize that much of the existing EV software, such as MS Project, uses the traditional notation. Table 14.1 summarizes the comparative core EV parameter dual terms.

It is important to note that both notations deal with the same concepts, so it is simply a matter of recognizing the dual naming convention. As the use of EV techniques continues to mature, there is a high probability of more new related status notation changes to emerge.

14.8 ROLE OF SOFTWARE IN THE PROCESS

The next stage of mechanics discussion involves converting a simple manual example into MS Project and letting the software take over the mechanical calculation details. With this step comes an important warning. *As is true with any reasonably complex software, if the user does not understand its internal assumptions the answers that are produced may be wrong.* This is certainly the case here. EV mechanics produced by MS Project must be performed with more discipline than most other processes to obtain both arithmetic and interpretative accuracy.

TABLE 14.1 EV Dual Notional Terms

New	Original	Meaning
PV	BCWS	Current baseline cost value to status point
EV	BCWP	Earned value calculation up to a status point (work performed)
AC	ACWP	Actual cost of planned work performed up to status point
BAC	BAC	Total direct work unit cost for the full project
EAC	EAC	Estimated direct cost of the total project at completion
ETC	ETC	Estimate of incremental direct costs from status point

Assuming that proper discipline is followed, the good news is that much of the raw EV calculation complexity can be taken over by software, and the parameters are essentially created as a by-product of basic project status mechanics that have already been discussed. A second significant advantage offered by utilizing a software approach is the efficiency of repeating the process at each project status point.

One of the early and maybe existing reasons for not producing EV measurements was the perceived mechanical overhead related to this process. That part of the argument is now no longer valid, but the process discipline required to generate the underlying data remains an issue.

14.8.1 Garbage In Garbage Out

There is a well-known adage regarding the use of computers in decision-making roles—"*To err is human, but to really screw things up requires a computer.*" This point is very relevant in this section. Recognize that MS Project has many built-in default assumptions that will be used to generate the resulting EV parameters. In some cases, these built-in assumptions may not be appropriate, so it is important to understand what the fundamental assumptions are. The basics of this issue will be outlined here, but local usage may create even other computational issues. Mishandling *the raw driver data will result in EV parameters that are inaccurate for the defined role.* When the derived data does not represent what happened, confidence in the value of the tool is compromised. To mitigate this issue, it is up to the project manager to make sure that the correct data and assumptions are used to produce the status and forecast parameters. Simply generating an EV parameter and saying that it always means something is erroneous.

Warning number two—MS Project is somewhat picky in the data input sequence leading up to producing EV parameters, and it will require some practice to generate accurate values. The mechanics outlined here are the most manageable and accurate for normal situations, but care must be taken not to just retrieve values blindly without manual review.

14.8.2 MS Project Calculation Steps

To generate an accurate set of EV parameter values, it is critical to follow the sequence outlined below. Consider this sequence of steps as a checklist to be followed in the order shown. The steps are as follows:

1. Be sure to produce a backup copy of the plan before starting this process.

2. The project plan should generate its internal work unit costs by allocating resources from the Resource Sheet into individual tasks. If planned costs are manually entered without linking them to work unit resources, a level of analysis granularity is lost.

3. Explicitly define the EV actual cost calculation option. Go to [File/Options/Schedule] in the Calculations options section; uncheck the box for "*Actual costs are always calculated by Project.*" This means that you are going to manage actual costs for each work unit in the plan—this simulates the accounting system delivering costs to the project. Second, review the [File/Options/Advanced] Earned Value section for measurement and baseline. The typical options here are %Complete for the EV method and a particular Baseline to be used for the comparison process. In this example, we will only have one Baseline but realize that this may not always be the case.

4. Make sure that the plan has a set baseline [Project/Set Baseline]—without this step, no EV parameters will be calculated since EV is %Complete times the Baseline value.

5. Open up data fields for desired output items—that is, BCWS/PV, BCWP/EV, CPI, SPI, and EAC. Recall the earlier note related to old versus new notation. If the new notation is preferred, the label titles can be edited by right-clicking on the column and selecting *field settings*. The normal practice would be to create a custom view and save that for reuse at each status point.

6. Set status date by [Project/Project Information]. Be sure to check the calendar year value carefully as projects often flow across year boundaries, and it is easy to set the wrong year—without a valid status date, no EV parameters will be calculated.

7. Set a red gridline at status date by [Format/Gridlines/Gridlines/Status Date]—if this line does not show on the Gantt bar sheet at the defined status point, check the Project Information Status box for date error or gridline format color. The typical error here is created by the wrong year, which is very easy to miss.

8. Use the Gantt Chart Wizard to define the initial bar formatting—one useful wizard option is [Other/Baseline: Style 3] as this shows plan, completion, and baseline on the same view.

9. From this starting format, use the standard bar format commands to add any other desired custom format features—that is, [Format/Format/Bar formats...]
10. Set actual durations for work accomplished task units and change any other future planned task durations as needed. This is essentially updating the plan to both observed and new forecast values.
11. Set %Complete measurement values for tasks in the status range.
12. Open the Actual Cost field and input actual cost data for each Task—this simulates data values from the external accounting system.
13. Remember that the EV driver data fields are Actual Cost, BCWP/EV, and BCWS/PV. Given the data entered above, EV parameters should now contain non-zero values. If no EV values appear, the most likely error sources are that no baseline or status date has been set. Also, check the location of the status gridline to confirm the status date is properly placed (all three of these items are common error sources).
14. Project status interpretation can now be determined from the derived EV parameters. Note in the output below that the same parameters shown earlier for the Gantt example are now available, not only for the whole project but all levels down to the lowest task.

Before moving forward to the next section, it would be worthwhile to review the mechanics outlined above. Also, it would be worthwhile to review the *Project N Motion* library video tutorials for more theory and mechanics on this topic (refer to Chapter 1 for details).

14.8.3 MS Project Gantt Example Setup (Checklist Steps 1 to 3)

The Gantt chart example from Figure 14.6 will be used to illustrate the mechanical steps described above. To map the Gantt example into MSP, the only data item needed is the addition of a generic "Worker" with a standard rate of $1000/day. That worker is defined in the Resource Sheet view and then allocated to each task to create the desired cost. This single resource and rate are designed to make the resulting task costing arithmetic easier to follow. All values shown in Figure 14.8 should now be matched as closely as possible to the earlier graphical Gantt model. Do note that the software has a more sophisticated calendar than the manual Gantt chart,

	Task Name	Duration	Cost	Start	Finish
1	⊟ EV Start	40 days	$40,000.00	1/13	3/8
2	A	10 days	$10,000.00	1/13	1/26
3	B	20 days	$20,000.00	1/27	2/23
4	C	10 days	$10,000.00	2/24	3/8

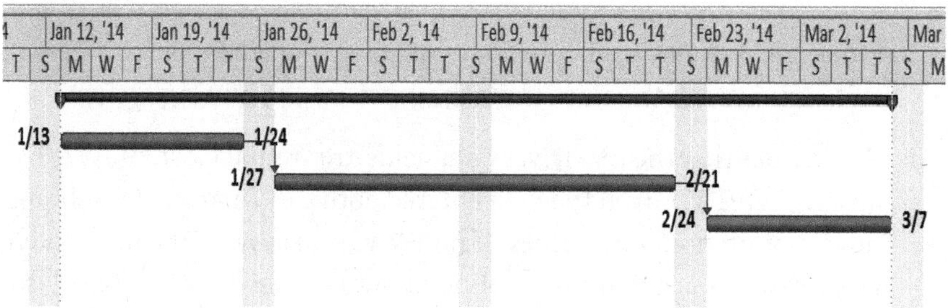

FIGURE 14.8 EV setup (steps 1 to 3).

so there is now an additional reference to planned calendar dates based on the resource's work schedule. In this case, the project is defined to start on January 13 with a standard five-day-per-week work schedule. Task duration is the number of workdays shown for the task, with a cost of $1000 for each day. Figure 14.8 shows the initial output split into two separate segments (table and Gantt bar views). This view essentially mirrors the earlier manual Gantt model.

14.8.4 Generating the EV Parameters (Steps 4 to 5)

This set of steps sets up the desired table format to make the desired EV parameters visible (Figure 14.9).

	Task Name	Duration	Cost	Start	Finish
1	EV Start	40 days	$40,000.00	1/13	3/6
2	A	10 days	$10,000.00	1/13	1/24
3	B	20 days	$20,000.00	1/27	2/21
4	C	10 days	$10,000.00	2/24	3/6

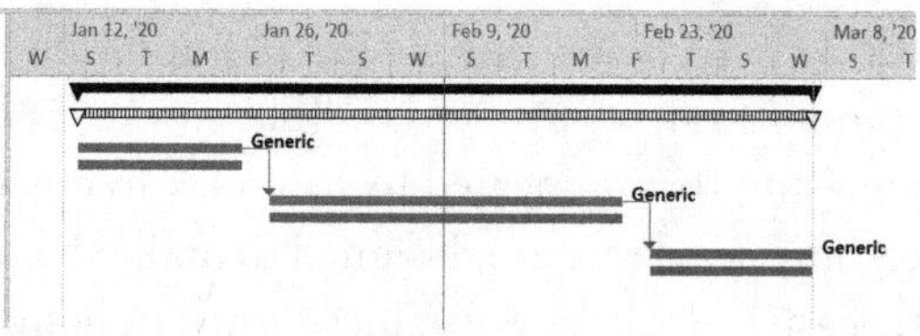

FIGURE 14.9 EV parameters (steps 4 to 5).

To generate this view, the following instructions were executed from the initial base plan view:

Step 4. Set project baseline—[Project/Set Baseline].

Step 5. The Gantt Chart Wizard was used to format the bars.

14.8.5 Generating the EV Parameters (Steps 6 to 9)

Column formatting for selected EV parameters. Header titles were edited manually using the *Field Settings* option.

Step 6. Set the status date to 2/8/2026 at [Project/Project Information].

Step 7. Display gridline at status date by [Format/Gridlines/Gridlines/Status Date]; Enter line Type (solid) and Color (red). The vertical line should now appear at the defined status point. If it does not, you have made an error in either the status date or gridline definition.

Step 8. This is a cosmetic step to convert the default bars into a desired format. The recommended sequence for this is to first use the Gantt Chart Wizard to initiate the formatting process. This icon can be found in the Commands section. See the MS Project checklist (Appendix D) for more details regarding how to make the wizard option visible on the menu ribbon panel. Use the Gantt wizard option "Other/Baseline Style 3" for a starting bar format point. Then go to the [Format/Format/Bar Styles] menu to finish any other item formats desired for the display—that is, dates, colors, bar sizes.

Note that no EV parameters or Actual Cost data are shown at this point in Figure 14.10. Remember, EV calculations are driven by measuring

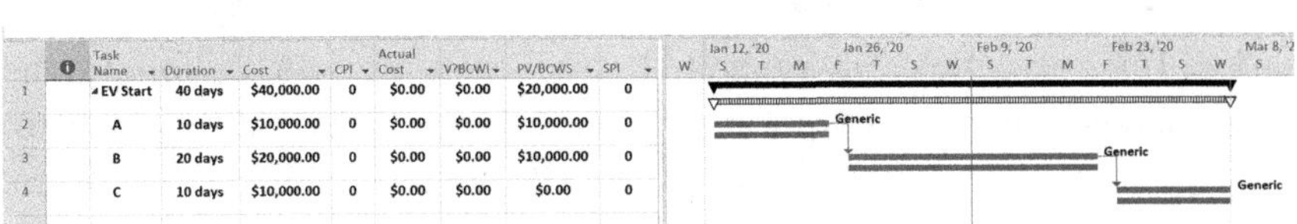

FIGURE 14.10 EV parameters (steps 6 to 9).

	Task Name	% Compl	Baseline Duration	Duration	Baseline Cost	Cost	CPI	Actual Cost	V?BCWI	PV/BCWS	SPI	EAC
1	EV Start	52%	40 days	47 days	$40,000	$47,000	0.82	$24,500	$16,400	$20,000	0.82	$48,781
2 ✓	A	100%	10 days	12 days	$10,000	$12,000	0.83	$12,000	$10,000	$10,000	1	$12,000
3	B	50%	20 days	25 days	$20,000	$25,000	0.8	$12,500	$6,400	$10,000	0.64	$25,000
4	C	0%	10 days	10 days	$10,000	$10,000	0	$0	$0	$0	0	$10,000

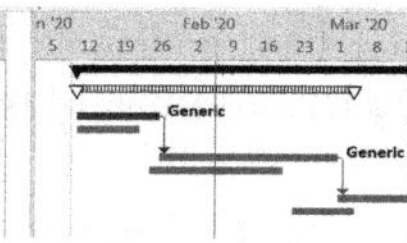

FIGURE 14.11 EV calculated parameters (Option A—Effort driven).

performance, and we have not provided %Complete data yet. The next section will add the accomplishment data.

14.8.6 Measuring Status with Effort-Driven Logic (Steps 9 to 13)

Figure 14.11 shows steps 9 to 13, and this represents the highest EV calculation complexity area of the process. The resulting EV parameter values will vary depending on how actual cost data and measurement values are generated. For example, in Option A, we will use the MS Project default approach for producing actual task costs, which is called effort-driven. *Realize that this means the check box for letting MS Project calculate Actual Cost is left checked (See [File/Options/Advanced]).* In Option B of this example, the box will be checked to show the impact that it has on the calculation. As subtle as this point is, this checkbox is a major key to getting accurate EV values and is also important to conceptually understand.

Note in the table data values shown in Figure 14.11 indicate that both tasks are overrunning baseline planned time. As a result of the default, MS Project effort-driven cost calculation assumption task costs are calculated as overrunning at the same rate as the duration. In this example, Task A overruns from 10 to 12 days, and the projection for Task B is for it to overrun from 20 to 25 days. Both revised duration values are shown in the Duration column. The cost of each task overrun is dictated by the effort-driven assumption. Second, %Complete values are added for Tasks A and B. These decisions generally mirror the original manual Gantt example. A revised table and Gantt bar result for this can be seen in Figure 14.11.

14.8.7 Option B—EV Status with Manual Entry of AC (Steps 9 to 13)

Figure 14.12 uses manually generated actual costs. This option follows the same calculation path as Option A; however, in this case, specific control of actual cost values used is external to MS Project defaults. This means that the "MS Project will Calculate Actual Costs" box is unchecked. Note that the Actual Cost for task B is now shown as $33,500 for task B and $12,000 for task A (these were manually entered with the hypothetical

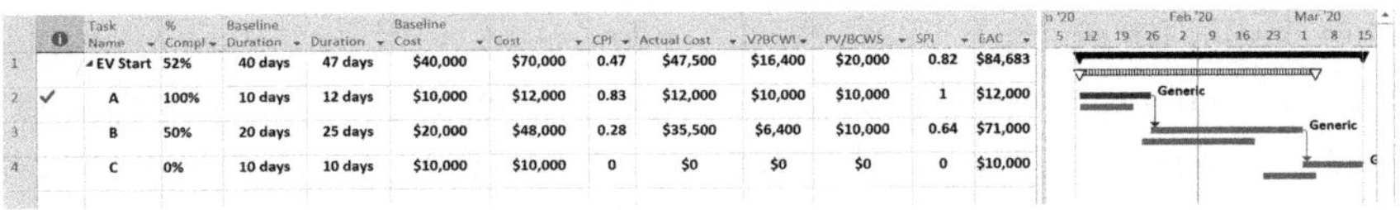

		Task Name	% Compl	Baseline Duration	Duration	Baseline Cost	Cost	CPI	Actual Cost	V?BCWI	PV/BCWS	SPI	EAC
1		EV Start	52%	40 days	47 days	$40,000	$70,000	0.47	$47,500	$16,400	$20,000	0.82	$84,683
2	✓	A	100%	10 days	12 days	$10,000	$12,000	0.83	$12,000	$10,000	$10,000	1	$12,000
3		B	50%	20 days	25 days	$20,000	$48,000	0.28	$35,500	$6,400	$10,000	0.64	$71,000
4		C	0%	10 days	10 days	$10,000	$10,000	0	$0	$0	$0	0	$10,000

FIGURE 14.12 EV calculated parameters (manual actual costs entered).

values retrieved from actual expenditures). These values will potentially change the previous values for CPI and SPI. *Note: Compare the EV calculation values shown in options A and B to see the significance of the step 3 checkbox and its impact on the output values.* Figure 14.12 summarizes the basic EV parameter set for option B. At this point, all of the column headings and data meanings should be understood.

In comparing option A output values versus option B, we find essentially the same values for SPI, but quite different values for CPI. *This is the key point to remember!* Inspection of the table data shows where this difference is produced. MS Project decided via the effort-driven assumption that the project's actual cost was $24,500 and not the $47,500 defined in the manual example. As a side note, the normal result of this is just the opposite, given that time expansion does not necessarily mean that resources are being expended through the expansion period. If the assumption is that these tasks are overrun because the resources could not finish them on schedule, then the effort-driven assumption might be reasonable. There are many similar reasons why generating duration cost with the effort-driven assumption is not valid, so the calculation rule stressed here is not to let MS Project automatically calculate task cost based simply on the expanded duration of the task. To accomplish this, always set the calculation option as described in checklist step 3.

Figure 14.12 clearly shows how actual progress varies from the baseline. One added analysis point is whether we can anticipate the rest of the project to suddenly start fitting the same trend. Research says that projects tend to develop a culture early in the cycle, and this behavior continues for the rest of the life cycle. Once that issue is decided, it is time to either modify the rest of the plan accordingly or assume that the future will go as planned from this point forward. If the same trend continues, one must start dealing with how to forecast completion values. In any case, the active project plan should represent the latest views on both current and future status. As the periodic analysis repeats, forecast value ranges will become better defined.

It would be an instructive reader exercise to reproduce this example and manually follow it through the steps to observe the software behavior as data values are added. This will provide more insight than seeing the static views shown here. It would even be instructive to change some of these sequence steps and see what happens to the calculated values.

14.9 EVALUATING MS PROJECT EV PARAMETERS

The final step in this example process is to interpret the various derived status parameters. Option B has accurately recreated the manual output originally shown for the manual Gantt example. More importantly, the required set of steps to produce that answer has been shown. This same set of steps will produce EV parameter output for any project plan with minimal additional data entry.

The basic status EV parameters are all included in the standard MS Project data set. The only data label ambiguity is in the classic definitions of AC (ACWP), EV (BCWP), and PV (BCWS). However, two EV status parameters are not native to MS Project. That is, ETC and EAC(time) parameters are not automatically calculated by the software. ETC can be easily produced by creating a custom formula to show the difference between EAC and AC. Also, one must be careful to recognize that there are more than one possible forecast technique for EAC, and this may not agree with the software version. The EAC (time) calculation is particularly vulnerable to error, as described earlier in the chapter.

The major warning concerning EV computer output accuracy is to recognize that some of the calculation processes embedded may not be correct and, for this reason, should be checked. The recommendation on this front is to move the core data out of MS Project, where it can be more easily manipulated to show specifically what is intended. More on this philosophy is upcoming in a later section.

14.10 DEVELOPING EV PARAMETERS USING EXCEL

Industry data suggests that many project plans are created using traditional spreadsheets. One typical argument from this segment is that the EV calculation process will not work for that model. This example will show that the same mechanics will work, but admittedly with a higher-level status view. The spreadsheet shown in Figure 14.13 illustrates a summary-level project plan that could be extracted from MS Project or just created at that level without going through higher-level software. The key with all EV calculations is to follow the basic steps outlined. Note that the

	A	B	C	D	E	F	G	H	I	J	K	L	M	N	O	P
1	Activity	Jan	Feb	Mar	Apr	May	Jun	Jul	Aug	Sep	Oct	Nov	Dec	PV	% Comp.	EV
13	Monthly Planned Value (PV)	4,000	10,000	10,000	10,000	12,000	16,000	10,000	6,000	8,000	4,000	5,000	5,000	100,000		42,000
14	Cumulative Planned Value (PV)	4,000	14,000	24,000	34,000	46,000	62,000	72,000	78,000	86,000	90,000	95,000	100,000			
15	Monthly Actual Cost (AC)	4,000	11,000	11,000	12,000	15,000										
16	Cumulative Actual Cost (AC)	4,000	15,000	26,000	38,000	53,000										
17	Monthly Earned Value (EV)	4,000	10,000	10,000	10,000	10,000										
18	Cumulative Earned Value (EV)	4,000	14,000	24,000	34,000	44,000										
19	Project EV as of May 31	44,000														
20	Project PV as of May 31	42,000														
21	Project AC as of May 31	$ 53,000														
22	CV=EV-AC	$ (9,000)														
23	SV=EV-PV	$ 2,000														
24	CPI=EV/AC	83%														
25	SPI=EV/PV	105%														
26	Estimate at Completion (EAC)	$ 120,455	(original plan of $100,000 divided by CPI of 83%)													
27	Estimated time to complete	11.45	(original plan of 12 months divided by SPI of 96%)													
28																

FIGURE 14.13 Excel spreadsheet for producing aggregate EV parameters.

major spreadsheet project plan segments are scheduled into monthly time boxes.

The first setup step required is to map this data format into the EV model. Note that the plan structure shows planned values (PV) by summary activity level and month grouping (Jan to Dec). The solid vertical line at the end of May signifies that it is the status point for this evaluation. Column "N" contains the cumulative PV data for each work unit. This is calculated as a row total. Column "O" contains %Complete measurements for active work elements up to the status point. So, the product of columns "N" and "O" computes a value of EV for each active work unit. From these data sources, it is possible to derive the project values for PV ($46,000) and EV ($42,000).

A good conceptual question to ask at this point is "What additional data is now needed" to produce the core EV parameters? If you remember the AEP magic box structure, we need actual costs to complete the required data. Let's assume that the aggregate AC value is $53,000 as delivered from the accounting system. Using these data values, one can define the values for CPI, SPI, BAC, EAC, and VAC. All required elements are available in the spreadsheet cells if you know where to look. It would be a good exercise to review rows 19 through 27 to see if you can retrieve the data used to calculate these values.

This spreadsheet format illustrates how the various EV data items can be captured in a different format and then used to provide at least a high-level status view. In some situations, this approach is easier to manage and a preferred EV starting option to get familiar with the data issues. Review this format until you are comfortable that all the core parameters are either shown or could be developed from this set of data. With Excel's graphing capabilities, it is also possible to use this same data source to create a visual view of the status. One major disadvantage of this high-level approach is the loss of a task-level view in evaluating individual work unit

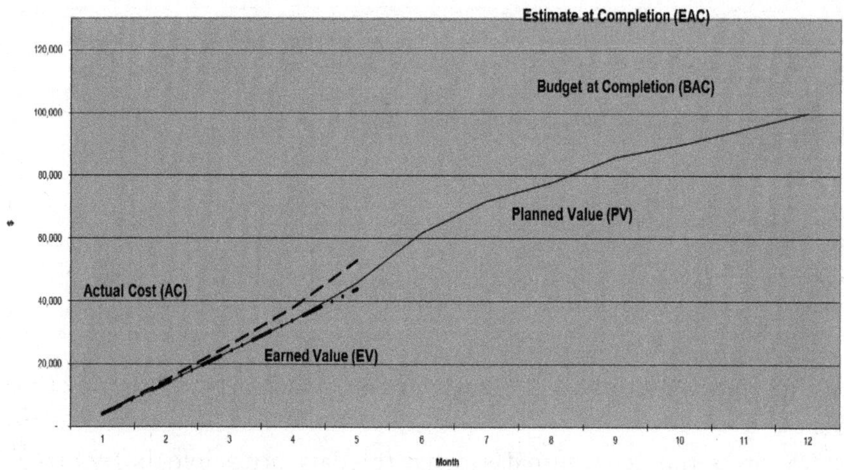

FIGURE 14.14 Graphical status view of earned value.

performance; however, recognize that this is still a worthwhile starting approach for any project. Also, if status values were saved at each monthly status point, it is possible to create various time-oriented trend views, which add visual insight into whether the project is improving or declining. Figure 14.14 illustrates the communication value for a graphical view.

14.11 EV EARNING RULES

There are several options for recording WP accomplishment. The list below summarizes the implications of the most used measurement techniques.

- 50/50—This rule gives a 50% credit once the WP is started and the last 50% is withheld until the unit is completed.
- 0/100—This rule penalizes work in progress since no credit is given until the unit is completed.
- Percentage of work completed—This is the most accurate mathematical calculation of accomplishment if one can assume that the measurement can be done accurately and honestly.
- Level of effort—This class of resource charge is normally either allocated to the WP as defined in the support agreements or as billed to the project based on actual charges. In the former option, the work unit should be taken out of the mix for performance measurement purposes.

- Previous examples used the term *%Complete* to represent work accomplished, and this value then drives the EV calculation. This is the most controversial and problematic variable in the EV model. Recall that the design goal is to measure what has been accomplished, and experience suggests that individuals are not always accurate in creating this measure.

14.11.1 Measurement Factors That Can Distort EV Accuracy

Several data and process areas in EV calculation impact the appropriate selection of a measurement approach. A few operational rules to consider when making this selection are summarized below:

Work package size—A general rule of thumb for a WP size is "2/80," meaning that it will be sized in the general range of two weeks elapsed time and 80 hours of associated work effort. Other values can be used as target guidelines, but it is important to keep the size moderate since large WPs can hide overruns. Errors in a large WP can distort the overall assessment and are harder to uncover in the short run.

Level of effort (LOE)—Tasks in this class are essentially recurring service-oriented activities not directly related to the project deliverables and are often billed at fixed values. Including data from his class of WP can cloud the project status analysis since related resource charges are independent of work accomplishment. Productivity studies should be done without these included in the mix.

Material costs—The process of developing material costs is quite complex and hard to isolate to a single WP. Including allocations from this resource category can also hide labor productivity and variance trends. If material costs represent a significant percentage of the total work unit cost, they should be extracted from the analysis of work-related performance.

Work unit measurement rules—Pros and cons for the major measurement options are summarized in Table 14.2.

As one can see from the pros and cons listing, each method has both advantages and disadvantages. The goal is to select the method that best measures work accomplishment. This selection should be documented as part of the project control management plan.

TABLE 14.2 Measurement Option Pros and Cons

Approach	Pros	Cons
%Complete	Easy to implement; fits calculation model	Human bias can affect accuracy; errors here distort results
Units produced	Accurate measurement	May not be granular enough
Weighted milestones	Visible measure	May create inaccurate measures, i.e., 0/100 granularity
Apportioned	Typical for support	Does not measure accomplishment
LOE	Work unit is often support oriented and hard to link to direct output	Another method should be used

Other techniques that may be appropriate to consider are as follows:

- Internal work unit milestones—key events within the work unit
- Weighted milestones—allocating completion credit for various key events
- Units complete—if the work produces physical items, measurement of these could be used for %Completion. As an example, drawings completed in the work unit could be counted as a measurement unit.

14.11.2 Labor Efficiency and Rate Variation

Stratton suggests that EVM for labor-intensive projects should also include an analysis of labor related to the work activities (Stratton 2007). The relationships shown below can be used for such analysis and can help reveal additional insights into project performance.

Cost variances (CV) can be broken down into two components to compute rate and use variances—also known as efficiency (Shulte). The bid and rate variance formulas follow:

Bid rate = EV Cumulative Hours/EV Cumulative Dollars

Actual rate = AC Cumulative Hours/AC Cumulative Dollars

Rate variance = (Bid rate – Actual rate) × AC (cumulative hours)

Use variance = (EV hours – AC hours) × Bid rate

Cost Variance = Rate Variance + Use Variance

TABLE 14.3 Example Work Unit Labor Data

Status Item	Earned Value (EV)	Actual Cost (AC)	Cost Variance (CV)
Cumulative Hours	680	960	−280
Cumulative $	75,480	90,000	−14,520

Table 14.3 shows example hours and cost data for a WBS element at a specific point in time. The earned and actual hours support rate and use variance calculations as illustrated. In this example, the computed CV value was −$14,520, which implies that the budget is overrun by the team. Let's see what else might be involved here.

Recall that a negative CV indicates that work is more expensive than planned. Calculation of a labor CV provides an additional view as to the separation of this value into two components: labor rate and productivity variances. The process outlined below shows how the estimated labor rate versus actual rate variance can affect the overall CV value and result in a quite different interpretation for that factor. Think of a bid rate as the initially planned resource rate, either for internal or external labor. Using the sample data in the example above, a labor rate analysis can be calculated as follows:

Bid rate = ($75,480/680hrs) = $111.00/hr

Actual rate = ($90,000/960hrs) = $93.75/hr (actual rate is cheaper than bid)

Rate variance = (111.00 – 93.75) × 960 = $16,560

Use variance = (680 – 960) × 111.00 = −$31,080

CV = $16,560 – $31,080 = −$14,520

These calculations show that the labor rate (rate variance) is cheaper (positive) than originally estimated (111.00 – 93.75) and the productivity (use variance) is below the planned value (negative). So, about half of the aggregate CV value is related to productivity loss, and half is gained back by a favorable lower labor rate. The actual labor rate values are less than forecast, so this has a positive impact on CV status variables. However, the opposite can also be true. In some situations, one might find a perfect work forecast indicating a negative variance when in reality the only issue causing this is the rate of the resources rather than hours. Recognition of the rate and use factor variance calculations is important to review since the team rates are often external to project control. Also, this type of analysis can help isolate resource issues and root cause factors. Understanding the interrelationship of this type of data can support an interesting conversation with resource providers.

In addition to examining the rate and use variances for a single period, trend charts can be used to examine how the parameters change over time at various stages of the project, by work unit, or organizational skill group.

14.12 EARNED SCHEDULE

Walt Lipke, who is one of the outstanding pioneering project management conceptual thinkers, recognized the inherent errors in EV time forecasting. To improve that, he derived an insightful contemporary model to evaluate project schedule status. This technique is labeled Earned Schedule (ES), and it is a time-based view of EV (Lipke 2009), rather than the class EV cost-based view. ES is now being broadly accepted in the industry, and the only operational complexity with it is the lack of a simple formula; however, there are software add-ins available to assist in this calculation. Figure 14.15 shows a simple schematic diagram that illustrates the concept

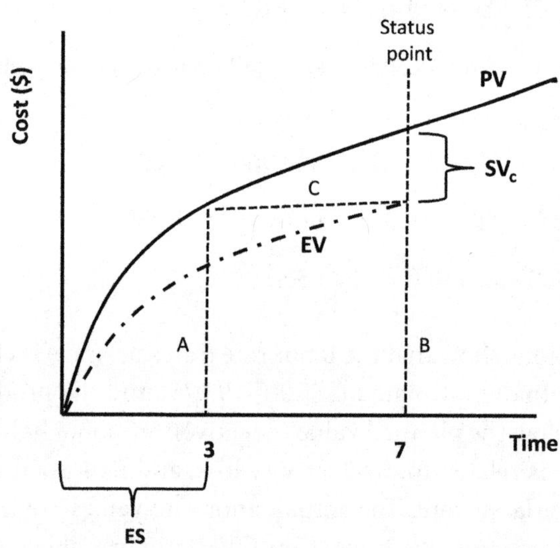

FIGURE 14.15 Earned schedule schematic.

Source: Adapted from Lipke, Walt. "Earned Schedule—An Extension to Earned Value Management," Accessed at www.earnedschedule.com.

of ES. The concept is much easier to visualize in graphical form rather than trying to understand it algebraically, so that is the way it will be described here. The interested reader should pursue some of the Lipke sources readily found through web searches.

Recall the earlier statement that the SPI parameter formula will fail as a forecasting metric around the 70th percentile point in the project life cycle based on its formula definitional structure (i.e., EV approaches 1.0 at the end of the project). ES offers a general method to overcome that shortcoming and will work throughout the life cycle. It is simply defined by reviewing the project's PMB S-shaped curve and evaluating the EV at a particular status point, then comparing a common point on the PMB. The difference in these two values represents the true time status for the project. As Mr. Lipke says, *"The idea is to determine the time at which the EV accrued should have occurred."*

ES is determined by projecting the current EV status horizontally to the equivalent PMB baseline curve. For this example, the project status point is period 7. The current schedule performance gap can be determined by moving from the current EV value at period 7 and evaluating at what period the PMB/PV has that same value. The horizontal line labeled C connects these two points. Moving horizontally from the EV curve along line C intersects the PMB curve at period 3. This indicates that the current planned EV should have been reached at period 3 rather than 7. So, the project has a current time variance of 4 time units (7 – 3) and an ES value of 3. Both values should be obvious from the schematic.

Figure 14.10 also graphically shows two methods for defining schedule variance. The traditional EV cost-based view is labeled SV as the gap between the PV and EV curves, while the newer ES time-based view is represented by the ES value on the X-axis. During project execution, both measures should be evaluated and tracked as a check and balance review.

14.12.1 TCPI

This parameter is designed to measure the likelihood that the project can be completed on schedule given its status. TCPI stands for "To Complete Performance Index" and is a third-level EVM forecasting tool. This relatively new parameter is designed to help organizations evaluate the potential that a project can be completed within budget, given current EV parameter values. The basic formula for this term is as follows:

TCPI = (Remaining Work)/(Remaining Funds)

The definition of remaining work is BAC – EV, which is a straightforward concept, given that this is the planned level of work remaining at this point. The remaining funds would be the increment between EAC (planned forecast) and the actual funds expended at this point. From this definition, there are two options to calculate the parameter, which are as follows:

Option 1. Using planned budget value

Remaining Funds = BAC – AC

Option 2. Assuming an overrun (EAC) budget value

Remaining Funds = EAC – AC

As can be seen from the basic formula, if there is more work remaining than funds remaining, the project is going to have to perform better than the plan to finish on budget. So, if the value of TCPI is 1.2, the project would have to operate 20% better than the plan for the rest of the way to finish at BAC, or the same interpretation if EAC is used. Every project manager feels that they can accelerate a plan, but the question is, how much is reasonable? That interpretation is where the value of TCPI emerges.

At this stage of usage, more research is needed to verify the interpretation rule, but there is growing evidence that projects do not recover budgets easily. Lipke has evaluated this process somewhat synthetically and concluded that project TCPI values greater than 1.1 would have a poor chance of budget recovery (Lipke 2009). Other researchers are coming to the same relative conclusion. Certainly, one should not expect projects with values above 1.2 to be able to recover their budget. When one looks at this parameter in this way, we see a cross-check on the project EAC forecast. Using measures of this type, we are starting to see evaluations of the project forecast reasonableness, yet another dimension of the EV model. Figure 14.16 shows a sample calculation for TCPI compared to CPI.

Using the TCPI value, should one expect the project to meet the baseline plan? According to industry experts, the answer to this would be no!

14.13 EV PARAMETER VALUE DISTORTION ITEMS

The data shown in Table 14.4 will be used to illustrate how misuse of data can distort the value of EV in status and forecast roles.

$$\text{CPI} = \frac{\text{EV}}{\text{AC}} = \frac{\$3600}{\$4800} = \underline{0.75}$$

$$\underset{\text{(BAC)}}{\text{TCPI}} = \text{cost efficiency to complete on budget}$$

$$= \frac{\text{Work Remaining}}{\text{Funds Remaining}} = \frac{\text{BAC} - \text{EV}}{\text{BAC} - \text{AC}}$$

$$\text{TCPI (BAC)} = \frac{\$10000 - \$3600}{\$10000 - \$4800} = \frac{\$6400}{\$5200} = \underline{1.23}$$

69

FIGURE 14.16 TCPI calculation.

TABLE 14.4 Sample EV Control Account Raw Data

Budget Area	AC	EV	Baseline
Labor	2,400	1,000	2,000
Material	1,500	1,000	2,000
$	1,200	1,000	2,000
Totals	**5,100**	**3,000**	**6,000**

Assume that this work unit is evaluated at 50% complete. The respective component EV parameter calculations are shown below. We will use this granular view of data to show what happens when we look at each resource separately and then compare that to the traditional total work unit EV equivalent value. The granular calculations follow:

CPI (Total) = EV/AC = 3000/5100 = 0.59

CPI (Labor) = 1000/2400 = 0.42

CPI (Material) = 1000/1500 =0.67

CPI ($) = 1000/1200 = 0.83

Each of the internal items indicates a cost performance status for that specific segment, so reviewing each individually gives a measure of performance for that component. However, if the goal is to forecast EAC for the project, is it correct to say that the proper formula would be EAC = BAC/CPI (Total), or 6,000/0.59 = $10,169? This forecast is valid if the CPI parameter is static for the remainder of the project. However, if there is reason to believe that the various components might have different levels of stability, each could be forecast independently. Using this logic, the labor portion would be estimated as 6,000/0.42 = \$14,286, and the same logic could be used for forecasting material and dollar costs. The key point in this decomposition approach is to recognize that not all components in the WP have the same behavior profile. Depending upon how the trends are anticipated for material and dollar expenses, a quite different result could occur when examined separately. Remember, the CPI parameter measures the profile of the entire WP, so assuming all pieces in that package behave the same may not be valid.

14.13.1 EAC Calculation

The standard EAC formula discussion described multiple ways of forecasting EAC, and these concepts are also relevant to this discussion. For this example, the assumption is made that the various component trends will continue through the life cycle as these parameters indicate. Given this assumption, a combined EAC calculation could be developed as follows:

$$\begin{aligned}\text{EAC (Total)} &= \$14,286 \text{ (labor)} + 1000/0.67 \text{ (matl)} + 1000/0.83 \text{ (\$)}\\ &= \$19,681\end{aligned}$$

Note that this forecast is nearly double the original forecast using a single bundled view. This simply highlights the sensitivity regarding what the forecast assumption can do to the resulting estimate if one does not understand the implications of the various resource types.

Four other alternatives for forecasting the project EAC are described below:

- Standard budget forecast for EAC—Used when *future progress is predicted to be like previous history.* The estimation formula for this assumption would be

  EAC = BAC/CPI.

- Flawed estimating forecast—Used when past performance is not believed to represent plans *and the existing plans are not felt to be valid.* In this case,

  EAC = (Current actual costs) + (A new estimate for all remaining work).

- Poor startup—Used when current actual cost variances are seen as atypical, but the project management team expects that *the remaining plan is valid* and similar variances will not occur in the future. In this case,

  EAC = (Current actual costs) + (Remaining planned work unit budget)

- Critical ratio estimate—This offers insight into the interaction of these two parameters. If both values are less than 1.0, this can indicate a more serious overrun than the standard EAC formula would indicate. Even if one of the parameters is significantly below 1.0 and the other is above 1.0, there might be reason to question the existing plan structure. More typically, both indices are less than 1.0, which means major issues with the current plan. From this view, a pessimistic estimate can be generated from the following formula:

  AC = BAC/(CPI * SPI)

14.13.2 Forecasting considerations

Included below are a few resource rules to consider in using EV for team productivity analysis or overall project outcome forecasting:

- Material variances do not impact either team performance or labor forecasting. This segment of the work unit should be reviewed separately to interpret material forecasting outcomes.
- The dollar resource category has the same interpretation as material regarding team productivity and outcome forecasting.
- LOE and recurring activities are typically fixed allocations regarding %Complete and possibly actual cost allocation. These typically should be removed from team analysis and forecasting parameters if these resource allocations are of significant size.
- Contractual labor charges should be reviewed for proper allocation in team productivity or outcome forecasting activities. The normal rule would suggest that this be dealt with as a separate analysis category.

- Task padding clouds both the team productivity and EAC forecasting calculation as described in this chapter. Proper metrics can only be achieved when the work unit estimates are based on actual forecasts and not padded values.

Once the various resource groups are segregated, decisions related to how each group should be manipulated follow the guidelines outlined earlier. The key aspect of this decision is to evaluate the likelihood that the current trends will continue. If so, the standard formulas can be used to evaluate status and forecast future outcomes. If not, other choices would be more appropriate. From a data viewpoint recognize that MS Project does not easily extract the resource groups into the categories supporting this discussion. It is left for the reader to evaluate a specific situation and decide on the best analytical techniques for that scenario.

14.14 CONCLUSION

This chapter has highlighted the analytical value of EV, which is often considered to be the best status measurement metric available. For this to be operationally true, one must understand its underlying model to achieve this promise. Modern automated project status tracking tools makes use of this technology more feasible than in the past. However, failure to understand the internal WP resource characteristics in this analysis can lead to erroneous use of the automated produced parameters. For instance, labor resource performance provides insight into team productivity, while material and dollar expense components do not directly link to productivity. Earned value analysis is based on previous project performance trends that are developed by comparing current actual project accomplishments to the approved baseline. To accurately evaluate project performance, the project manager must understand not only what created the past variances but also the assumptions regarding what will drive future predictions and challenges. One objective of Earned Value is to assist in the process of understanding project quantitative variances; however, strictly focusing on past performance does not necessarily yield the future picture. The EV model defines key data elements needed to evaluate aggregate project performance, but more value granularity of these evaluation parameters is often needed for detailed root cause analysis. In addition to the historical evaluation role, derived EV status parameters can be used to create completion forecasts. A third phase of this discussion illustrated how to use work unit data for analyzing rate and work efficiency.

By utilizing this suite of status evaluation techniques, the project manager is provided with very sophisticated tools for use in examining project performance. Collectively, these methods contain some of the most sophisticated quantification analyses available for status reporting, evaluating work unit progress, and identifying variance sources. The ability to isolate specific casual issues for performance variance is a fundamental success criterion for the project manager. Through the identification of work unit time and cost variance, the project team can better identify how to influence the project back toward its baseline direction.

The examples used here have been very simple to focus on the underlying mechanics of the processes. Larger projects would create additional data manipulation and interpretation complexity for this process.

The following list contains a summary of the EV process as outlined in the chapter.

1. The EV "magic box" is a good way to structure EV core driver data into a simple format that easily leads to the core parameter formulas.
2. MS Project can produce numerical EV data as a by-product of its basic status tracking, but one must be very careful about the process steps used to achieve the accuracy of the derived results because of the software tool's internal calculation rules. Also, be aware of how the parameters are created and use them appropriately.
3. The MS Project EV checklist supplied in this chapter should be used in executing the EV calculation process.
4. A spreadsheet summary-level plan can also be used as a high-level method to generate summary task-level EV parameters. This is a reasonable method to use for the initial testing of the concept since the data values used can be more easily controlled.
5. Exercise care in using CPI and SPI parameters for cost and time forecasting. The appropriate method depends upon assumptions about future work compared to the past. A standard CPI formula is only valid if it is assumed that the future will progress in the same way as the past. Also, note other possible assumptions provided in the chapter.
6. The standard SPI formula is *invalid* as a time status predictor after around the 70th percentile of the life cycle. Use only the ES formula

after this point and always compare the standard SPI value with ES for consistency. Monitor trends in this forecast category for further validation.

7. At 20% through the project execution life cycle, the CPI will normally not change more than 10% (Christensen).

8. When a project falls behind the baseline plan by 15%, it will not recover. TCPI parameter values above 1.15 indicate a low probability of budget recovery.

9. Organizations with higher process maturity can deal with the EV process more cost-effectively than those with lower maturity because of their existing support processes.

10. If a project has significant material costs embedded in the work structure, those cost components should be separated for team productivity and completion forecasts. Also, realize that LOE, labor rates, and overhead elements may distort EV analysis.

11. Review labor data for rate and efficiency variations. The rate variance is often outside the project manager's control and needs to be assessed as to its impact on the CPI calculation.

12. It may be worthwhile to move the MS Project data to a spreadsheet for analysis purposes. In this format, the data can be sorted and analyzed effectively.

QUESTIONS

1. If a current project's actual cost expenditure is less than the baseline plan at the status point, is it valid to assume that the project is doing well? Why or why not?

2. After reviewing the material in this chapter, why do you feel it took so long for the EV concept to catch on in general project management practice?

3. What are the key driver variables for EV? Draw the EV "magic box" and show how this leads to the core formulas.

4. Is EV calculated from the Work Package or Cost Account data?

5. If CPI is 0.5, what is your expectation for future project performance? Why might this not be the case?

6. If SPI is 0.9 at the 80th percentile stage of the project, should you assume that this project will finish within 10% of its baseline schedule? Why or why not?
7. What does the Earned Schedule model add to the EV parameter analysis set?
8. What is the role of the Critical Ratio (CR)?
9. List the key steps necessary for using MS Project to generate reasonably accurate EV parameters.
10. What is the 20% rule for CPI?
11. What is the major EV issue with %Completion?
12. What can rate and use variance help with?
13. If accurate raw data is available, why might an EAC forecast be wrong even if it is assumed that the project will continue as it has?

PROBLEMS

1. The simple Gantt chart plan shown below contains three activities. Planned values (PV) for each activity are shown in bold font on each bar. Task completion estimates are shown as a percentage, and period two is the status point. The accounting system has collected actual costs for the three activities: 6, 12, and 6, respectively. Compute the EV parameters for this model. What is your prediction for the cost and time completion of this project (ignoring the SPI formula flaw)?

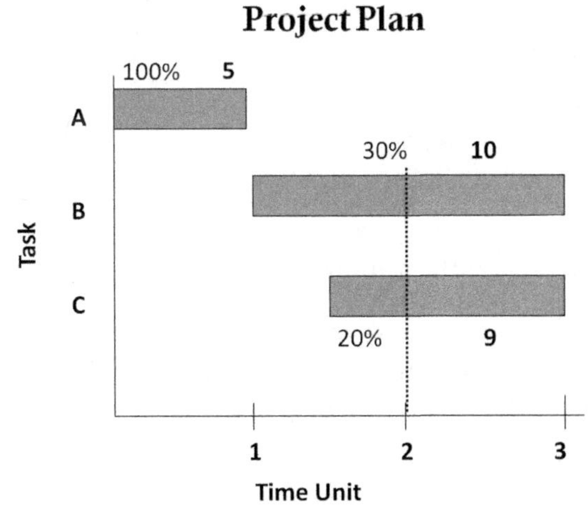

2. EV Dog Pen Exercise. Assume that you are the project manager for the Dog Pen project. The goal is to build a four-sided dog pen. Each side is budgeted to cost $200 to complete and requires one day of your time. As the project progresses, the following results are recorded:

   Day 1: Side A completed; spent $200

   Day 2: Side B completed; spent $240

   Day 3: Side C 50% completed; spent $120

   Day 4: Side D has not started; spent $0

   a. By inspection and without resorting to EV formulas, what is our perception regarding the general status of this project? Estimate cost and schedule percentage completion by inspection of the raw data.
   b. Calculate the following EV parameters and compare your answer with part A:

      PV =

      EV =

      AC =

      BAC =

      CV =

      SV =

      EAC =

      ETC =

   c. Use your knowledge of Earned Value to interpret the status of this project.
3. What are the essential steps in producing valid EV parameters using MS Project?
4. If a project is approaching the 70th percentile point, what method should be used to evaluate schedule status?
5. Name three task cost element types that can distort a project team's efficiency analysis.

TABLE 14.5 Car Problem Status

Task Name	% Complete	Duration (days)
Car Problem First Cut	39	283
Planning & Project Initiation	88	70
Charter	67	15
Stakeholders Identification	50	10
Scope Definition	100	15
Schedule Development	100	15
Risk Assessment	100	5
Budget Definition	100	15
Management Charter Approval	100	5
Set Baseline	100	5
Engineering	40	147.5
Body/Engine Draft Design	100	22.5
Initial Draft	100	14.5
Integration analysis	100	5
Mechanical Engineering	87	124
Engine Design	100	104.5
Mechanics Design	65	60
Electrical Engineering	0	77.5

6. Table 14.5 contains sample output from MS Project. The columns labeled Baseline Duration and Baseline Cost are equivalent to planned values for time and cost in the EV notation. The dashed line appearing after the Task B Gantt bar reflects the status point (January 1). Manually compute the EV parameters for this table view. It would be a good test of your understanding to duplicate this output in MSP using the sequence rules provided in the chapter.

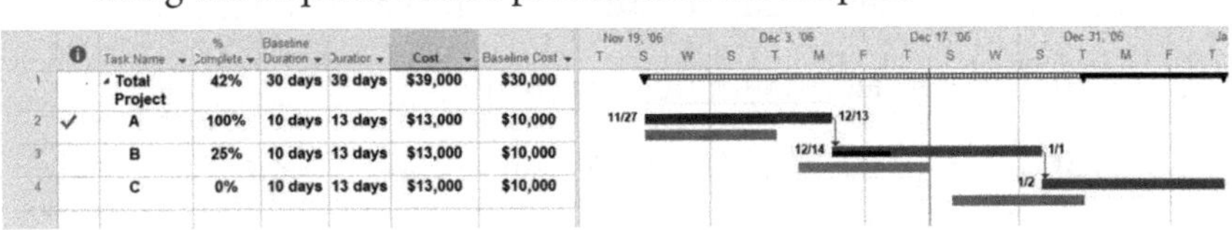

	Task Name	% Complete	Baseline Duration	Duration	Cost	Baseline Cost
1	Total Project	42%	30 days	39 days	$39,000	$30,000
2	A	100%	10 days	13 days	$13,000	$10,000
3	B	25%	10 days	13 days	$13,000	$10,000
4	C	0%	10 days	13 days	$13,000	$10,000

What is the planned project cost and duration? How does this compare to the computed EV equivalents?

What are CPI and SPI for this model?

What are the EV parameter values for project cost and schedule at completion?

7. One of the more common methods to compute EV parameters is through the use of a spreadsheet. Create a 12-month spreadsheet

summary-level plan with planned cost for each month (see format sample in this chapter example). From this base, add the following data parameters:

a. Set the status date at the end of month six.

b. Provide %Completion values for all active groups.

c. Add actual cost information for each month.

Using your EV knowledge, answer the following questions.

a. Translate this table data into core EV status parameters.

b. What is the condition of the project based on these metrics?

c. What is the estimated cost and schedule at completion?

d. Use the Excel trend graphing utility described in the chapter, and create a trend chart for CPI and SPI (make up the data for the chart).

Advanced Exercise

Extract file *14 Car Project Start.mpp* from the publisher's website (refer to Chapter 1 for access details). This file contains an MS Project baseline plan. Use the current %Complete and Duration status information shown in Table 14.5 to update the project status. From the results generated by MSP, answer the EV-related questions outlined below.

Evaluating Results

1. How is this project doing regarding planned cost and time?
2. What is the projected duration overrun?
3. What is the EV predicted cost overrun at completion?
4. Why are the values of SPI and CPI the same? Do you see a potential flaw in this calculation?
5. What is the MS Project's predicted date of completion, and how does this compare to what you would estimate from the EV EAC calculation? What would cause this difference?

REFERENCES

Christensen, David S. (1998). “The Cost and Benefits of the Earned Value Management Process.” http://www.dau.mil/pubs/arq/98arq/chrisevm.pdf

Fleming, Quentin W. & Koppelman, Joel M. (June 16, 2006). Start with ‘Simple’ Earned Value on All Your Projects, *Crosstalk*, www.stsc.hill.af.mil/crosstalk (Accessed October 17, 2008).

Lipke, Walter H. (2009). *Earned Schedule*. Lulu Publishing, ISBN 978-0-557-17738-7

National Defense Industrial Association (NDIA). (April 2014). Earned Value Management Systems EIA-748-C Intent Guide, https://www.ndia.org/-/media/sites/ndia/divisions/ipmd/division-guides-and-resources/ndia_ipmd_intent_guide_ver_d_aug282018 (Accessed September 2, 2025).

Stratton, Ray W. (May 14, 2012). The EVM Newsletter, “Usage of EVM Grows Worldwide,” Management Technologies. Arington, VA. https://www.mgmt-technologies.com/ (Accessed September 2, 2025).

Stratton, Ray W. (February 18, 2007). PM Boulevard, Project Management Institute, http://www.pmboulevard.com/Default.aspx?page=View/Content&cid=2205“Management”

CHAPTER 15

Resource Leveling

15.1 INTRODUCTION

PMs face many challenges when leading a project. One of the most significant challenges is labor resources because PMs do not have unlimited labor and monetary resources to perform the defined project work. There are often mismatches between the plan and actual resource availability. Therefore, complex adjustments to the original plan are frequently necessary based on various decision factors. This will usually result in sacrificing scope, schedule, or cost goals.

Before diving deeper into this issue, let us first revisit part of Chapter 8 by discussing the term *overallocation*. In the project environment, this term means allocating a specific resource beyond what is available in that period. For example, assigning 300 weekly work hours to a single resource is impossible. The question for this chapter is, "What are the appropriate management actions that can be brought to this class of problem?" When the overallocation condition occurs, the project team must reach *resource level*, which essentially means balancing the plan versus actual available resources.

Recognize that many resource capacity problems can be resolved by shrinking scope, stretching the schedule, or adding resources (e.g., task crashing) to overallocated areas. However, the correct way to make resource decisions is to understand the degree of flexibility within the project resources and the project priorities, such as whether the project can be delayed or if more resources need to be assigned if a deliverable does not have a flexible completion date. Beyond dealing with the specific

DOI: 10.1201/9781003532989-18

solution choices, various other management considerations are also involved.

15.2 HANDLING THE OVERALLOCATION DECISION

The following outlines standard decision options and sample trade-offs that accompany the capacity resolution option:

1. *Move constrained resource types from slack tasks to overallocated tasks*—Recognize that this should have been done as part of the initial planning process. This option can resolve the capacity problem with the least external visibility and minimal impact on the budget.
2. *Obtain underutilized organizational resources external to the project*—Appropriately applied to the critical path, this new allocation of resources should shorten the duration. However, it would typically add resource costs to the effort. Here, we are trading extra resources to achieve time and scope goals.
3. *Add third-party contract resources*—Similar result options as discussed above in item number 2.
4. *Change the sequence of tasks to remove the resource conflict.*
5. *Cut scope in areas with constrained resources*—This has the potential to cut the schedule and budget but decreases the project functionality in the areas impacted.
6. *Work the team overtime*—In a salaried environment, this is essentially a zero incremental resource cost; however, this is not a recommended long-term solution because of the decrease in team morale and productivity.

Properly recognizing and dealing with resource overallocation is a significant project success activity (Ashayeri & Selen, 2005). Unfortunately, each project is unique in the timing and allocation of resources needed (Vidal & Marle, 2008). For this reason and the complex interaction between work tasks, schedules, and resources, the problem does not lend itself to a cookbook-type solution. This chapter aims to sensitize the reader to the issue of resource overallocation and present some of the tools embedded in Microsoft Project (MSP) to help resolve this situation. Nevertheless, recognize that the resource requirement is often dynamic based on task overruns, time delays, changing resource availability, etc. As a result, the associated resource status can also change through the life cycle.

As indicated above, the most common method for handling resource mismatches is to add overtime hours to the existing team to bridge the gap between the plan and actual resource availability. This is a way to stretch the defined resources, but this approach can have negative consequences despite often being a universal solution. While this may be a viable temporary solution, there are better long-term solutions. Excessive overtime eventually decreases productivity and signifies the need for a better solution using one of the above-defined decision options.

15.3 MANAGING RESOURCE GAPS

This section aims to use MSP-embedded views and tool options to manage resource gaps. It may be helpful for the reader to download the "Car Problem File for Chapter 15" MSP file from the student resources folder of the publisher's website. Figure 15.1 displays the MSP Gantt Chart view of this file, where the "red men" can first be seen. Please note that textbook images are in black and white, but that MSP will show "red men." These red men were first introduced in Chapter 8 as a signal for resource overallocation. A red man icon indicates that defined resource levels cannot support the identified task. Figure 15.1 provides a sample of the car tasks, specifically 1.2.1.1, 1.2.1.2, 1.2.2.1, etc., affiliated with the red men icon, which has overallocated resources. The complete status can be reviewed by extracting the file from the publisher's web library to view all tasks, including those with resource overallocation dilemmas. When the cursor is placed over a red man icon, an information message box appears to explain that resources affiliated with the task have been overallocated (see Figure 15.2). In this view, the tasks involved in resource overallocation are identified, but little is known about how best to resolve the situation. For instance, the tasks linked to an overallocation may be affected by other tasks. So, it is necessary to look at various views of the problem to resolve this. As one goes through this process, it is much like peeling an onion. Each step opens new levels of problem understanding (i.e., breadth and depth). With this perspective, we move on to the next layer of dealing with this problem.

Poor capacity planning and lack of required capacity are the root causes of scheduling problems (Ashayeri & Selen, 2005). Improved results can be

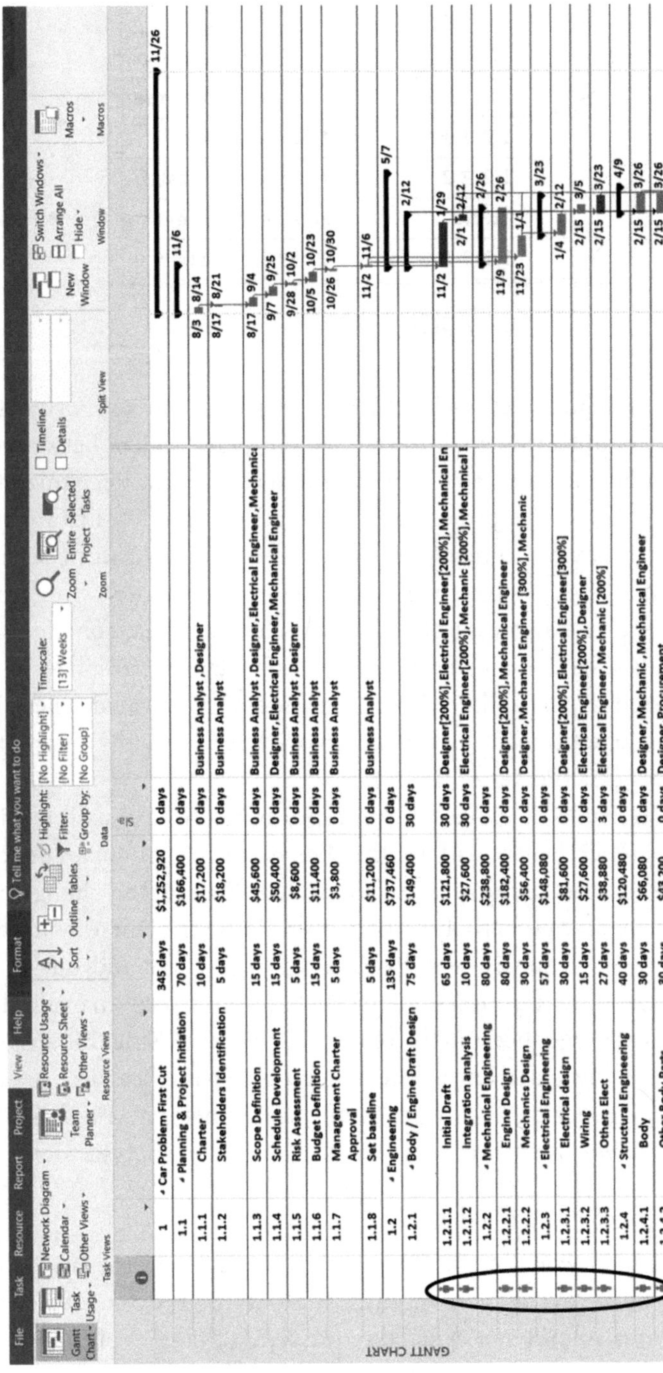

FIGURE 15.1 Gantt chart view of red men.

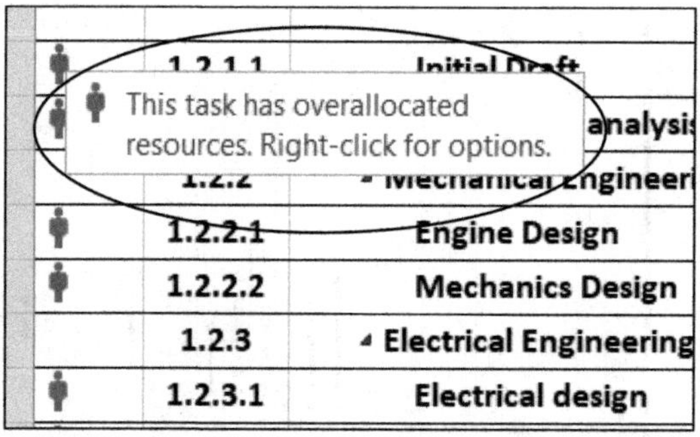

FIGURE 15.2 Red man cursor message.

obtained through more exact scheduling of project resources (Porter et al., 1996). To accomplish this goal, there must be tight control between the project plan and the associated task resource allocation. Recognize that almost everything in the project plan depends on resources (Kalyani & Sahoo, 2011). It is vital to recognize how project results can be severely impacted when this management area is not adequately monitored (Shetach, 2010). It is also essential to understand that the complexity of this problem is such that an automated solution wizard to resolve all situations is beyond the capability of even the most sophisticated scheduling tool, and most practitioners have given up on the idea of fully automating the leveling process (Porter et al., 1996).

A simple example of dynamic overallocation can be illustrated in a construction project where bad weather has made it impossible for some planned tasks to be performed as scheduled. This event changes the resource picture and potentially creates overallocation. One way to mitigate this issue is to move the active tasks under cover and out of the weather and use other resources during that period. This is an example of changing the task sequence and the resource schedule. Using this strategy, the schedule may be kept using a different resource plan. Resource leveling should be viewed as a balancing act in all these situations. The following section will present samples of this process using MSP analysis tools.

15.4 MSP OVERALLOCATION TOOLS

Once the red men icons in the ID column are detected, the magnitude and location of the overallocation must be assessed. The first step of this process is evaluating the involved resource types, the level of overallocation, and the associated status. Phase one of this process examines various resource views in the existing plan. Multiple MSP views are available to aid in reviewing the status of planned project resources. A brief overview and comment on these analysis views follow:

- *Gantt or tracking view*—These plan views show the task calendar location of the overallocation, and in some cases, that will be enough to diagnose the severity of the problem. One of the most essential status issues in this view is to assess whether the overallocation involves critical path tasks.
- *Resource sheet*—This view provides a high-level indication of overallocated resources but not the calendar location. If more capacity of that resource is readily available, simply adding resources to the plan will solve the problem.
- *Resource graph*—This view graphically shows each resource's quantified total staffing requirement over time and the periods in which the resource is overallocated. This view does not show the specific tasks involved, but this information may resolve the problem if the source of overallocation is obvious.
- *Resource usage*—This view offers detailed data regarding specific time frames, tasks involved, and resource overallocation amounts. When more straightforward solutions have failed, this will be the playing field for defining a detailed resolution.

We have a word of warning before we start to describe this process. Please realize that this is often an iterative process. Therefore, it is essential to use backup copies of the plan before formally committing any changes to the formal plan. So, please save a copy of your initial MSP file before making changes to the formal plan. This is extremely important because resolving one resource problem often creates issues in other areas that need correction. Sometimes, you must abandon your changes and start over from your initial plan when trying different ways to resolve resource issues.

The first step toward resolution should be to identify simple solutions and deal with those first. The two preferred methods to accomplish this are to add resources to the plan or to cut your project scope. When that phase is exhausted, moving on to the higher-level supporting tools available in the MSP will be necessary. The warning here is that using the internal automated MSP solution options will usually be unsuccessful because the model solution tends to increase task times until the resource availability fits. That often has the effect of stretching out the schedule beyond an acceptable range.

15.5 MSP TOOL EXAMPLES

This section will describe the mechanics related to the MSP utilities that aid in dealing with more complex overallocation issues. There are different ways to view information regarding resources. We will begin by clicking on "Team Planner" and then clicking on "more views," which will show many different options, such as "Resource Sheet" (see Figures 15.3 and 15.4). Alternatively, your MSP may already be set up to show a shortcut on the menu ribbon to go directly to some of these views.

15.5.1 Resource Sheet View

A resource sheet view is displayed in Figure 15.5. Overallocated resources are highlighted in italics and red in the eBook, along with an information note that will be attached, indicating that the associated resource is overallocated. In this example, the Designer, Tester, Mechanical Engineer, and Mechanic resources are shown as being overallocated. When the cursor is placed over the resource, a pop-up message is displayed, stating that the resource has been overallocated and needs to be leveled. The term *leveled* means an imbalance between the planned and actual available resources.

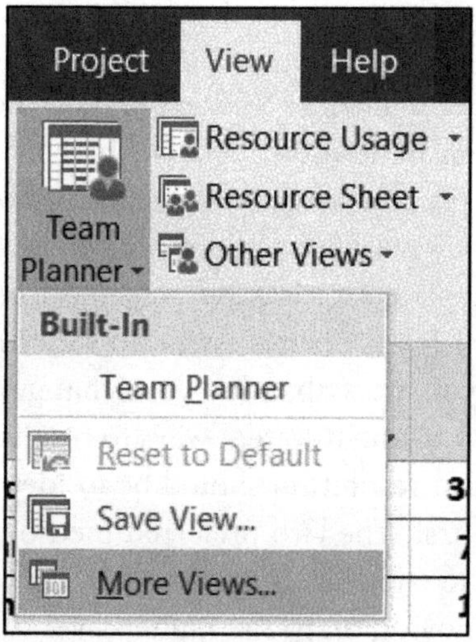

FIGURE 15.3 Team planner view options.

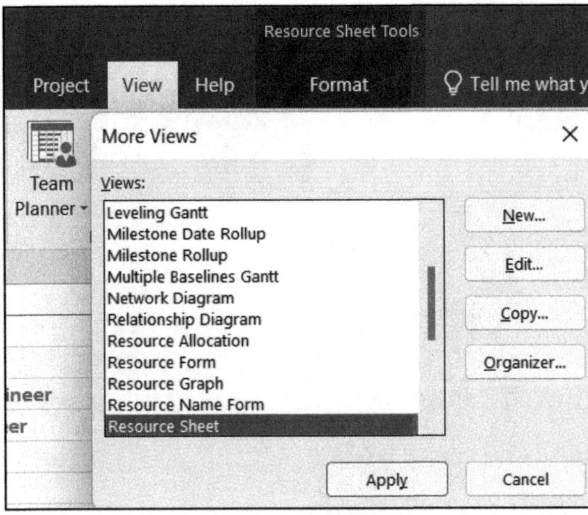

FIGURE 15.4 Team planner more view options.

		Resource Name	Type	Material	Initials	Group	Max.	Std. Rate	Ovt. Rate	Cost/Use	Accrue	Base
1		Designer	Work		DES	DES	400%	$75/hr	$75/hr	$0	Prorated	Standard
2		Tester	Work		TEST	TEST	200%	$45/hr	$45/hr	$0	Prorated	Standard
3		Business Analyst	Work		BA	BA	500%	$50/hr	$50/hr	$0	Prorated	Standard
4		Mechanical Engineer	Work		ME	ME	400%	$40/hr	$40/hr	$0	Prorated	Standard
5		Electrical Engineer	Work		EE	EE	200%	$35/hr	$35/hr	$0	Prorated	Standard
6		Procurement	Work		PROC	PROC	500%	$25/hr	$25/hr	$0	Prorated	Standard
7		Mechanic	Work		MECH	MECH	200%	$40/hr	$40/hr	$0	Prorated	Standard
8		Driver	Work		DRIVE	INV	100%	$100/hr	$0/hr	$0	Prorated	Standard
9		Fiberglass	Material		GLASS	INV		$225		$0	Prorated	
10		Tires	Material		TIRE	INV		$50		$0	Prorated	
11		Wires	Material		WIRE	INV		$300		$0	Prorated	
12		Metal	Material		METALS	INV		$0		$0	Prorated	

FIGURE 15.5 Resource sheet view.

One simple mechanic to fix the Designer overallocation is to adjust the "MAX" column value until the red man icon disappears. This says we are adding that number of resources to support the project. Unfortunately, this simple fix is usually the most difficult because it requires allocating more resources (i.e., hiring, contracting, or internal movement within the organization). In practice, the desired answer will likely require more creativity from the PM. Nevertheless, for the first stage of this review, the Resource Sheet defines the offending resources to deal with. However, nothing else is known about the actual magnitude or location of the problem. To pursue this needed information, we move to the second stage of the evaluation found on the Resource Graph view.

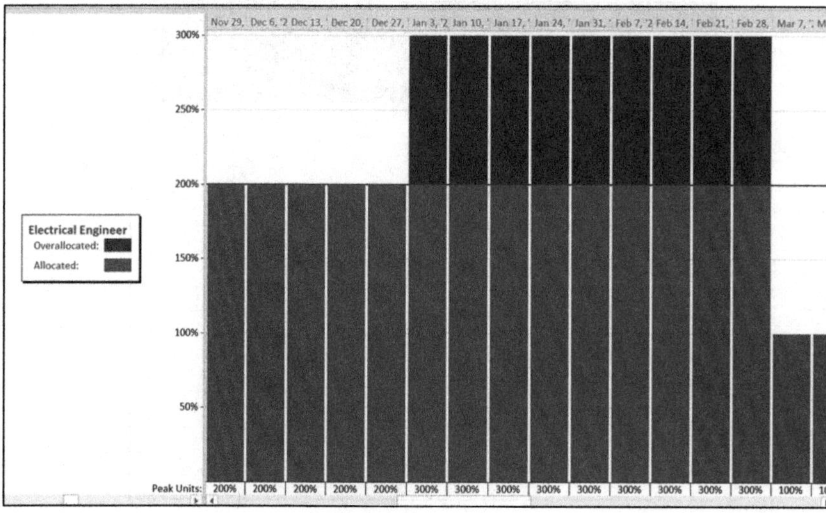

FIGURE 15.6 Electrical engineer resource graph.

15.5.2 Resource Graph View

A sample Resource Graph view is displayed in Figure 15.6 with the "Units" field set to weeks.

You can find this view by repeating the steps in Figures 15.3 and 15.4. Note that this view shows planned capacity and overallocation levels for the named resource through the project life cycle. By scrolling through the plan resource name groups, the graph view shows where the overallocation occurs and to what degree. Each graph view is for a single resource. The critical concern is the red resource periods and associated tasks that consume those resources. The *Car Problem Example NOT Leveled First Cut* Example project plan shown in Figure 15.6 shows the periods in which the Electrical Engineer requirements exceed the available capacity. The graph shows an overallocation of one resource (100%) starting in early January and continuing through February. The management problem for this period is deciding how best to handle this resource gap. As stated previously, this problem can be resolved by allocating an additional person in the periods shown on the graph. However, the PM must deal with other areas in the life cycle schedule where additional resources may be needed. A quick way to evaluate the full life cycle is to right-click on the timeline and zoom in on the entire project. This will compress the timeline view to show the full life cycle picture.

The Resource Graph view provides essential information as to the timing and magnitude of the problem but lacks low-level detail granularity regarding which tasks and exact dates are involved. Also, there is little data to aid in finding how best to resolve the problem. One crucial data item is the amount of slack time the related tasks have. If the target tasks were not on the critical path, it may be possible to expand the task time and stretch out the resource requirement to stay within constraints. Also, if more than one task is involved, the problem analysis further increases complexity. To examine additional lower-level details, we move to the Resource Usage view.

15.5.3 Resource Usage View

The Resource Usage chart provides detailed allocation data at the task level through the life cycle. This can be navigated using the steps displayed in Figures 15.3 and 15.4. It displays a segment of the view of the Electrical Engineer resource life cycle. The following details are shown in this view (see Figure 15.7):

- WBS task codes involved
- task dates involved (added to the default view)
- critical path flag (added to the default view)
- work hour requirement by period
- overallocation totals

WBS	Resource Name	Start	Details	M	T	W	T	F	S	S	M	T	W	T	F
	Electrical Engineer	8/17	Work	24h	24h	24h	24h	24h			24h	24h	24h	24h	24h
			Overalloc.	8h	8h	8h	8h	8h			8h	8h	8h	8h	8h
1.1.3	Scope Definition	8/17	Work												
			Overalloc.												
1.1.4	Schedule Develo	9/7	Work												
			Overalloc.												
1.2.1.1	Initial Draft	11/2	Work												
			Overalloc.												
1.2.1.2	Integration analy	2/1	Work												
			Overalloc.												
1.2.3.1	Electrical design	1/4	Work	24h	24h	24h	24h	24h			24h	24h	24h	24h	24h
			Overalloc.	8h	8h	8h	8h	8h			8h	8h	8h	8h	8h
1.2.3.2	Wiring	2/15	Work												
			Overalloc.												
1.2.3.3	Others Elect	2/15	Work												
			Overalloc.												
1.4.2	Station Electric	7/20	Work												
			Overalloc.												
1.4.3	Finishing	7/27	Work												
			Overalloc.												

FIGURE 15.7 Resource usage view.

With this background data collection, the challenge is to decide what best fits the plan goals and available options. A sample set of these resolution decisions was summarized earlier in this chapter (i.e., hiring, contracting, scheduling overtime, expanding task durations, cutting scope, moving resources from non-critical paths). Proper resolution of specific issues requires an in-depth understanding of the project situation. Recognize a complex linkage of underlying related data and an equally wide variety of decision options, collectively making this problem complex.

Please note that the data shown in this view is grouped by resource (i.e., Electrical Engineer) and then indicates work hours required for all the tasks associated with that resource. The timeline scale is arbitrarily set to "Weeks," but this could be adjusted up or down in granularity. This shows that task 1.2.3.1 is the culprit for six weeks because overallocation occurs once a resource is over 40 hours weekly. This fits the one-person short value previously shown above in the resource graph. However, now we know the problem detail level by associated tasks, hours, and timeframe. Matching this data to the Gantt view adds one more critical input to understanding the situation. From the Gantt view, Task 1.2.3.1 is found to be non-critical, with a total slack of 138 days. This means the task can be stretched without compromising the schedule, lowering the resource need within that period. Stretching a task while keeping the exactly defined work estimate will reduce the period requirement proportionately. Because this task also uses other resources, it is essential to consider how this change impacts each resource. Executing this change on the sample MSP plan using a work file would be instructive. Increase the task duration value until the red man icon disappears and positive Total Slack remains. This illustrates one way to solve the simple overallocation for a task having available slack time.

15.5.4 Team Planner View

The team planner view offers yet another view of the overallocation situation. This view, shown in Figure 15.8, is similar to the Resource Usage view, except that the related tasks are similar to Gantt bar chart labels with no detailed data. There are red outlines around the tasks involving resource allocation challenges.

The added advantage of this view is that it is easy to see which tasks involve individual resources. Overallocation areas are set in italics and shown in red in the eBook. In this view, tasks can be manually moved with the cursor, and the resulting allocation status is visual. A key point to

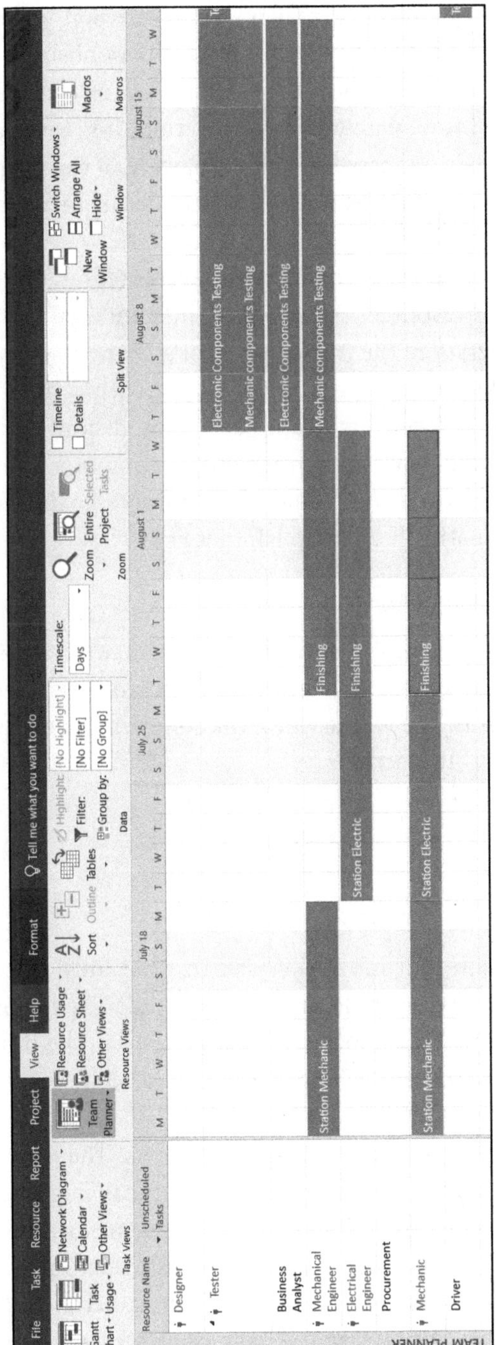

FIGURE 15.8 Team planner view.

recognize in this view is that fixing one overallocation may also affect other linked resources. From the reader's perspective, it would be best to open up a copy of the actual MSP plan from the publisher's library and move tasks around to observe how the resource status changes. One conclusion will surface immediately from this action: it is very easy to make matters worse with this process if you are unaware of the overall task status upfront. Also, note that resource allocations can be changed within this view.

The Task Planner view provides a high-level integrated view between tasks, schedules, and resources. However, significant confusion can result if one does not understand the implications of a change, so always do this process with a backup copy, as suggested earlier within this chapter, and not the master plan. Given this example, it should now be evident that solving this type of problem is not amenable to computer automation but rather involves balancing time, scope, and resource goals. No computer algorithm can automatically accomplish that goal.

15.5.5 Task Inspector View

However, another MSP utility is available to deal with a resource conflict. This is the Task Inspector view. Right-clicking on a red man task enters this utility from the Gantt view. The following two repair options are shown on the pop-up menu:

1. Reschedule to the available date
2. Fix in Task Inspector

The reschedule option will move the task in time until the overallocation disappears. This is a simple task if a schedule expansion is not an issue. However, if exploring another option is desired, the other choice is to select "Fix in Task Inspector." Figures 15.9 and 15.10 show the pop-up menu for option two and WBS 1.2.1.1.

This menu offers several background data items. The top of the menu shows that the associated Designer and Mechanical Engineer resources are overallocated, and it provides to fix both problems using two methods (option one and Team Planner). Also, below this section are other data items that can affect the solution decision: schedule method, predecessors, and related calendars. If the previous task analysis has been performed through the basic status views, the Task Inspector view may be the easiest

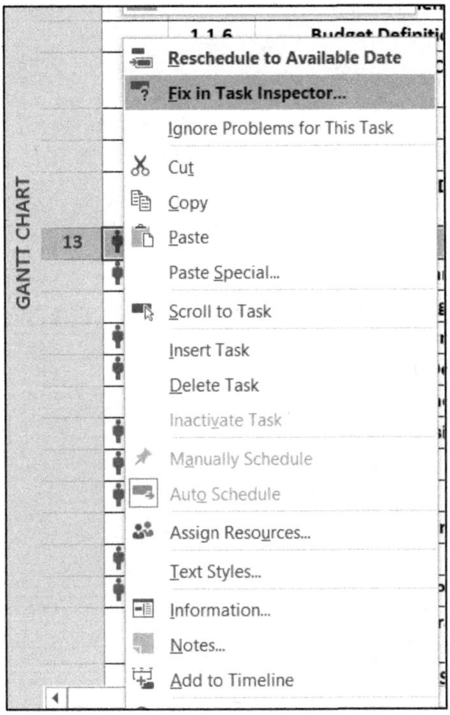

FIGURE 15.9 Right-click on red man in Gantt view.

way to implement the fix. A collection of all views described thus far adds insight into the status, but none offer a global fix.

15.5.6 Leveling Wizard

For capacity management, Nirvana needs an automated option to resolve resource overallocation. The goal is to turn the problem over to the computer, which will resolve it. No more fuss. There is a somewhat modified, famous expression to remember here. To err is human, but to screw up requires a computer (author translation). That adage fits here. In deference to the software engineers who developed this wizard, the recommendation is to use it carefully with a backup copy and then be prepared to start over manually. There are three automatic leveling options available from the Resource menu. These are as follows:

1. Level resources—shows the overallocated resources and offers to level them.

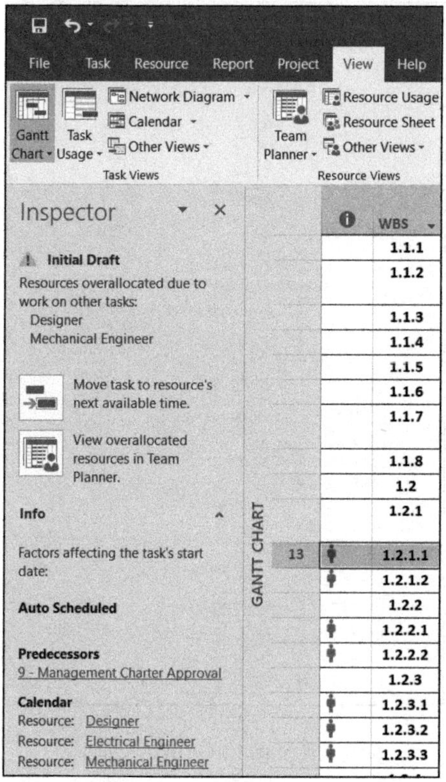

FIGURE 15.10 Task Inspector for WBS 1.2.1.1.

2. Level all—attempts to level the entire plan automatically.
3. Leveling options—offers a wider variety of controls, but essentially the goal of automatically leveling.

Invoking either option 1 or 2 changes the plan, which is worth reviewing to see the effect on schedule and general task fragmentation. If the result is usable, it can be kept. However, what occurs in most cases, and in this sample plan, is unresolved remaining red men. This means the system could not resolve those task issues, and you are back to the manual resolution mechanics.

Figure 15.11 shows the pop-up menu for "Resource Leveling." Please note that the granularity options are more significant for this method and provide more control. The best technique here is to limit the work areas and keep the results separate from the production plan. This option is a

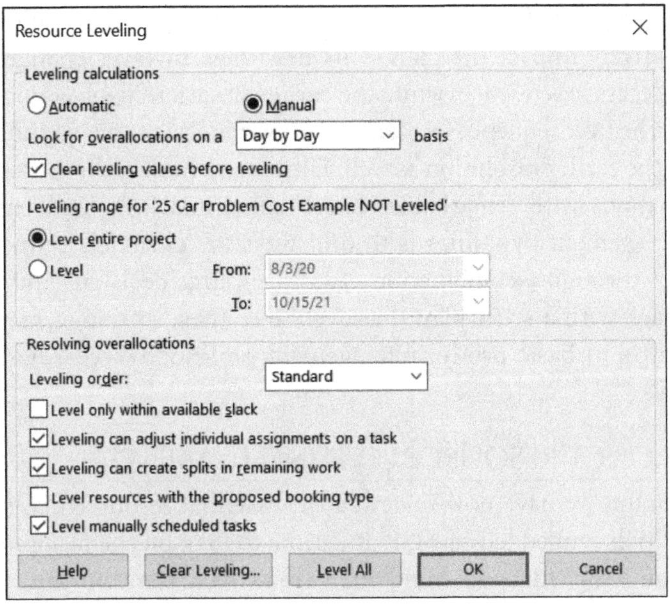

FIGURE 15.11 Leveling option wizard menu.

reasonable first solution attempt after one has gone through the fundamental analysis and solved the easy problems described earlier. Then, the Leveling Option can be attempted on selected resources and time frames to observe the results. Use the library project plan as a reading exercise and play with various utilities and decision strategies. Observe the impact of the solutions regarding schedule and task slack. This lab-type explorative exercise is the best learning method for showing the interaction of resources on the plan.

15.6 AN EXTERNAL REALLOCATION SCENARIO

Not all resource allocation issues are identified by red men in the plan. One example of this is illustrated in a weather-related example, where some part of the plan cannot be worked on owing to external factors such as rain, cold, and heat. In this case, the planned tasks and resources must be reallocated around the external problem. In a classic construction example, when the weather outside is too poor to work, a good strategy would be to focus on other tasks inside. Good PMs will look for ways to keep the overall effort moving forward rather than just stopping work and waiting for the event to clear. Failure to evaluate the impact of this class of project resource analysis violates the tenets of good project management in the same manner that failure to deal with the lack of appropriate resources will negatively impact the plan. This new view involves changes to the task predecessor sequence, while the earlier situation involved more dealing with the task sequence as planned. However, this same method would apply to the earlier resolution as well. Ultimately, both are resource allocation situations with remarkable conceptual similarities. The mantra for plan management dynamics is to find ways to "positively influence the outcome" through a combination of task/resource decisions and not just sitting back with a static plan that overruns. These examples make good case studies in basic project management philosophy, regardless of the root cause.

15.7 PROJECT RESOURCE LEVELING EXAMPLES

At this point, we have now reviewed the essential resource management analysis tools embedded in MSP. It is time to take on the task more logically, simulating what one would do in a real-world environment. This section will show more detailed decision examples to illustrate how the status analysis leads to corrective action using the *Car Problem First Cut* plan (see the publisher's website for access to the Car Problem for Chapter 15 file). The root-level mechanics for each example will use the MSP resource tools and apply them to specific decision examples. These examples will illustrate the following:

1. Adding resources to resolve the capacity gap.
2. Reducing resources and extending durations.
3. Resolution through decreasing task scope or using worker overtime.

Beyond what is illustrated here, these same methods can be used equally well in more advanced decision processes.

EXAMPLE 1: Adding Resources

In this first example, the challenge is to deal with the overallocated Designer resource as described in the resource usage chart shown in Figure 15.12.

The Resource Graph view shows two overallocation situations for the Designer from late November through mid-December. Note that the resource gap for this period is 100% or one person. We assume we can add an additional "Designer" as needed for this segment. The gap is two persons in the second overallocation segment from January into early February. In both cases, the decision is to add internal and contractor resources. We can make those changes to the resource sheet and the task allocation. The red men icons should go away. Problem solved!

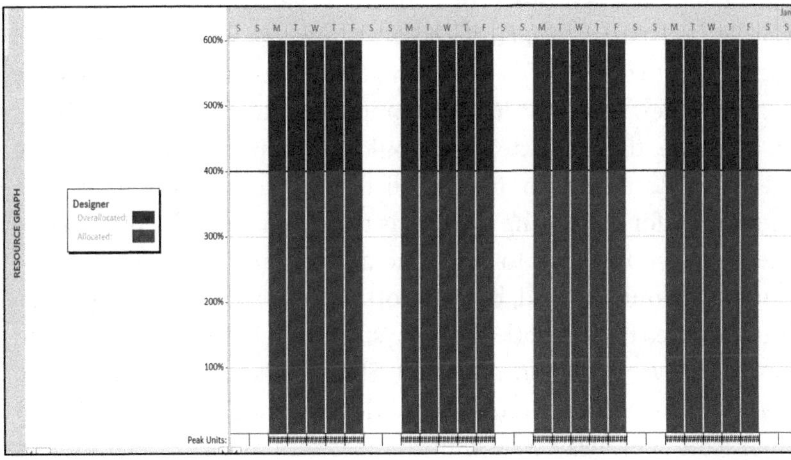

FIGURE 15.12 Designer resource graph.

EXAMPLE 2: Reducing Resources and Extending Durations

This example illustrates the reverse of typical overallocation. In this case, the project team lost one resource in January due to a transfer, illness, or other reasons. The related assigned tasks will need to be adjusted accordingly. In this example scenario within the car project plan, there is only one affected task to deal with: task 1.2.3.1, which is "electrical design." Assume that one of the three allocated Electrical Engineers will be off work for surgery during January and February. Reviewing the Resource Usage chart shows the overallocation for both the Designer and Engineer during this period. At this point, multiple variables come into play:

1. Extending task 1.2.3.1 will move an overallocation into a follow-on task 1.2.3.2 (see resource usage view).

2. Both tasks involved have high levels of slack (see Gantt view).
3. The overallocated designer from the example above is involved in this task, so a decision here could affect another resource.

The interaction of these elements highlights the complexity of making overallocation decisions.

The high levels of slack for both tasks make extending the duration of both attractive, so the first test change would be to extend these tasks to see if the Electrical Engineer issue is resolved. If so, the second review would evaluate this impact on the Designer's decision. In this case, the task extension would work on the surface, and the Designer decision can be modified to accommodate that. The mechanics for expanding the task is to first change the allocation of the Electrical Engineer to 200% and use the smart tag to say that the duration should expand, but the work estimate would stay the same. This changes the schedule to 45 days, and Slack drops to 123, which is still an overallocation condition. Step two is to right-click on the task and select the option "Reschedule to an available date." The red man icon goes away with this, and the task shows 60 days and a remaining slack of 108 days. Even though stretching out durations and lowering resource requirements is simple, doing this within the planning environment will take some practice. Another mechanical issue also surfaced here. That is, the resource pool value only changed during one period. To properly handle this, you would need to change the resource calendar schedule for the Electrical Engineer and schedule him as another source. Administrative overhead such as this is just one of the reasons why a rigorous allocation of resources into the plan is often not followed.

EXAMPLE 3: Working Overtime

The Car Problem First Cut plan is like the previous one but illustrates another more complex option. In this case, the target resource is Mechanic for the February to April period. Resource Usage view for the Mechanic shows WEEKLY resource overallocation of 40h to

160h between February and April. The tasks linked to the overallocation are 1.2.5.1 and 1.2.5.4 (see Figure 15.13). Both overallocated tasks are on the critical path; therefore, technically, resolving this through overtime would require a clear understanding of how much overtime is reasonable for the February to March period.

The overtime load requirement would be 40 hours per week, distributed over the three resources, so approximately two hours per day per person, or add Saturdays to the schedule. The level of gap time shown for April cannot be met by overtime. Working each Saturday through this period would consume approximately 24 hours of overallocation, leaving a considerable resource gap. Also, because both tasks are on the critical path, more discussion of alternatives is needed; however, extra resources will be required if the schedule and scope are to be maintained. It is essential to recognize that only the mechanic resource is involved in this overallocation, so it will only be necessary to deal with resolving that singular resource issue. In many other overallocation cases, multiple resources are involved in a single task, so it is impossible to deal with one. The Team Planner may be the best tool for numerous resources because it offers a more global insight (see Figure 15.8).

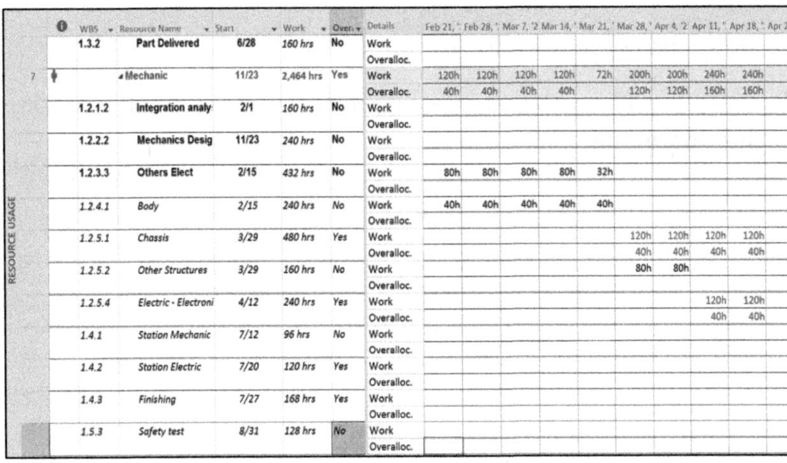

	WBS	Resource Name	Start	Work	Over...	Details	Feb 21, '	Feb 28, '	Mar 7, '2	Mar 14, '	Mar 21, '	Mar 28, '	Apr 4, '2	Apr 11, '	Apr 18, '	Apr 2
	1.3.2	Part Delivered	6/28	160 hrs	No	Work										
						Overalloc.										
7		Mechanic	11/23	2,464 hrs	Yes	Work	120h	120h	120h	120h	72h	200h	200h	240h	240h	
						Overalloc.	40h	40h	40h	40h		120h	120h	160h	160h	
	1.2.1.2	Integration analy	2/1	160 hrs	No	Work										
						Overalloc.										
	1.2.2.2	Mechanics Desig	11/23	240 hrs	No	Work										
						Overalloc.										
	1.2.3.3	Others Elect	2/15	432 hrs	No	Work	80h	80h	80h	80h	32h					
						Overalloc.										
	1.2.4.1	Body	2/15	240 hrs	No	Work	40h	40h	40h	40h	40h					
						Overalloc.										
	1.2.5.1	Chassis	3/29	480 hrs	Yes	Work						120h	120h	120h	120h	
						Overalloc.						40h	40h	40h	40h	
	1.2.5.2	Other Structures	3/29	160 hrs	No	Work						80h	80h			
						Overalloc.										
	1.2.5.4	Electric - Electroni	4/12	240 hrs	Yes	Work								120h	120h	
						Overalloc.								40h	40h	
	1.4.1	Station Mechanic	7/12	96 hrs	No	Work										
						Overalloc.										
	1.4.2	Station Electric	7/20	120 hrs	Yes	Work										
						Overalloc.										
	1.4.3	Finishing	7/27	168 hrs	Yes	Work										
						Overalloc.										
	1.5.3	Safety test	8/31	128 hrs	No	Work										
						Overalloc.										

FIGURE 15.13 Mechanic resource usage view.

15.8 DIAGNOSTICS SELF-STUDY

Use the file *New Product Development* to test your knowledge and ability to use MSP resource tools. Use the toolset to answer the following eleven resource status questions:

1. What is the life cycle date range for this project?
2. What resources are overallocated? Which view shows this?
3. Define the periods of overallocation for Product Engineering (using the Resource Graph utility).
4. Repeat this period analysis process for the other three resource overallocations.
5. On the Resource Usage view, add columns for WBS, "overallocated," and Start (see Figure 15.14).
6. Which tasks are involved in the overallocation of Product Engineering, and to what degree?
7. Are these overallocated tasks on the critical path? How do you know?
8. Change the timeline on the Resource Usage table view to weeks.
9. Used the Resource Usage table to define which tasks are linked to Legal.
10. How much overallocation is indicated for Marketing?
11. How would you propose resolving the identified resource overallocation?

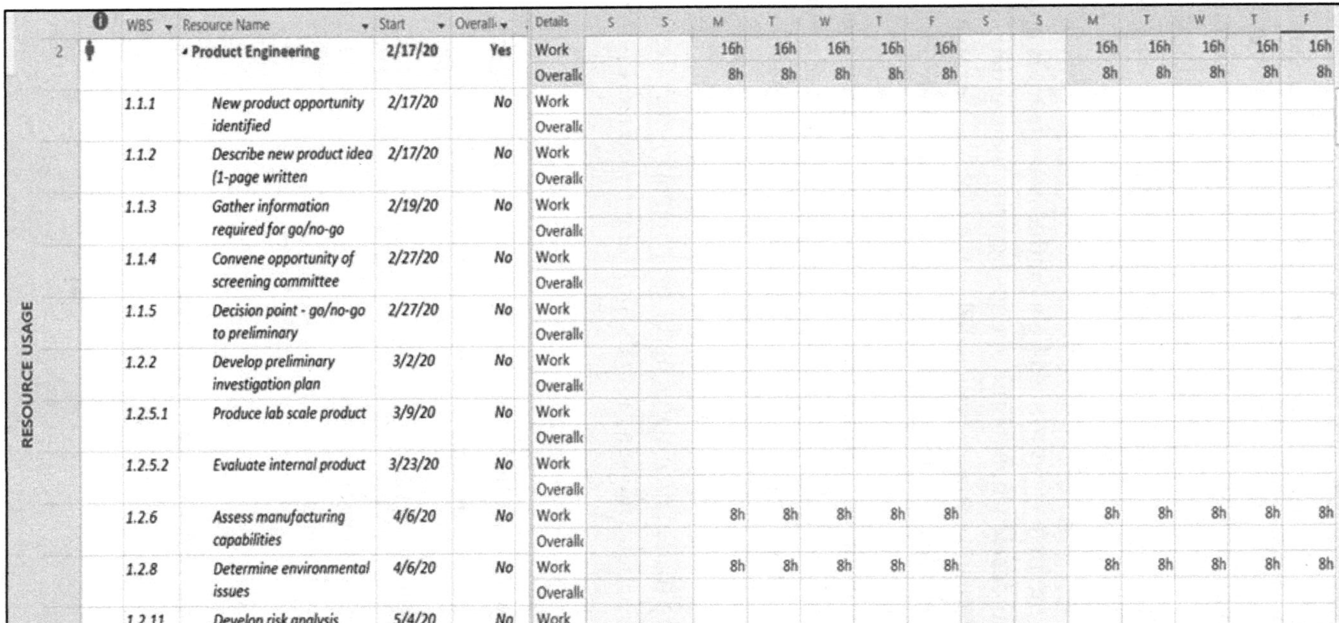

RESOURCE USAGE

	WBS	Resource Name	Start	Overalloc	Details	S	S	M	T	W	T	F	S	S	M	T	W	T	F
2		◢ Product Engineering	2/17/20	Yes	Work			16h	16h	16h	16h	16h			16h	16h	16h	16h	16h
					Overallo			8h	8h	8h	8h	8h			8h	8h	8h	8h	8h
	1.1.1	New product opportunity identified	2/17/20	No	Work														
					Overallo														
	1.1.2	Describe new product idea (1-page written	2/17/20	No	Work														
					Overallo														
	1.1.3	Gather information required for go/no-go	2/19/20	No	Work														
					Overallo														
	1.1.4	Convene opportunity of screening committee	2/27/20	No	Work														
					Overallo														
	1.1.5	Decision point - go/no-go to preliminary	2/27/20	No	Work														
					Overallo														
	1.2.2	Develop preliminary investigation plan	3/2/20	No	Work														
					Overallo														
	1.2.5.1	Produce lab scale product	3/9/20	No	Work														
					Overallo														
	1.2.5.2	Evaluate internal product	3/23/20	No	Work														
					Overallo														
	1.2.6	Assess manufacturing capabilities	4/6/20	No	Work			8h	8h	8h	8h	8h			8h	8h	8h	8h	8h
					Overallo														
	1.2.8	Determine environmental issues	4/6/20	No	Work			8h	8h	8h	8h	8h			8h	8h	8h	8h	8h
					Overallo														
	1.2.11	Develop risk analysis	5/4/20	No	Work														

FIGURE 15.14 Resource usage.

15.9 CONCLUSION

This chapter has explored situations where planned resources are not available. To staff the defined plan. MSP highlights this situation clearly with red men icons by the affected tasks, but the proper resolution for the problem could be more straightforward and more amenable to computer automation. The technical manager must understand the various trade-offs that can be made between time, resources, and project scope. If infinite resources were available, the problem would be easy, but the options would become more political and complex without them.

MSP offers multiple views to shed light on this issue, but it is up to the project team to decide which of the options best fits their situation. Even though an automation wizard is included in the tool, the warning is that this problem is typically more complex than a computer utility can handle.

Recognize that the mechanics illustrated here are an introduction to this problem and are designed to sensitize you to the impact of an out-of-equilibrium situation. The dynamic management of resources is one of the most challenging tasks the PM faces. If this problem still needs to be resolved, it leads to schedule overruns, even with a perfect plan. Keeping a constant eye on the red man status in the evolving plan can offer insights into corrective action and help move the project forward more orderly.

QUESTIONS

Student Review Questions

1. What is the fundamental issue of resource overallocation?
2. What is resource leveling?
3. Name the primary methods for resolving resource constraints. List some of the criteria for selecting each of these.

In-Class Question

1. Use the Internet to identify and describe the functionality of two software tools available to support the project resource leveling process.

REFERENCES

Ashayeri, J., & Selen, W. 2005. An Application of a Unified Capacity Planning System. *International Journal of Operations & Production Management* 25(9), 917–937.

Kalyani, M., & Sahoo, M.P. 2011. Human Resource Strategy: A Tool of Managing Change for Organizational Excellence. *International Journal of Business and Management* 6(8), 280–286.

Porter, K., Little, D., & Kenworthy, J. 1996. Finite Capacity Scheduling Tools: Observations of Installations Offer Some Lessons. *Integrated Manufacturing Systems* 7(4), 34–38.

Shetach, A. 2010. Obstacles to Successful Management of Projects and Decision and Tips for Coping with Them. *Team Performance Management* 16(7/8), 329–342.

Vidal, L.-A., & Marle, F. 2008. Understanding Project Complexity: Implications on Project Management, *Kybernetes* 37(8), 1094–1110.

IV

Support Processes

CHAPTER 16

Risk Management

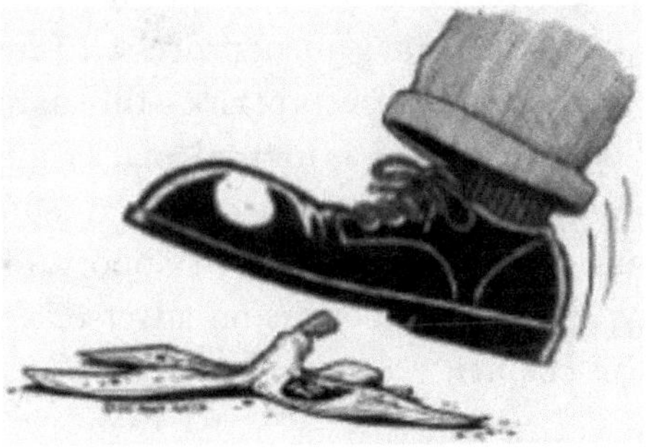

Project N Motion Videos for This Chapter

15R1	Risk Management	Risk Management Overview

Risk is defined by PMI (2021) as "an uncertain event or condition that, if it occurs, has a positive or negative effect on one or more project objectives" (p. 248). Managing risk is one of the project team's most complex and error-prone processes. This area tends to be more of an art than science. In recent years, understanding the consequences of a failed risk process has been more recognized but still not well practiced. One of the fundamental mechanical problems with risk assessment is that it is technically not part of the scope of work defined to produce the project goal. Risk factors lurk in the dark and emerge probabilistically during the later stages.

DOI: 10.1201/9781003532989-20

16.1 INTRODUCTION

Technically, this class of project work is labeled unknown/unknowns, which means that they are hard to anticipate and even harder to specify when and if they will occur. Because these are not tasks to be executed, they are not part of the official work plan. However, they do represent future work in some ill-defined form. Some formal recognition of these events is needed in a category defined as a contingency reserve. More will be described about this segment of the project plan later.

The basic goal of risk management is to execute a process that will reduce the negative impact of risk events on a project. Also, when an unplanned event occurs, the project should have a mechanic to handle the event without it being overrun. From this vague definition, one can begin to see the complexity of the management process. To compound the definition even more, there are two aspects of risk—threats and opportunities. The typical layperson's view of risk is to think of it as only being the threat side of the equation, and, in fact, that seems to dominate the project world. However, one should not forget that positive opportunities could benefit the project. The typical project focus is on adverse events, which is the primary focus of this chapter.

16.1.1 Working Definitions

A starting point for examining risk is to define some key working definitions. The first view of this is to recognize that a risk (threat) event is some unplanned event in the project life cycle that could occur but is not shown in the plan as a defined task because it may not happen. Recognize that a risk event could also be either a surprise event or a previously identified potential issue. A final definitional term is the idea of risk *triggers*. The idea of a risk event is unknown until it *is triggered*. The term *trigger* can be defined as an early warning sign that a risk event has occurred or is about to occur.

Different management actions can be performed on the inactive potential, but a quick and effective reaction is needed when the event occurs. Risk is an unusual concept in project management. Because it is often viewed as an invisible event in contrast to an overabundance of "real" visible work. As a result, some ignore the probabilistic invisible portion until it happens, while others spend much time trying to mitigate and minimize its impact through an analysis process. The ignoring strategy is dangerous and not recommended. However, attempting to plan risk out of the project is also an incomplete exercise. Regardless of the approach pursued, some level of risk events will occur, so a risk event handling process must be part of the recognized project functions. This activity lies external to the defined scope of work. In many ways, one can view risk handling as a firefighter waiting for a fire. Similarly, the project team needs to be able to handle these unplanned fires. Think of this as a supplemental process to help ensure that the project has minimal negative impact from the event.

Humans handle personal risk in various ways. Some ignore it and wait for it to happen. Driving a car with a spare tire is an example of personal risk mitigation; it does not stop the tire from going flat but does minimize the native impact of this. An interesting question is how many know how to use spare tires. The author's experience with this on a rural road with a slashed tire suggests it will not. Just as in our personal lives, the topic of risk is often incomplete or misjudged. These same behaviors are found in the project environment.

Many things in the project world can unexpectedly go "bump in the night." Good project teams consider these events and develop a "risk culture." That means a sensitivity to them and a watchful eye. In many cases, the tricultural viewpoint is even better than formal risk planning efforts as the human expert is more tuned to the dynamics of these events as they trigger.

From the definitional views outlined above, the basic objective of the risk management process is to identify and plan methods that decrease the probability of these adverse events. The process must also include handling events after the planning process.

It is essential to recognize that some risks are inherent in the project goal and can never be avoided entirely—as a radical example of the NASA project to send crews to Mars. On the other hand, some negative issues can be minimized with prior thought. Recognizing that a car could have a flat tire late at night could be mitigated by ensuring the spare was in good shape with support tools to handle the task, even in the dark. That is called mitigation. The risk event did not disappear, but prior planning made it less onerous. We will see more examples of this process later.

16.2 MANAGING RISK

This chapter aims to describe sample tools and techniques for the project team to utilize in dealing with the planning and execution activities related to risk. This material will focus only on the threat side of this equation because the positive opportunity examples are very local in origin. The ultimate vision of risk management is to understand its potential impact on the project and how these events can affect project success. Once again, it is essential to realize that risk has not happened yet and is not shown in the baseline execution plan.

Delivering complex project deliverables successfully within cost, time, and quality does not typically occur without effective risk management, so the project plan development process must incorporate risk considerations. Eaton and Little (2011) discuss how risks are ever-present, and although they can be significantly reduced, even the best risk management process cannot bring the total negative impact to zero. Risks are divided into two categories, known risks and unknown risks, which are further described below (Richardson & Jackson, 2025):

- ***Known risks*** are logically expected to occur, and some general probabilities and impacts can be estimated. These risks can be handled through risk management techniques and minimized by following those techniques.
- ***Unknown risks*** are not predictable events or generally anticipated in the formal risk evaluation process. Another radical example is a meteor falling on a building.

The analysis of risk as described here deals with the known category, although all triggered risk events will have to be handled during the execution phase.

16.3 PROJECT RISK MANAGEMENT

Envision a project without risk. The stakeholders have perfectly defined the scope, and all the required work tasks have been defined. Task estimates are also perfect regarding duration and cost. The project plans illustrated in earlier chapters perfectly define what needs to be done. How complex can project management be? We will ignore the apparent scope changes that occur because of newly defined requirements but now add risk events that are added to that view. In the perfect world, project management would be renamed "checker." Because that would be the only task—checking off perfectly planned tasks—adding the variability of scope and risk to the view is what makes the work activity what it is. Plan variability is what makes project management so technically challenging, and unplanned risk events are a significant part of that variability. Recognizing this, one might rename project management "risk management" because that comes closer to defining the work process.

The scope of risk management includes processes in all project areas, and a formal risk assessment and control model covers much of the project life cycle. The steps include planning risk management, identifying risk, performing qualitative risk analysis, performing quantitative risk analysis, planning risk responses, implementing risk responses, and monitoring and controlling risks. The Risk Management Plan is "a component of the project, program, or portfolio management plan that describes how risk management activities will be structured and performed" (PMI, 2021, p. 248). The PM and project teams will brainstorm risk events that are likely to affect the project and the characteristics of each and document these in the project Risk Register. PMI defines the Risk Register as "a repository in which outputs of risk management processes are recorded" (p. 248). A risk report can be created, which is defined by PMI as "a project document that summarizes information on individual project risks and the level of overall project risk" (p. 248). Different analyses can be conducted, such as identifying the probability and impact of defined risks, often shown as a Risk Matrix (described later) or a bubble chart, which is part of a qualitative analysis. From a quantitative risk analysis, a dollar impact on a risk event and numerical probability is difficult, but a numeric ranking system can be used to grade the events. Furthermore, how the risk will be dealt with involves response options and actions for the defined events. Also, from this activity, a *contingency reserve* is defined as "time or money allocated in the schedule or cost baseline for known risks with active response strategies," which is helpful should the risk occur (PMI, 2021, p. 237). The team can then implement the risk response. Once a risk event triggers, the project team would consult the risk register and the risk owner to decide how to move forward. The defined change is handled as a formal project plan change when that decision is made. The money to handle the change theoretically comes from the contingency reserve, which is why that reserve is so important for smooth operations. This should not be viewed as a budget overrun but as a probabilistically planned work event. The PM and team should continuously track the status of identified risks, execute risk response plans, monitor residual risks from previous fixes, identify newly identified risks, and evaluate the effectiveness of the process throughout the project cycle. The main outputs of this process are recommended corrective and preventive actions, requested changes, updates to the risk register, and a revised project management plan.

Note that four new supporting tools have been introduced here, which are as follows:

Risk plan—The implication for this item is to understand that not all risks are the same. Some organizations are willing to accept more risk than others, and this document guides the analysis process. In theory, more time spent analyzing the situation for risk should do a better job of dealing with the process of minimizing future adverse events. In comparison, less time spent would imply that a higher level of risk triggers later is acceptable. These decisions are made by senior management, which also has a fundamental role in risk management.

Risk Register—It is important to catalog planning decisions made regarding risk events and to capture future events. In both cases, the organizations will learn more about this phenomenon in their local environment, which will help them deal with it better downstream.

Risk matrix—When evaluating potential risk events, the management process will need to prioritize which events require additional time. High-risk probability and high-impact items will be prioritized over lesser ones. The matrix helps to sort out the evaluation process into evaluation groups.

Risk handling—When a risk potential is identified during the planning analysis step, the question becomes how to handle it. More on these decagons will be described later in the chapter.

At this point, we have summarized the key vocabulary and the concept of formal risk management. From here, we will move into more detailed levels of this process. From the definitional point of view, it is important to understand that the risk process is logically separate from deciding how to execute the planned scope, even though it can affect those decisions and the defense project work.

16.4 PLAN RISK MANAGEMENT

Time spent planning for risk management up front allows those involved to be exposed to less chaos later in the life cycle. Individual organizations have differing perceptions and tolerances regarding risk exposure.

These differing views are influenced by concerns regarding safety, project success, company image, environment, legal, etc. As a result of these priority areas, there is a wide variety of ways in which senior management prioritizes this subject and the required level of effort in analyzing the project goal. Beasley et al. (2010) suggest that there is a growing recognition that organizations are not spending adequate time on identifying, assessing, and managing the most important risks or later being blindsided by unknown risks, which then results in project failures.

16.4.1 Public Risk Events

One of the best ways to sensitize one to the negative side of project risk is to see public articles about major adverse events such as unfinished buildings collapsing, dams breaking, and nuclear power plants being shut down because of location issues. The Boeing Company, which has long been viewed as a mature producer of safe and high-quality airplanes, has had two public risk events that offer an interesting background (Wang, 2024). Both involve the highly successful 737 model. In 2016, the 737 MAX version of this plane was introduced. The MAX model was based on multiple smaller iterations of the same design. All this background is characteristic of improved risk and knowledge typically gained from each iteration. Then, in 2018 and 2019, two crashes occurred, and the entire commercial fleet was grounded as a result. Subsequent analysis of the root cause was found in a poorly defined software system that was not tested with the new version. It worked well with previous versions, but the new size parameters of the new model changed the characteristics of its operation. The plane would become unstable if the pilot did not know how to intercept this error. The 737 MAX was grounded all over the world from March 2019 to December 2020 and then again in 2024. The technical background of this story is too detailed to enumerate here. However, the essence is that some risk aspects of this project resulted in a significant loss of company reputation, legal issues, lost sales, and significant disruption of the international operation for the model. One could argue that the problem was caused by inadequate testing before release, poor acumination of the software, insufficient pilot training on the system, poor procurement strategy by the customer (only one sensor), or even the declining company risk culture that led to all of this. All these factors matter in the result. Each of the items listed here could have been corrected theoretically. However, the nature of risk is that it is always easier to see in the rear-view mirror than when examining the situation during planning.

As a learning exercise, this case study is worthwhile to explore in greater detail further, not to try to find fault for the problem but more to be better sensitized to the potential impact of risk and recognize the difficulty in anticipating where a risk event lies in the bowels of a project. If one had defined the root source of this issue, most of the related factors could have been resolved without any crashes. In this case, there was every reason to believe the existing navigation system was adequate for the job and previously well-tested, but that was not a valid assumption. Everything involved in this case highlights the complexity of risk assessment in planning. Boeing's senior management seems to have recognized its decline in risk culture and is working to rebuild it.

16.5 RISK IDENTIFICATION

Risk identification involves identifying potential risks associated with a project. These risks can be known risks for a potential project, such as looking at company historical records to see which risk factors have occurred before on a related project. However, brainstorming is a support process to learn more about any other risks that may have never occurred before but still pose a threat or potential risk to the project. Vargas-Hernadez (2011) suggests categorizing risks using group headings such as environment, technical, resources, integration, management, marketing, and strategy to help focus on those areas. These categories are grouped into a sample risk breakdown structure in Figure 16.1.

Identifying the root cause of potential risk events will also help an organization understand the potential for occurrence, identify additional risks, and then make an informed decision to accept, avoid, mitigate, or transfer these risk(s). Let us review a hypothetical example of a system on the U.S. Gulf coastline where hurricanes often hit. Given previous

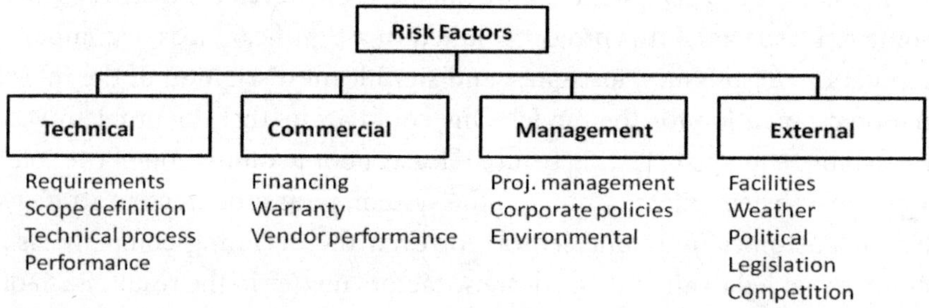

FIGURE 16.1 Risk breakdown structure.

Source: Richardson and Jackson (2025).

experience, there is a defined probability that any particular section of this area will experience a hurricane of some undefined magnitude. Looking at historical records would provide insight into a method to help mitigate the negative consequences of this class of risk. Clear evidence of this event would alert the project to ensure a safe backup system to handle any temporary downtime from this risk event. That could even include moving the team elsewhere during the recovery period. The question at this point is whether to continue enumerating more potential risk events or terminate this group. One lesson Mother Nature often teaches is that the past may not predict the future well, which is the case with weather data.

16.5.1 Identification Tools

One of the classic identification tools is called the Ishikawa or Fishbone Diagram (also known as the Cause-and-Effect Diagram). This is a schematic diagram that is used to provide a logical method to list potential risk factors that may affect a problem or a process outcome. Figure 16.2 shows the basic template structure of this diagram and displays an example of a populated diagram relating to a manufacturing problem.

Note: No standard set or number of "bone" categories exists, and the essential areas shown can be modified to fit the problem.

This type of analysis helps to graphically identify and detail the possible causes of a problem related to significant areas of the process or environment. The box to the far right of the diagram is the "effect" or problem area

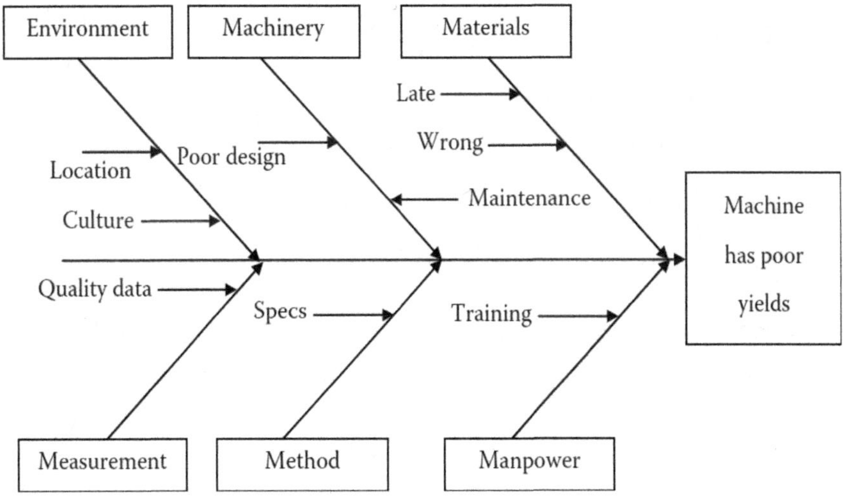

FIGURE 16.2 Fishbone diagram (major cause categories).

being analyzed, while everything to the left of the box is a possible group area and element leading to that effect. Possible causes are identified by subject matter experts (SMEs) from each technical area that come together to *brainstorm* the potential risk or process problem. This class of problem diagramming helps organize root causes that impact the target.

One of the strategies for identifying risk events is using a checklist covering pre-defined risk categories. One interesting example of this format is a checklist designed by the State of TX Information Resources organization containing 77 targeted assessment questions divided into the following 14 major risk categories:

1. Mission and Goals
2. Project Management
3. Decision Drivers
4. Organizational Management
5. Customers/Users
6. Project Characteristics
7. Product Content
8. Deployment
9. Development Process
10. Development Environment
11. Project Management
12. Team Members
13. Technology
14. Maintenance and Support

Table 16.1 shows an example fragment of a risk factor checklist that evaluates ten mission goal risk areas and illustrates the associated risk grading logic.

The outcome of a potential risk group grading process can provide both targeted and group assessment of risk. This can also provide guidance on the level and magnitude of potential risk areas in the project. The more

TABLE 16.1 Risk Checklist Fragment Example (Sample Category)

Factor ID	Risk Factors	Low-Risk	Medium Risk	High-Risk	L	M	H
			Mission and Goals				
1	Project Fit to Customer Organization	Directly supply Ports customer organization mission and/or goals	Indirectly impacts one or more goals of the customer	Does not support or relate to customer organization mission or goals			
2	Project Fit to Provider Organization	Directly supports the provider organization's mission and/or goals	Indirectly impacts one or more goals of the provider	Does not support or relate to the provider organization's mission or goals			
4	Workflow	Little or no change to the workflow	Will change some aspects or have a negligible effect on workflow	Significantly changes the workflow or method of organization			
6	Resource Conflict	Projects within the program share resources without any conflict	Projects within the program schedule resources carefully to avoid conflict	Projects within the program often need the same resources at the same time (or compete for the same budget)			
7	Customer Conflict	Multiple customers of the program have everyday needs	Multiple customers of the program have different needs but do not conflict	Multiple customers of the program are trying to drive it in very different directions			
9	Program Manager Experience	The program manager has deep experience in the domain	The program manager has some experience in the domain, can leverage subject matter experts	The program manager is new to the domain			
10	Definition of the Program	The program is well-defined, with a manageable scope by this organization	The program is well-defined but unlikely to be handled by this organization	The program is not well-defined or carries conflicting objectives in the scope			

details that can be obtained about the project at the early planning stage, the better the ability to define the project scope, prepare a viable plan, and anticipate related risk events.

A formal risk assessment process requires that all identified risks be assigned a probability of occurrence and impact magnitude. A more rigorous analysis level is required when a risk event is graded with a value higher than the defined tolerance level. As indicated, the defined tolerance level is a management-driven variable and will be variable across entities. For example, NASA deals with very high-risk projects. To say that they will construct a rocket that never fails is not reasonable, as history has shown, but that does not mean the absence of a formal risk assessment process. It simply says that the level of inherent risk in their mission has risks that cannot be controlled entirely. Even here, the history of their major adverse events is worthwhile background reading for this topic. In some cases, the risk assessment phase reveals a level of risk beyond the tolerance level constraint, and the project might be canceled despite positively defined benefits.

Years ago, a large U.S. metro area required a new roadway to be added to what was historically a rural area. The need for the venture was clear, but the early risk identification review indicated that obtaining access to the required land right away might be an issue. Management decided to proceed with the project, and it moved forward to a point where a group of wealthy citizens held up the extension through imminent legal domain hearings. This continued for a considerable time, so the project was costly and completed very late. This resulted in a poor tactical allocation of resources in the interim. This same scenario has been observed with the discovery of archeological remains found mid-project.

To be practical, tools in the checklist category require the following general design guidelines:

- The risk question list must be created and continually monitored based on the current project experience.
- A risk-scoring system needs to be included.
- The significant risk groups need to be assessed for current validity.
- When completed, the items identified need to be discussed for final handling decisions.
- Actively rank the top ten risks and have appropriate risk-handling plans in place for these items.

TABLE 16.2 Sample Risk Register Format

No.	Rank	Risk	Descr.	Category	Root Cause	Triggers	Potential Response	Risk Owner	Probability	Impact	Status
R6											
R43											
R21											

Using a risk checklist can also assist in developing a quantification scheme that provides a global assessment of overall risk. This grading scheme would help assess areas of the project that needed further review and help decide on the overall breadth of risk.

16.5.2 Risk Register

The *risk register* is the primary documentation source for capturing and documenting risk events. This document is a formal repository for capturing project knowledge regarding risks and their status (Richardson & Jackson, 2025). Table 16.2 displays a fragment sample risk register to document defined risk events and related information.

The name of each risk is listed under the risk column, along with a description. The formal name for each item in the risk register is a *Risk Event.* Each line item should have a formal Risk Owner assigned. This is the key person responsible for assessing the ongoing status of the event through the life cycle.

A risk may be linked to a particular category, such as technology, resources, and facilities. Other information on the specific risk will incrementally be added as the analysis process unfolds. Realize that not every item on the list will deserve further action. One important management variable in this area is the concept of a risk trigger. The term *trigger* is defined as an early warning event that signals the emergence of a risk event from possible to occurring. Lastly, the status column allows free-form status notes regarding the item. Realize that the more modern method of capturing risk data would be through an online database. However, the same data elements as shown here would be relevant.

16.6 QUALITATIVE RISK ANALYSIS

Once a risk event has been identified, the next step in the formal evaluation process is to evaluate it regarding the probability of occurrence and impact level. These two related grades are designed to provide a "fuzzy

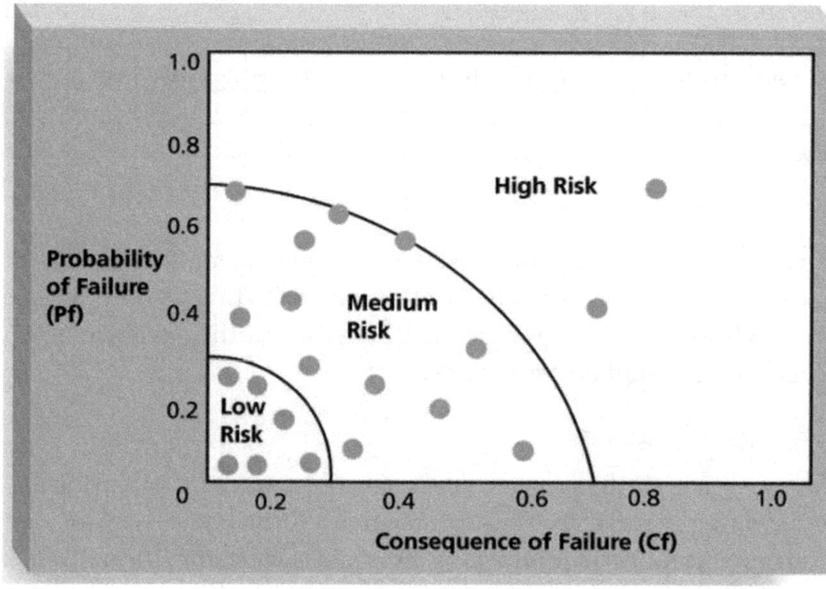

FIGURE 16.3 Risk matrix.

math" measure of quantification of the identified event. Other than providing fuzzy quantification, the operational role of this level is to segment the identified risk events into priority groups. There are multiple ways of doing this. One of the more typical schemes is to group each event into an HML grade. Other schemes are more numeric, with grades of 0–10. The net result of this exercise can be shown schematically by the sample *Risk Matrix* in Figure 16.3, where the axis scales are 0 to 1.0.

Note in this figure that the various events are arrayed into three broad groups equivalent to HML. From this high-level scoring system, the next step involves members of the prioritized groups and further examination. This is probably the time to admit one of the "dirty" realities of this process. *Not every risk event will be selected for further examination.* On the one hand, it is not economically feasible to do this, but on the other side, the risk assessment process is often curtailed when the time allocated runs out. This area is where the human element begins to emerge. Selecting the events to analyze further is done not just by category but also with insight from internal expertise.

16.7 QUANTITATIVE ASSESSMENT

The assessment model runs into a real problem in the process.

16.7.1 Formal Risk Impact Equation

From a purely mathematical point of view, the impact of a risk event is the probability of the event times the impact value. So, if an event were judged to occur with a probability of 10% and have a negative impact of $100,000, the "expected value" of that event would be $10,000. If the same data were available for all risk items, the sum would be the total probabilistic risk impact to the project. There are just a couple of minor issues with this theory. First, the quantified impact and occurrence values are not obtainable. Second, there is no logical reason to believe that this is the total list of all future events. We have a model with no way to execute a formal solution from these two shortcomings. Calculating a total risk impact will have to come in different forms.

Some additional risk assessment tools are needed to bridge this gap between pure math and empirical estimates. Three such tools are decision trees, sensitivity analysis, and simulation modeling.

16.7.2 Decision Tree Analysis

Decision tree analysis is more focused and quantitative. It provides an expected value probabilistic evaluation of an event in which the expected outcomes can be modeled. The technique traces different outcomes through defined decision options and projects' statistical evaluations of each pathway. From this perspective, one can judge the appropriate action and see the calculated expected value. One must define each potential decision or result branch to construct this model. From this, probabilities and impacts are estimated for each path. Mechanics to quantify each branch use is called Expected Monetary Value (EMV). An example of these calculations is shown in Figure 16.4.

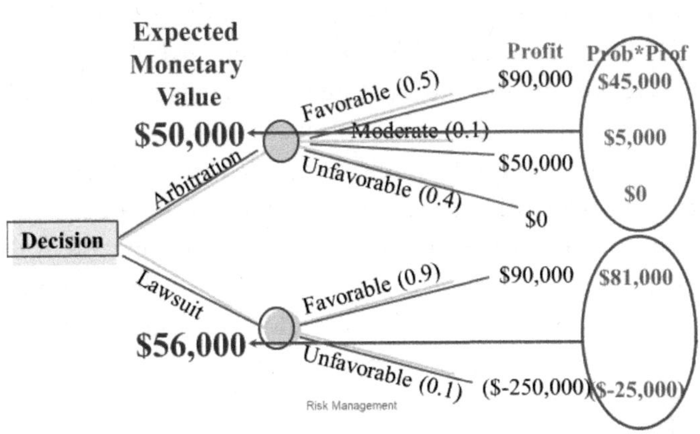

FIGURE 16.4 Sample decision tree analysis.

This decision tree example compares a strategy of pursuing a lawsuit versus using arbitration to settle a dispute. Each branch outlines probabilistic result options and the corresponding probability of each outcome. At the end of each branch, the monetary result is displayed. The value of a decision option is calculated by multiplying the probability of a branch event by the outcome value and then summing the decision branches. For example, the lawsuit decision branch has three outcomes. Multiply the value of each possible outcome by the probability and then sum the result. This yields an arbitration EMV of $50,000 compared to the $56,000 value for the lawsuit decision branch. Quantitatively, this suggests that pursuing a lawsuit has the most significant value. However, note that the option also has the highest potential for loss at $250,000. In some cases, such as this, the decision maker will avoid significant *regret* if the difference between the two decision options is marginal. Using that rationale, this scenario is an example of choosing the lower expected value to avoid a potentially harmful result. Notice that the arbitration option has no negative result. Sometimes, avoiding a significant loss is more desirable than achieving maximum return. The idea of avoiding a highly adverse event falls into the category of sensitivity analysis.

Details on simulation modeling are deferred to Chapter 20 as it is considered more of a general-purpose analysis tool. One of its significant roles in the risk area is estimating the size of contingency reserves for both schedule and cost.

16.8 RISK RESPONSE PLANNING

At this stage of the examination process, a subset of identified risks has been chosen for further review. The last step in the formal risk planning stage is called Risk Response Planning, and the goal of this is to develop action plans regarding how the identified risk item is to be handled. Let us start with an example. A critical server has been identified as a potential operational risk event. The team has concluded that this server needs to be online, essentially full-time, and that data loss would be significant. So, the decision to handle the situation is to have a backup system that mirrors the primary. A sophisticated data backup plan is also proposed. These decisions add cost to the project plan and add work units. Even though this is couched until the risk theme, this new work is melded into the project's operational plan. Note that the decision made for this event is to avoid the risk event by adding something to the plan. Risk events not dealt with this way would fall into another contingency category explained below.

Four risk event decision strategies can be made in the risk-handling stage. These are Accept *Risk*, *Avoid Risk*, *Mitigate Risk* (minimize), and *Transfer Risk*. Sheehan (2010) suggests that a risk event should be accepted without further action if the likelihood of the risk occurring and the financial impact on the project are both low. If the likelihood of the risk occurring and the financial impact on the project are high, a more careful analysis is required. The selection of a handling strategy is closely tied to the character of the event itself. If the net impact of an event is high, some action is required, even if it is only to assign careful tracking by a defined *Risk Owner*. Recognize that each industry, company, and even project is unique, so the decision for one may not be appropriate for the other. This is true for projects even within the same company, as their goals may differ. The sections below offer brief comments on what each of the four alternative handling diction actions involves.

16.8.1 Accept Risk

Accepting a risk becomes a potential choice when the impact is low and an easy fix is known should it occur. In this case, a logical option is to monitor the event. A less obvious example could be the situation of using new software technology. This new version looks promising with new desirable features; however, history has taught us that unexpected negatives can occur from new versions. For instance, new software may have unknown bugs that could emerge later and thus could negatively impact the project in some way. Also, user training might be an additional cost for this version. The decision is based on assessing the version benefits and the known experience of usage. Some organizations will not accept a new version until the next one is released, but that only hints at the historical nature of new versions.

A third risk-handling example is the weather-related risk for a community fair. This situation is outside the team's control, and the only options are to accept the problem or have a backup plan to avoid the adverse weather. One option is to accept this without mitigation and cancel the event as the weather dictates. Recognize that accepting a risky event is no different than what most people do daily as they drive a car through heavy traffic. Yes, the vehicle has seat belts and airbags to mitigate the impact of an adverse event, but that does not remove the total risk, including the risk of death. It is important to note that millions move into this hazardous risk pool daily without much thought as to the consequence, thus risk acceptance. Logically, we may psych ourselves into believing that driving a car in urban traffic is a low-risk event, but statistics would suggest otherwise.

16.8.2 Avoid Risk

Avoiding a risk occurs when it is decided that the situation is not tolerable to leave alone. In this case, the goal is to identify a plan of action to avoid risk. In the case of the new software mentioned above, if that were to be avoided, the apparent strategy would be to see about using an older version with known characteristics and trained users. In the case of the weather event, canceling would have a significant negative impact that could result in the search for an alternative method to allow it to proceed (i.e., new location, tents).

16.8.3 Mitigate Risk

The concept of mitigation means that the risk is reduced to an acceptable level through some action. The above example regarding a new software version and deciding to move down to an older version is an example of mitigating and avoidance. It avoids the stated risk of the new version, but it does not possibly get rid of a new software risk of some type. So, moving to the new software version creates a secondary risk issue to evaluate. The primary risk was mitigated. Mitigating a risk occurs when a project team decides that a particular situation is not tolerable as defined. In that situation, the goal then dictates the development of a plan of action to reduce the risk impact. One way of describing the process of mitigating risk is to envision deciding to lower the negative impact of a defined risk event. One of the best memory examples we see is in office buildings. There is an obvious fire threat (risk event). Depending upon the organization's view of this, one can find various mitigation strategies:

- Forbid smoking
- Install a sprinkler and fire alarm system
- Perform evacuation drills
- Assign local fire marshals

With these in place, we can classify the effort as mitigating the probability of a fire (maybe). These decisions improve the chances of employees getting safely out of the building if a fire occurs. These decisions do not help

TABLE 16.3 Risk Mitigation Plan

Rank	**Content**
Title	Reference name (short).
Description	A full description of the risk and its impact on the project.
Contingency plan	Describe the plan of action that will be implemented if the risk occurs.
Status	Update the status by continuously monitoring the risk throughout the project's life cycle and assign a category to the status, such as execute contingency, watch, hold, or closed.
Probability that the risk will occur	Use a consistent scale from 1% to signify an extremely unlikely occurrence to 99% to represent an almost inevitable occurrence.
Impact the risk will have on the project if it materializes	Use a consistent scale such as 1 = very low, 2 = low, 3 = moderate, 4 = high, 5 = very high.
Earliest date the risk impact could materialize	Identify the month/day/year or use the term *BOP* to signify the beginning of a project.
Latest date the risk impact could materialize	Identify the month/day/year or use the term *EOP* to signify the end of a project.
Departments or program areas that would be impacted by the risk	Identify department or project areas affected by the risk.

avoid a tornado hitting the building or any other risk to the employee. Thus, the fuzzy nature of this phenomenon.

In some cases, a mitigation decision involves specifying a contingency action to be performed after the risk has been triggered. The example above was proactive and included mitigation factors to avoid the event. Table 16.3 displays a mitigation plan template that could be used to document a contingency action plan to be executed once the target risk event has been triggered.

16.8.4 Transfer Risk

The transfer of risk option is viable when another party assumes the negative outcome consequences. For example, purchasing car insurance transfers the damage cost incurred by an accident to another party that would assume the funding consequences. The contingency fund would pay the third party as insurance against the loss in this case. Any similar example of paying a third party to mitigate potential issues would fall under the transfer strategy category. This class of response is typically insurance-like, but paying a third party to host a backup site that may never be needed would be similar. Partnering with a third party to mitigate risk could be viewed as mitigating or transferring risk.

At this point, we hit a critical operational problem with the risk methodology. If it were possible to assign numeric values to each event regarding probability and dollar impact, the mathematics for risk would be reasonable. However, that is not a typical real-world situation. Let us demonstrate the ideal assessment process if solid quantitative values were known. If a risk event was estimated to occur once in the project with a probability of 10% and if it happened, the impact would be $100,000, then we could model the expected impact of that event at $10,000 (i.e., 0.10 × 100,000). This is analogous to the actuarial process that an insurance company would use to calculate customer premiums. In both cases, the goal is to set aside sufficient funds to cover the probabilistic risk event (i.e., customer death or project negative). In this sample case, we would set aside a contingency fund of $10,000 to cover the expected impact of this known/unknown event. It is always confusing to those exposed to this idea for the first time about how $10,000 can cover a $100,000 event. The short answer to this is that we have a lot of such probabilistic events, and all of them will not happen. In mathematical terms, if all the estimates made had been arithmetically correct, we would have set aside sufficient funds to cover their resolution. The contingency fund would be exhausted at the end of the project with zero remaining. Given our rough quantitative method of defining probability and impact, the resulting math is inaccurate. Also, there is no assurance that we have even captured all the events that could occur. One could say that risk management is risky! Regardless, this is the current model process used to estimate the size of the project contingency, along with other historical experience data.

16.9 CONTINGENCY RESERVES

Resources must be set aside to cover active risk events as a byproduct of the risk decisions. Up to this point in the text, the plans related to schedule and cost did not include the impact of risk. There is no standard method for showing risk in a plan, but some indication of a risk contingency should be shown (authors' opinion). A contingency fund needs to be established to fund the resolution of risk events that eventually occur during the execution process.

16.10 RISK MONITORING AND CONTROL

The risk monitoring and control process occurs throughout the life cycle. This consists of tracking previously identified risks and implementing response strategies as necessary so that the project impact is minimized. Risks previously identified and ongoing identified risk strategy need to be actively monitored. A PM should not leave the risk monitoring and controlling adrift from a management view. Instead, a specific time must be set aside for this activity (Benta et al., 2011). Therefore, this activity should not be considered once in the life of the project as that is not enough. As indicated, developing the monitoring process for the team to carry out is important. This process should include, as a minimum, the following tasks:

- Review current risks and explore how these risks could be reduced, eliminated, or no longer need coverage.
- Track systematically the identified risks in a *risk tracking log* similar to the sample shown in Table 16.4.

TABLE 16.4 Risk Tracking Log

Risk ID #	Content
Risk Category	Vendor, software, hardware, resources, etc.
Description	Describe the risk.
Impact on Project	Describe what will happen to the project if this risk materializes.
Probability of Risk Occurring	Use a consistent scale such as 1% to signify an extremely unlikely occurrence to 99% to signify an almost inevitable occurrence.
Range of Time Risk could occur	Identify the range in terms of month/day/year or use the term *BOP to EOP* to signify the beginning of the project till the end.
Status (Execute Contingency, Watch, Hold, or Closed)	Update the status by continuously monitoring the risk throughout the life cycle of the project and assign a category to the status, such as execute contingency, watch, hold, or close.
Action Currently Taken	What is being done to monitor or control the risk, such as extra meetings with the customer or hiring additional resources?
Assigned Team Member(s)	Team member(s) assigned to the project.

- Identify new risks continuously throughout the life cycle of the project.
- Produce lessons learned for future risk assessment and allocation efforts.

After the risk identification process is complete and events analyzed, the project team should set aside enough reserves to support handling these future events (Richardson & Jackson, 2025).

16.11 RISK PATH REVIEW

One of the techniques not often seen in projects is a method of mapping risk events to the plan. This can be done simply by evaluating each work task according to some risk assessment. In this case, HML will be used. The method shown here is crude but seems worthy of showing for communication purposes, at a minimum. Assume the goal was to highlight tasks by risk level in the planned periods. To produce a risk map, assign a grade value to each task in the plan. This grade assessment could come from the earlier assessment exercise or some less sophisticated local process. Regardless of the method used to define a task risk level, the mechanics are to sort the plan by risk grade and display it in that format—those on the project plan.

For this example, the risks are coded into H, M, and L groups, plus one other code used to remove summary tasks from the listing (they could also be coded if desired). To set this problem up in Microsoft Project (MSP), open a free column such as Text20 (it could also be any other free variable). Figure 16.5 shows a fragment of the car project plan. Note the manually entered risk codes shown in the Text20 field.

The coding scale shown here is arbitrary and is only used to group the results into the desired risk groups. From this point, the regrouping process is relatively simple. The MSP mechanics are as follows:

1. View/Sort/Sort/Sort by Text20 ascending; uncheck the outline structure box; the Text20 column should now be in numeric code order by Text20
2. View/ Group/New Group/ group by Text20
3. Format the display to show the desired columns
4. A fragment of the total result is shown in Figure 16.6

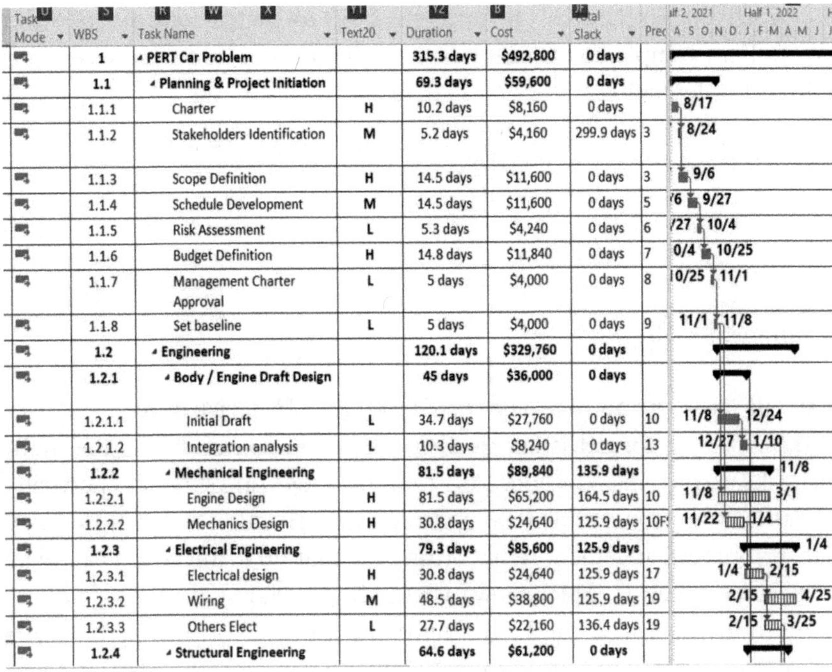

Task Mode	WBS	Task Name	Text20	Duration	Cost	Total Slack	Prec
	1	PERT Car Problem		315.3 days	$492,800	0 days	
	1.1	Planning & Project Initiation		69.3 days	$59,600	0 days	
	1.1.1	Charter	H	10.2 days	$8,160	0 days	
	1.1.2	Stakeholders Identification	M	5.2 days	$4,160	299.9 days	3
	1.1.3	Scope Definition	H	14.5 days	$11,600	0 days	3
	1.1.4	Schedule Development	M	14.5 days	$11,600	0 days	5
	1.1.5	Risk Assessment	L	5.3 days	$4,240	0 days	6
	1.1.6	Budget Definition	H	14.8 days	$11,840	0 days	7
	1.1.7	Management Charter Approval	L	5 days	$4,000	0 days	8
	1.1.8	Set baseline	L	5 days	$4,000	0 days	9
	1.2	Engineering		120.1 days	$329,760	0 days	
	1.2.1	Body / Engine Draft Design		45 days	$36,000	0 days	
	1.2.1.1	Initial Draft	L	34.7 days	$27,760	0 days	10
	1.2.1.2	Integration analysis	L	10.3 days	$8,240	0 days	13
	1.2.2	Mechanical Engineering		81.5 days	$89,840	135.9 days	
	1.2.2.1	Engine Design	H	81.5 days	$65,200	164.5 days	10
	1.2.2.2	Mechanics Design	H	30.8 days	$24,640	125.9 days	10FS
	1.2.3	Electrical Engineering		79.3 days	$85,600	125.9 days	
	1.2.3.1	Electrical design	H	30.8 days	$24,640	125.9 days	17
	1.2.3.2	Wiring	M	48.5 days	$38,800	125.9 days	19
	1.2.3.3	Others Elect	L	27.7 days	$22,160	136.4 days	19
	1.2.4	Structural Engineering		64.6 days	$61,200	0 days	

FIGURE 16.5 Modifying the plan to examine risk.

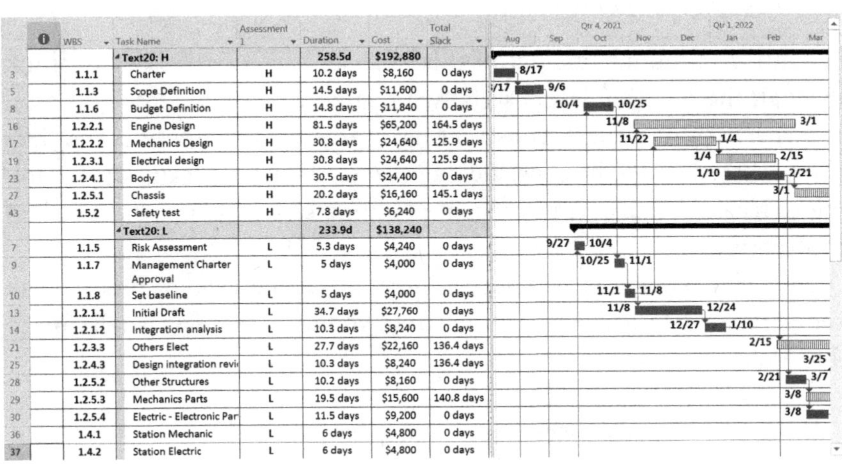

	WBS	Task Name	Assessment 1	Duration	Cost	Total Slack
		Text20: H		258.5d	$192,880	
3	1.1.1	Charter	H	10.2 days	$8,160	0 days
5	1.1.3	Scope Definition	H	14.5 days	$11,600	0 days
8	1.1.6	Budget Definition	H	14.8 days	$11,840	0 days
16	1.2.2.1	Engine Design	H	81.5 days	$65,200	164.5 days
17	1.2.2.2	Mechanics Design	H	30.8 days	$24,640	125.9 days
19	1.2.3.1	Electrical design	H	30.8 days	$24,640	125.9 days
23	1.2.4.1	Body	H	30.5 days	$24,400	0 days
27	1.2.5.1	Chassis	H	20.2 days	$16,160	145.1 days
43	1.5.2	Safety test	H	7.8 days	$6,240	0 days
		Text20: L		233.9d	$138,240	
7	1.1.5	Risk Assessment	L	5.3 days	$4,240	0 days
9	1.1.7	Management Charter Approval	L	5 days	$4,000	0 days
10	1.1.8	Set baseline	L	5 days	$4,000	0 days
13	1.2.1.1	Initial Draft	L	34.7 days	$27,760	0 days
14	1.2.1.2	Integration analysis	L	10.3 days	$8,240	0 days
21	1.2.3.3	Others Elect	L	27.7 days	$22,160	136.4 days
25	1.2.4.3	Design integration revi	L	10.3 days	$8,240	136.4 days
28	1.2.5.2	Other Structures	L	10.2 days	$8,160	0 days
29	1.2.5.3	Mechanics Parts	L	19.5 days	$15,600	140.8 days
30	1.2.5.4	Electric - Electronic Par	L	11.5 days	$9,200	0 days
36	1.4.1	Station Mechanic	L	6 days	$4,800	0 days
37	1.4.2	Station Electric	L	6 days	$4,800	0 days

FIGURE 16.6 Task risk grouped output.

Space limitation does not allow the complete listing to be shown here. However, note that the code = H (high-risk) tasks are grouped at the top, followed by the code = L in the second grouping. The other groups are not shown from a presentation format view. It would make more logical sense to convert the alpha codes to numeric ones to better control the order shown. For example, to translate the risk alpha codes into a more controllable sort order, convert the Text20 data codes as follows:

H = 100

M = 70

L = 30

Summary = 10

This will cause the output to be sorted into the HML risk order.

The risk map format is an excellent format to communicate the main risk tasks within a plan period. Note that some tasks are shown as noncritical paths, a key concept here. Sometimes, the critical path is not the only high-priority management target. This presentation format is an easy way to convert the standard plan view to better highlight risk, and thus, it offers a way to focus on risk rather than just the schedule critical path. Given coding freedom, all critical path tasks could be made high risk if that was desired. Think of this technique as a risk communication strategy tool.

16.12 A RISK CASE STUDY

To gain maximum value out of this case study, it is necessary to think through the question layers serially. Imagine you are the lead technical manager on this project, constructing a large nuclear power plant. This plant is scheduled to be constructed in Japan near the ocean, where cooling water is available. A significant planning step is to assess the risk of this venture.

Risk management is of concern here, given the type of process and comparable historical events. There are two major risk events relevant to this project:

Three Mile Island—1979 in the United States

Chernobyl—1986 in Russia

Both events are well-researched and provide significant insights into the potential threats related to their operation. A nuclear power plant is a complex integration of mechanical and nuclear elements that can potentially be a significant environmental, human, or financial threat if something goes wrong. One prominent risk event is a nuclear core meltdown, which could lead to explosions and radiation exposure over a wide area; the Three Mile Island plant failure event did not create a major environmental issue but was significant enough to deter continued nuclear plant construction in the United States. This was likely somewhat related to the perceived threat and the memory of the Japanese atomic bomb damage in 1945. The Chornobyl event that occurred some years later was characterized as a disaster, with long-term radiation exposure and population displacement issues. These two events were sufficient to highlight the need to perform a rigorous risk analysis on the case study project.

Post audits of the two events highlighted somewhat similar root cause characteristics, although the details are complex. The Three Mile Island failure was traced to ineffective processes in place to handle unexpected events, such as a failed valve and the operators not handling it properly. One might call this process immaturity at both the hardware and process levels. In some ways, the Chernobyl event followed a similar path. A fire broke out in a turbine that was under repair (non-nuclear). From this, a switch failed, and the operator did not handle it well. Major fire, explosions, and radiation exposure followed. The surrounding area became inhabitable, forcing citizens to abandon a nearby city. The half-life of the associated radiation was 30 years. By any measure, these scenarios represent a mandatory risk assessment for the new project.

16.12.1 Mitigation Strategy

Based on a review of previous nuclear plants, the following core mitigation strategies seem obvious:

1. Establish known mature operational practices and emergency procedures for a nuclear core overheating.
2. Install redundant backup for electrical and water, which would be needed in case of an emergency shutdown.
3. Construct a 35-foot seawall to protect the plant from heavy waves, which were known to occur.
4. Establish an evacuation plan for the local population in case of radiation emission.

This class of strategies would be called classic mitigation factors for a high-risk target. The key question here is: what event could change the status of these items? The value of this review would be improved if you stopped reading and asked what could go wrong to minimize these legitimate strategies.

16.12.2 The Actual Triggered Risk Event

All the above risk analysis and planning steps were undertaken before the Fukushima nuclear plant moved into production successfully. However, in March 2011, a defined risk event was triggered when a devastating tsunami spawned from a large nearby earthquake, delivering a wall of water estimated at 50 feet, swamping the plant, and causing a core meltdown as support services failed. The effects were catastrophic on the power plant operation and the nearby area. Over the next few days, the power plant failed with explosions and radiation emissions to the atmosphere. As a result, the population in the area had to be displaced because of radiation emissions, and the plant and soil cleanup began. More than ten years later, the plant is still out of service, and the cleanup costs are now estimated at nearly $200 billion. Water stored in the tanks will not be drinkable for 100 years, and the half-life of the soil radiation is 30 years. These examples illustrate the impact that an unmanaged risk event can have even when the risk potential is recognized. Also, note that the failure of this project occurred years after the physical project was completed "successfully." The interested reader can find a plethora of literature available on this accident. So, now the question is, what do we learn about risk planning and mitigation from this real-world example?

16.12.3 Risk Mitigation Lesson

There have been many successful nuclear projects over the years, most of which have had no visible problems. Procedures in place for those would make good design guidelines for the new project. Learn from others! Without doubt, three key factors emerge from this review:

1. Provide robust backup facilities to handle emergency shutdown of the reactor.
2. Provide mature training programs for the personnel.
3. Provide active control processes to evaluate the online status of the operation.

So, if the Fukushima team knew all of this, what went wrong? That is not an easy question to answer. A post-audit of the two deviant sample projects uncovered operational shortcomings. There was some evidence that the operators were not adequately trained to handle unusual events, and even when operators did know what to do, the failure of backup capabilities for electricity and water wiped out the rest of the risk plan. Having a plan on paper does not equal a working plan. Reviewing events of this type provides a great learning opportunity for understanding the risk environment, but this example offers more insights than that. Here is an additional thought to make one ponder the process a little deeper. First, this location was known to be a frequent earthquake area, which often triggered tsunamis. Knowledge of that led to the design of a sea wall to protect the plant. This linkage of factors is a unique risk view. The sea wall was the only defense against the potential of high waves and, thus, a single point of system failure! How high should the wall be? Decisions of this class are not just minor blips. They are system failure events that represent mega-risk. Serious consideration should be given to avoiding this decision if possible. In this case, how can the wave risk be removed from the equation? Build an extremely high wall or move the plant away from the water area. Surprisingly, the post-audit did not mention the decision to move the plant.

High-risk project ventures such as this highlight one of the significant roles that risk management plays in the planning process and future success. However, the underlying message here is that some risk events can be catastrophic, hard to forecast, and may differ from previous similar projects. Failure to properly deal with this aspect of project planning is a major contributor to future failure. Looking back at a situation and seeing what should have been considered is always easier. The Fukushima project is a good example to keep as a reminder of this.

16.13 CONCLUSION

This chapter has illustrated a model approach to analyzing and controlling project risks using quantitative and qualitative assessment techniques. The process results in identifying risks and evaluating ways to decrease their negative impact. One of the resulting artifacts is produced in a risk register repository that is used to show global status for this analysis. The operational step involves monitoring and controlling defined and new risks throughout the life of the project. Several risk-centric tools and techniques related to risk management were illustrated.

Failure to recognize the impact of risk on project success will negatively impact the result. It is important to recognize that this process involves the recognition part of the risk equation and the ongoing monitoring and control aspects. It is important to recognize both. Even though the risk identification process commences at the beginning of the project, new risks may be uncovered throughout the project's life cycle. It is important to assess these as they are recognized continuously.

The final section of this chapter highlighted the need to communicate areas of risk related to defined work tasks. This was illustrated using a simplified risk mapping technique embedded inside MSP. Managers who develop an understanding of risk in the overall scheme of the project will be more successful than those who focus on the defined project scope. The PM must create an environment whereby risk is recognized as an important element, even though it has not yet occurred and may occur at any time. This is called a risk culture and a key part of the learning organization.

ASSESSMENT QUESTIONS

1. Describe and differentiate between quantitative and qualitative risk assessment.
2. Describe the risk response strategies.
3. Explain the importance of risk monitoring and controlling processes.
4. Give three examples of risk assessment tools used by companies.
5. Create a risk scorecard appropriate for your university or company when rolling out a new system.
6. Name and briefly describe the project risk management process.
7. Describe how risks for a project can be identified and categorized.
8. Explain the value of documentation in the risk management process.
9. Select a specific industry, such as healthcare or banking, and brainstorm risks that may apply to implementing a new system within the selected industry.

RISK EXERCISE

1. Using the Car Case Study file, produce hypothetical examples of the following risk artifacts:

   a. **Risk Planning**—Develop a Risk Management Plan that the Formula Two Project Team can use throughout their project.

      (Sources for *Risk Management Plans can be found in Appendix C).*

   b. **Risk Assessment and Identification**—Identify five risks associated with the Car project. Use a sample Risk Log by conducting an online search to record your answers.

   c. **Qualitative Analysis**—Develop a Risk Matrix that categorizes the defined risks associated with the Car Case Study regarding the likelihood of a risk event occurring and the probability of the risk event's impact on the project.

   d. **Risk Mitigation**—Using the risk log, develop a Risk Mitigation Plan for each of the five defined risks. The plan should include the definitional components listed in the Risk and Mitigation Plan previously shown in Table 16.3.

   e. **Risk Response**—Assume probability and impact values for each of the five risk events and develop a contingency reserve for this data.

   f. **Risk Monitoring**—Develop a process for your team to monitor high-level risks. The process developed is not limited to but should include the five sample risk events.

REFERENCES

Beasley, M.S. Branson, B.C., & Hancock, B.V. (2010). Are You Identifying Your Most Significant Risks? *Strategic Finance November*, 29–35.

Benta, D., Podean, I.M., & Mircean, C. (2011). On Best Practices for Risk Management in Complex Projects, *Informatics Economica* 15(2), 142–152.

Eaton, G.H., & Little, D.E. (2011). Assessing and Mitigating to Deliver Sustainable Safety Performance. *Professional Safety* July, 35–41.

Project Management Institute (PMI). (2021). *A Guide to the Project Management Body of Knowledge*, 7th edition, Newtown Square, PA.

Richardson, G.L., & Jackson, B.M. (2025). *Project Management Theory and Practice*, 4th edition, CRC Press Taylor & Francis Group: New York.

Sheehan, N. (2010). A Risk-based Approach to Strategy Execution, *Journal of Business Strategy* 31(5), 25–37.

Vargas-Hernadez, J.G. (2011). Modeling Risk and Innovation Management, *Advances in Competitiveness Research* 19(3/4), 45–57.

Wang, B. (2024, August 25). *Summary of Boeing Problems*. Next Big Future. https://www.nextbigfuture.com/2024/08/summary-of-boeing-problems.html

CHAPTER 17

Project Control Concepts

This chapter explores the theoretical side of evaluating project status and the associated management control actions that result from identified performance gaps. Two critical components are related to the project control process—identifying a delivery status gap and taking effective management action to deal with it. This chapter will present a theory-oriented overview of project status measurement and the associated management process.

17.1 INTRODUCTION

The simplistic traditional planned versus actual approach to status tracking is described here as generally meaningless when examined from a theoretical standpoint. These techniques are based on the wrong target to be meaningful, assuming the goal is to measure a project's goal vision perspective-based performance. Instead, most traditional calculated measures focus on resource consumption but not a quantitative assessment of the goal that spawned the project. This statement is made from the standpoint of the conventional project plan components and not one that could be constructed more accurately to reflect the future goal. To support this harsh thesis, various project performance gap scenarios will illustrate how traditional techniques do not show the current resource status.

More operational details describing project performance from other standpoints and assumptions will be described in Chapters 19 through 22.

DOI: 10.1201/9781003532989-21

Each of the future chapters will focus on a single aspect of status. Also, Section V of the text will look into less-recognized ideas about viewing the project from a more reality-oriented perspective.

From these multiple views, one might be tempted to ask, Why can a project's performance not be accurately measured using the example plans described in the text? One of the issues that has already surfaced is the gap between the initial project requirements and how future events change that view as the project unfolds. This dynamic view means that the approved project, which is the basis for execution, has changed in some unplanned manner that is not reflected in the plan. Then, the mythical laws of Murphy and Parkinson emerge to create further unplanned turbulence, and we are left with a chaotic picture to deal with in this Matrix-like dream state movie. A project execution movie must contain a video scene in a chamber filled with snakes, spiders, and vampires, all swirling in the abstract plot. That is a better picture of the world for which we are trying to produce a status view. Various project management model concepts have been defined in other chapters of this text. Each focused on methods to define and structure an environment, outlining how the project should unfold. However, there were subtle hints that some related process might go awry during the execution process. These deviations have been described as factors to explain why the success rate of projects is not stellar, and there must be a reason for that result. Various survey data indicate that between 50% and 70% of projects are unsuccessful or compromised. From a status viewpoint, projects drift away from their targets in the execution phase. A status control process goal is to identify deviations from plan targets and trigger actions by the PM to influence the project outcome. This is the essence of a project control process.

Note that the management objective of project control is to "influence" the output in a desired direction. That translates to management actions such as nudging, leaning on, motivating, making a decision, or doing whatever is needed to move the project back toward a planned value. There are many variables used to describe a project's performance. Trying to move one of these variables may require sacrificing one or more of the others. Thus, the control process prioritizes various outcomes in favor of lower-priority ones. A significant part of the decision process involves quantifying actual results and deciding what management action to take from that set of values.

17.1.1 Twelve Magic Success Indicators

We start this journey with a summary list of the 12 *magic* indications that signal a project going well. Each of the following is considered a positive indicator:

1. Planned milestone events are being met.
2. The budget is under control.
3. Quality control results are within specifications.
4. The control process indicates a reasonable level of change requests.
5. Project resources are supplied per schedule, and skill levels are adequate.
6. The project team appears to be cohesive, and morale is good.
7. Users seem satisfied with the progress of the work.
8. Top management is visibly supportive of the project goals.
9. Third-party vendors are delivering quality items on schedule.
10. Risk events are under control, and nothing unusual appears.
11. The project team training program is progressing according to plan.
12. Relationships with support groups appear to have no identifiable issues.

One might consider this outcome the PM's wish list, which will likely not be the actual state. However, this attribute set provides a control envelope of desirable attributes to evaluate project execution status. If one or more of these are not going well, it is time for the PM to initiate the influencing process. The key management question is what to do about an observed variance. The fundamental problem here is that all of the project variables are in play at this point, and many of them affect each of the 12 performance areas.

Operational control of the project involves a complex interaction with multiple sets of processes. A memory metaphor for this is to envision an airplane pilot navigating his plane through the fog and controlling the aircraft through a set of control knobs that could be turned to change the state of some operational element on the plane. Gauges represent status items, and the knobs represent management actions to change the current direction. The PM works with a similar arrangement where the knobs are various decisions that can be taken. The project influencing control knobs are related to the following core impact areas:

1. Scope
2. Resources
3. Team
4. Stakeholders
5. Sponsor
6. Risk
7. Procurement
8. Quality

Some management actions related to one of the control areas will change the project's direction. However, these are not magic knobs like the airplane that will instantly convert the trend back to a new direction. Many management decisions will have slow, visible status reactions and may not be predictable as to the actual result. Examples of a decision targeted at one factor can negatively impact other project parameters.

Before the project moves into execution, status tracking, and management actions, the project control activities have been "soft" because plans are still evolving and being formulated with frequent changes. As a result, the control activities are more approval-related than operational decision control. Technically, nothing has gone wrong at this point. One could describe the early project planning segment as more of a chaotic whirlwind of activity than a controlled environment. However, once the planning process is completed and approved, specific work delivery targets are specified, and associated resources are defined. Then, a formal control process to measure outcomes can be pursued. The traditional definition of control measurement compares actual parameter results to the approved plan values. Variances from these comparisons are triggers to signal actions by the PM and project team. Dealing with execution variances is a management artform more than a defined arithmetic process.

17.1.2 Parameter Tradeoffs

Chapter 6 described the project as composed of work units and the overall project as a three-dimensional box with sides of function, time, and cost. During the execution process, finding one of these parameters that varies from its planned value is common. For example, if the project schedule is overrun, there is a desire to "influence" it back toward the approved plan. The management question is, what control knobs might help with this? One decision knob that could be turned is adding more resources (cost) to the task. That may solve the scheduling problem but wait! Costs have now overrun the plan value. This is an example of the "rubber box." It is possible to move one of the sides of this box to deal with one of the other sides, but the overall capacity of the box remains fixed (in concept, at least). This sort of decision-making is called tradeoffs. In our knob metaphor, it is best to think of it as turning one knob and finding one or more of the others changing in some other direction. Functionality, time, and cost have an interlinked relationship. A change in one of these can affect the other two. There are very few project situations where one decision does not affect others, often negatively, as in the example above. The tradeoff relationship is a vital management equation to understand.

17.2 THE HUMAN INTERFACE

Resolving many of the problems that occur during execution involves interfacing with the project team members. Almost any decision made to correct a status variance would be passed to the team for their action. In some cases, corrective action is needed within the dynamic team itself. Corrective factors related to motivation, conflict resolution, skill development, and resource capacity are frequent management issues that must be resolved to address defined gap issues. Expertise to deal with this breadth of complex human interaction requires the best skills of a PM. Successfully dealing with problems in this area is a significant project success factor, and it is one of the key metaphorical variance decision control knobs.

17.3 STATUS TRIGGERS

A significant portion of management action is triggered by various status measures comparing plan versus actual performance. One of the interesting views involves how the status is transmitted. There are many formal and informal methods for receiving status. Some PMs walk around and observe, while other use more quantitative and formal-based methods.

The formal status production process is often in parallel with other data systems that have results available periodically. This could be daily, weekly, or monthly cycles. This data collection strategy is based on visual events, and the related status question is "compared to what." Data related to the volume of change requests, risk events that occurred, technical performance measures, test results, and others collectively help paint the overall status picture of the project. Beyond these more quantitative and mechanical values, unstructured interactions with team members, sponsors, users, and senior management add a richer level of understanding to the status view (sometimes timelier and more accurate).

17.3.1 Tracking Processes

Status tracking attempts to map how the project moves compared to an approved plan. Two critical execution and control concepts must be in place for the tracking process to work correctly. First, a well-defined plan with quantifiable objectives is necessary to establish comparison targets. Without this base, there is no method to compare ongoing results. Theoretically, if there is no goal (plan), there is no control. As we can see, even with a plan, the control process is complex. The second requirement for status tracking is to measure the actual performance of appropriate parameters representing goal status. That topic also has complexity woven into it, but for now, the best approach is to deal with the basic mechanics of the process. These two plans versus actual concepts are called the *Siamese twins* of project control.

The definition outlined above requires a planned measurement target to be defined. Model theory specifies that this is the initially approved plan, called the *baseline*. If the project never changes, this would be the perfectly logical base to judge; however, that is not reality, so this theory is worth exploring further. Every project has some level of scope change, usually not insignificant and sometimes not even measured. It is not uncommon to find that the level of project-defined work grows by 2% per month. Each project type and situation has variations, but zero scope growth is almost unheard of. One can proceed with the earlier definition of status comparison using baseline data, but the resulting values need interpretation. They are only status in the sense of saying the calculated comparisons quantify how the project is doing versus the approved plan, but that is not a true status of the project goal. Scope changes have been formally approved; therefore, they must be considered part of the revised status view. Unfortunately, variances created from baseline comparisons are often viewed as indicators of poor team performance. That is the fundamental flaw in traditional status-tracking methods. Let us see what this quantification process might create regarding management reaction to reported values. Six case scenarios illustrate how status metrics can lead to the wrong interpretation of the problem.

17.3.2 Project Status with Scope Change

If a project status calculation shows that the budget or schedule overruns the plan by 15%, one traditional interpretation is that the project team efforts have consumed that much more than the approved plan indicated. A variance of this amount might be sufficient to energize higher-level management to look into the problem. At this point, the negative focus is on the project team; however, the increased scope could be the reason for this variance. As the scope is added to the approved work set, that is not the same work set defined in the baseline plan. This is a calculation of apples against oranges. This result might be a measure of planned cash flow but not of project goal status. At least five more sample scenarios could result in erroneous status views based on the plan versus actual comparisons. These will be summarized in the following sections using the same status numeric values and the audits that followed.

Scope change. A project audit found a significant number of change requests that added additional work to the plan. There was not an actual variance in any of the task's estimates. This overrun is entirely caused by scope change.

Team resources. This audit indicated multiple issues with the project team. The main problem was the level of resources allocated compared to the plan. Actual resource allocations were 20% less than the plan specified. Also, the team members have not been trained. Most of the variability can be traced to these factors.

Project team performance. The audit shows that the project team results were, in fact, over the planned values. Task work estimates were unmet, and some material cost variance was added to the overrun. This is the worst-case scenario for the project team as it uncovers internal performance issues.

High-productivity team. This project team is composed of an all-star group of highly productive professionals. They execute tasks as fast as possible and produce more work than planned. Their schedule status looks good, but the budget comparison to the plan has a significant variance. They are consuming much more human and material resources than planned and producing much more work than the plan indicated. This is a very unusual profile but mathematically possible.

Mixed results. This audit yields a more typical result. Variances are found in baseline tasks, and some components of the variance are created by scope change. In addition, issues were identified in the project team composition, size, and skills.

17.3.3 Summary View

The key insight from these real-world status-based scenarios is that project variability is created by factors related to task estimating, resource allocation, scope change, work performance, risk, and a host of other harder-to-identify factors. Once these factors are understood, it is possible to design a status system that yields the type of indicators that help most with tracking the project. The following list contains some design thoughts:

1. Using the initial baseline plan only supports evaluating how the project is doing versus the original baseline view.
2. The concept of status should focus on factors such as work efficiency, resource issues, scope change, team performance, and the likelihood of meeting the desired completion budget and schedule.

If the plan could be changed with every approved scope, then a current plan versus an actual data approach might be helpful for tactical review. However, that is not feasible because measuring the work change is seldom definable except in a third-party contractual environment. The one element that does come through this screen is the concept of work required to deliver the results, whether that is baseline or new scope. Collecting plans and actual data on these items remains a core part of the status process.

17.4 HOLLOW PLANS

Often, a project plan does not represent the actual project scope, and many examples in this text will be guilty of that view (but we know better). To be a complete plan, it must show the full scope, even if some of that is not known yet. The direct core of a project consists of a collection of work units and deliverables called scope. However, other elements in the reality plan are implicitly part of that scope. From a management point of view, even though future scope change is not known at plan time, it exists. Similarly, risk events are not known, and they also exist. Various supporting processes are needed for the project, which are not typically recognized. Examples such as facilities, IT, and host systems are all part of this third grouping. So, in recognition of the impact on the project, these variables come into play in various ways. Project plans that do not show risk and scope imply that neither of these will occur. Not showing various supporting elements indicates that they will not incur any cost to the project. None of this represents reality. These are hollow plans in that regard.

If the goal is to define status, the plan must show the elements that dictate status beyond the initial work units or boxes in the Work Breakdown Structure (WBS).

Each of these items can be measured and compared but is relatively passive in terms of defining the status that spawned the project.

So, the concept of a "hollow plan" does not reflect the overall scope of the project's cost or time variables. Scope and risk are the main factors missing from most project plans (and most example plans shown in the text). The textbook plan examples strip the work to show only the core units and simplify the explanation, but from a status viewpoint, this is not a valid project plan.

17.5 TRADITIONAL STATUS-TRACKING APPROACH

The best approach for dealing with this subject would have been first to describe the traditional view and then critique its validity. However, the logic used here was to first look deeper into the theory of status measurement rather than describing a meaningless comparison of two numbers that may not reflect perceived status. The examples shown here clarify that status reporting should have more than a single value or three traditional values. The conventional model compares actual performance to the original approval plan. The question with that is, "What does this comparison reveal?" Given that the baseline plan does not reflect the final definition of the project owing to the scope and possibly other factors, the traditional plan parameters versus current value comparison only measure performance versus the initial plan view of the project. Nevertheless, this standard status-tracking process needs to be described and understood.

The formal planning steps in project evolution focus on the original vision of delivering something. The following stages, decisions, and processes were all theoretically focused on the work required to deliver that goal. An approved baseline plan represents a view at the end of the planning phase to define the work, schedule, resources, and cost to produce the defined deliverables. Following that initial milestone point, scope change, triggered unplanned risk events, and resource allocation variations emerge to impact that initial plan. Comparison of actual results to these various parameters looks like status factors and is considered that way in many situations. However, they are not the items that matter most from an operational viewpoint. A project is initially approved because it has a perceived functionality that aligns with organizational objectives and has better benefits than competitors. Future cost and schedule factors related to the project are essential status measures because the planned schedule of the deliverable is part of the benefits equation, and the associated cost represents a scarce organizational resource that could fund other initiatives.

The traditional approach to status measurement has three key components: Baseline Plan, Performance Measurement Baseline (PMB), and Technical Performance Measurement (TPM).

17.5.1 Measurement Concepts

The definition of project scope must include the total work required to produce the defined deliverables. Scope change and risk events represent two unknown factors that are difficult to anticipate but must be represented in a form to satisfy this requirement. Also, the related schedule variability that results from these additional work requirements and other internal work variability factors is inherent in the project. In other words, the PMB view will show the direct work and all other anticipated additions to that view. These undefined elements will be displayed as additional time and cost on the baseline plan. In this view, the schedule and budget exceed the defined baseline values.

A second aspect of a coherent plan is integrating the factors that produce the plan. For example, the scope definition must represent the total work required to produce the defined deliverables. Work definition parameters are then used to determine the duration and cost of the underlying scope components. This aggregation produces the total cost and supports a schedule creation based on available resources. From this description, the project plan represents the defined deliverables and their associated work requirements. Attached is the linkage to planned resources, which drives the calculation of planned project cost and schedule. This is the essence of the PMB concept. In this view, all of the plan elements are interconnected and have tradeoff implications between them. For example, decreasing the allocation of resources and the schedule could expand, or the scope could be reduced. The key point of this approach is that the project plan has substance to support the value shown.

At this point, we reach a crossroads in plan design. One approach is to ignore the scope change. In this mode, the plan becomes "hollow," as described earlier. An alternative structure is to add estimated values for the unknown future scope change, risk events, support elements, and schedule overruns. In this expanded format, the baseline plan comparison is more interesting as a status target, but the mechanics of handling the new work events are still complex. Most baseline plans seem to ignore this expanded latter view, so we are back to the first mentioned erroneous comparison. One visual for project plan design can be seen through an approved WBS with its companion Dictionary containing all the described elements.

Oftentimes, the management process gets lost in controlling the resource consumption and status comparisons, which shortens the reasons for the project. That is, a deliverables performance goal is implicit in every project. The new product is planned to have characteristics such as speed or weight. The new process is to improve throughput speed, handle a transaction, etc. So, the fourth status segment involves the functional performance of a project or work unit.

The full equation for status tracking is now uncovered. The integrated plan outlines the various work units and resources involved in the execution process (i.e., work scheduled, human resources, material resources, and calendar time/schedule). The TPM process matches this consumption-oriented view with work accomplished to derive a balance across this set of factors. The underlying theory here is that status is more than resource consumption. It needs to recognize what was accomplished as a result of this.

Earned Value is defined by theorists as the most robust measure of performance, yet only marginally used. Detailed operational mechanics related to this topic will be described in Chapter 22.

From the various views outlined here, the complexity of generating a simple statement of project status is unrealistic. The operative answer becomes "status of what?" Upcoming chapters (19 through 22) will focus on various theoretical status and control process concepts related to the execution phase of the life cycle.

17.5.2 Scope Definition—The Achilles' Heel

One of the common reasons why projects get stressed during execution is the negative impact that scope changes have on the plan and the project team. As these volumes grow, the complexity of work coordination and the related resource allocation defined in the plan increases. As the breadth of these changes occurs, the stress level becomes exponential. Ideally, the PM would like to avoid all changes in work requirements to keep the initially planned view smooth and stable, but that is seldom a viable option. The best compromise is establishing a formal and rigorous integrated change control (ICC) process to manage scope creep. ICC would not usually be considered a management control knob-turning process but rather a workflow issue for the overall process—this is the fire hose of volume. All of this must be integrated into the defined work. This activity is considered a status variable because it changes the baseline view and is a management control variable.

17.5.3 Third-Party Control

One of the control issues that is often difficult to deal with is the interaction across decision domains from other projects or external contractual relationships. For example, a third-party contractor's late delivery of an item can impact several internal project areas (i.e., schedule, budget, resources, and activities). Similarly, a team member asking a contractor for a "quick favor" by making an unauthorized and unapproved change can also negatively affect the plan's integrity. Also, conflicts across multiple projects can emerge from various sources. The most prevalent is competition for highly skilled individuals or general competition when resources are scarce. This management knob may be senior management, who adds their level of authority to the situation.

17.5.4 Evolving Target Environment

When explaining a proper method for producing project status, one continually seeks appropriate methods to deal with the dynamics in this arena. This last example of measuring status is seldom mentioned but is still relevant in the list of measurement conundrums. If the goal is to measure something against a plan representing a valid target, what happens if the valid target moves? In some cases, organizational objectives change before the project is completed. Similarly, some planned benefit factors are based on external economics or technology that can wipe out the benefit of a project if they change in the wrong direction. This suggests that another aspect of the status measurement process involves evaluating the external environment in which the future deliverable will exist. This again shows that a linkage to the future user and environment factors needs to be validated along the way, as the target project must be compatible with the operational environment. There are documented cases where a complex deliverable was "successfully" installed, only to find that the target environment had changed and the output was unusable. Comparing anything in the plan for this external view of status tracking would not uncover such an issue. If status tracking is designed to produce a successful project, this is another required view of the activity. The only valid comparison here is the future operational environment. A PM must be prepared to take radical action regarding the project when this external mismatch class of status variance is identified.

17.5.5 Turning the Management Control Knobs

Through all the various scenario paths described, there has been a variety of status measurement perspectives. Management action is often needed regardless of how these performance gaps have been identified. This corrective action is what is described here as "turning the knob" to influence the outcome. To complete this thought, envision a control panel with a list of decision areas as outlined earlier. Recognize that the knob-turning exercise is non-trivial because turning one knob could negatively affect others. Returning to the airplane pilot example, only part of the landing gear works. The plane will be badly out of control with only one wheel. The primary goal is to land the plane, so the decision is made to land without the landing gear. As a result, this decision damages the plane, but it achieves the primary goal. The key point of this example is to recognize that sometimes, a worthwhile component of project landing gear will have to be compromised for the overall good of the project. In a similar view, more resources (people or money) will need to be allocated to fill in the status gap of the project, and this will affect the functionality of the deliverable or the required schedule. These are the metaphoric knob-turning decisions for which all project status measures are not kept in place as planned. Management control knobs have priorities attached, and the PM must be aware of what those are from the project sponsor's view. Looking at final project success as meeting all of the various plan values is a dream but one seldom achieved. Prioritization is the operative term.

The pragmatic goal is to meet the most important items and know which order to tradeoff.

Because our management control knobs are metaphorical, we need to translate them into a more understandable view. First, recognize that this process is a management decision action triggered by some operational status variability. When a gap level reaches a decision point, this triggers action. Usually, this variance has caused the project to drift, or some related occurrence will cause the drift. If the target work unit variance is on the critical path, adding resources to bring the project back in line might be necessary. However, if achieving the budget goal is more critical than the schedule goal, the decision might be made to let the project schedule slip and meet the budget target. In this case, the observed variance triggered the decision to take no action. Realize that taking no decision is also an example of a decision.

A second knob example involves the area of human resources. Here, gaps are more challenging to quantify, and the decision response is equally complex to pursue. Assume that some conflict is identified in the project team. The decision rule for this is to react quickly, as this class of problem does not automatically correct itself. Here, the resource knob is turned, and some prescribed action to correct the problem is put in motion. This example may not be identified directly from a quantified work unit variance so this second-level status variance may be based on non-quantitative factors. Human conflicts are often complex to match to a work unit but rather come from more random sources. Regardless of the trigger, the management issue is to decide on a control action. The option, in this case, is linked to the management knob that has the PM's authority and influence attached. This is complex and has many characteristics beyond the available time and space for discussion. Review the team-related management concepts outlined in previous chapters that provided insights into the fuzzy nature of human management practices when a status gap is identified.

Based on the management decision knob metaphor described here, there are two very different triggers. One trigger is based on quantitative data that signals a need to take action, while the other is a more qualitative data collection process that only indicates broad areas of corrective action related to human issues. A "softer" human-based action is often more designed to influence behavior than deal with a measured work unit problem. All control scenarios aim to influence the project variables to a more desirable collective state (i.e., schedule, cost, scope, quality, morale).

Hopefully, the use of the physical airplane analogy to illustrate the use of a status measurement dashboard and related management control theory has not been too abstract. Let us try our luck with one more. Look at the PM as an orchestra maestro who waves his baton, and all the orchestra members follow the "plan." If the musical "plan" is well designed, the outcome can be beautiful based on the members' skills. This metaphor is correct up to a point but falls apart when we must admit that some of the players are not in sync. In that case, waving the baton will not correct the problem. This view shows the orchestra leader's control aspect very well when the overall organizational and staffing environment is mature. These metaphorical examples are still interesting enough to remember from a management control perspective. The goal is to develop the project team to the point where the relationship can be more of a hands-off maestro and less of a physical management corrective action control knob-turning-oriented environment. Experience suggests that an authority-driven and heavy control approach is not the most successful, but status measurement still has a role that helps to keep the project aligned toward the defined goal.

17.6 HUMAN RELATIONS STATUS DRIVERS

Beyond the general status-driven control-oriented activities outlined earlier, much of the PM's time is spent on daily decision-making and influencing activities involving various human resources, internal or external, for the project. PMI (2021) describes how poor communication, vaguely understood project goals and objectives, and poor leadership pave the way to project failure. From these and other expert views, we can conclude that communication failures are at the heart of many control-related issues. It has been previously stated that 90% of a PM's time should be spent on various forms of communication. This discussion will focus on the communications issues around the major human-related management processes dealing with team resource acquisition, development, and general team management.

17.6.1 Team Resource Acquisition

The basics of the team resource acquisition process are outlined in Chapter 3, and the primary item to add here is the recognition that the initial acquisition decision may not be permanent. A second iteration may occur when a team member does not deliver planned results, a key member leaves the project early, or the functional department does not deliver on promised commitments per the project plan. On the motivation side, failure to properly recognize positive performance is the most direct way to lose control of the project output.

If the original staffing estimates were accurate, but the actual staffing allocation rate was slow, it is reasonable to predict that at least the project schedule performance will drift. This same result can occur if the skill level commitment for staffing is not delivered as planned. Given the direct impact that staffing has on the project, a vital status metric is to track the actual allocation of project staff to the approved plan. The project performance gap can often be traced to this shortfall. Because this directly links to future performance, the PM must be prepared to show one of the significant sources of the performance gap outside the project team.

17.6.2 Team Development

Project team development relates to the processes involved in improving the productivity of the team and its members. This is both a short-term and long-term management activity. Short-term actions can enhance the process of the current project. However, long-term actions can transform the resource skills along a path of continuous career improvement that impacts future project performance and has positive value in overall motivation for the team members.

17.6.3 General Team Management Perspectives

When project teams are initially formed, there is typically some level of internal confusion as to roles and responsibilities. Later, with proper management, the team becomes a more cohesive unit, and productivity grows. Tuckerman (2001) defined a four-stage model for team development: forming, norming, storming, and performing (Wong, 2007). Each team member focuses on the leader for guidance and direction in forming. The relationship at this stage is polite but distant with the other team members. In the storming stage, team members become more concerned with their impact on the project. They want to be respected and often battle others with feelings of inadequacy, wondering who will support or undermine them, and above all, trying to prove their value to the team (Krawzcyhk, 2024). The third evolutionary stage moves the group into the resulting work structure groups (good or bad), where group feelings and cohesiveness develop. Performance standards evolve, and new roles are adopted. At this stage, personal opinions are expressed more freely (Tuckman, 2001). The fourth evolutionary stage is called performing. Here, less internal structure is needed because the team members understand their roles, and group energy is channeled into the task. Structural and relationship issues have been resolved, and the resulting team now generates the output (Krawzcyhk,2024). This evolutionary process for team dynamics does not mean every team has high productivity. This process can lead to that goal properly, but a poor outcome will follow if managed poorly. The process of forming a successful team is a true management art. There is no cookbook method to accomplish this, but clear guidelines for the process must be understood. PMs who can assemble a group of diverse and competitive individuals into a productive project team will have long careers. Failure to have this ability will yield the opposite career path.

It is important to understand why team development is a status variable. This is the instrument through which the work is produced. It is not a homogeneous static funnel but can be managed as a quantity measure. This is another knob to deal with that is on the dashboard. Think of this as fertilizing and watering the lawn to make the grass grow. If you do not do this, the lawn will die. Grass dying is another memory management metaphor for you!

17.6.4 Team Training

Too often, the assumption is made that the team members are trained professionals who do not need training. That is a flawed assumption if one wants to develop a productive team. There is more to the project work than raw technical skills. Roles and work processes within the team are also key activities. Specific training requirements vary by the project type and profiles of the individuals assigned. Sometimes, the requirement could be very mechanical and involve techniques to estimate work and establish plans. In other cases, change control could be proved accurate, including the logic for the approval plan, risk issues, and other targets.

Examples of typical management actions to improve team productivity are as follows:

1. Training programs to improve team and individual skills (Low, 2024)
2. Interpersonal activities to build morale and improve performance (birthday, recognition actions, social events, etc.)
3. Use visible recognition and reward actions to motivate positive performance

4. Developing an effective team communication environment
5. Creating a work climate conducive to high productivity
6. Providing appropriate leadership practices
7. Teaching and practicing effective conflict resolution
8. Providing timely team and individual performance feedback

Most items outlined above relate to behavioral-oriented activities outside of specific work unit-level factors. However, individual process training can occur when dealing with a team member who has not followed correct procedures or has created an internal team conflict over a controversial technology-oriented disagreement. Training does not have to occur in a classroom, but individual efforts of this type must be performed positively (Mind Tools, 2008). The team members need to understand why the discussion was held and be motivated not to repeat the adverse event. Telling the person they will be fired if they do this again does not fit the proper motivation technique. The method by which the PM manages the work environment of the project team will dictate many aspects of the future project outcome, for the project deliverables, team culture, and future skills of the participants.

17.6.5 Managing Team Performance

Evaluation of overall team performance can lead to various strategies for improving outcomes. This evaluation can determine tasks being frequently delivered late, not delivering what has been requested, making poor use of the tools and resources, not integrating work efforts with others, and a host of other such daily events. Each of these events represents visible signs of an improvement needed by the team. Edwards Deming, the father of Quality Management, stated that the root cause of poor performance was mainly related to poor processes, inadequate training, or poor management. Based on this classic observation, much of the team development productivity potential comes from project processes and team skills. This means that the PM must focus on the work processes and an individual strategy for improving workers' skills. If the individual does not view this evaluation as punitive, it can be a positive motivational event. The key to this process is to get every person to admit that there is some developmental need to help them move in that direction. Properly done, this can also be a strong individual motivator as well. The same statement can be made for every work process, so constant attention is needed for skill and process improvement areas. Skill changes can be dealt with through formal training or mentoring activities, while process issues would fall into a more mechanical corrective arena. A mature learning organization focuses on improvement in these areas beyond just pursuing the project work, and that is a cultural attitude as much as a physical set of targets.

17.7 CONCLUSION

This chapter has explored the theoretical side of project control. The trigger that defines a gap in project performance typically comes from some quantitative measurement process related to work unit performance. However, some status gaps are identified by a non-quantitative process, such as management walking around.

Discussions in this chapter outlined why the traditional practice of measuring actual performance against a typical approved baseline plan did not describe performance in a true sense. The operative goal of defining project status is to produce a forecast quantitative measure for future results based on the initial approved Charter rather than planned versus actual comparisons of the subsequent baseline plan produced in the planning cycle. An accurate status analysis must focus on work unit performance, resources, team issues, user evaluations, and the external delivery environment. Historical parameter comparisons do not show this breadth of view effectively.

Given the existing project culture in many organizations, the process of status tracking will remain as a baseline plan versus actual comparison. However, that simplified view needs to migrate more toward the perspectives outlined here.

DISCUSSION QUESTIONS

1. What are the key factors in assessing the general status of a project?
2. What is the flaw in comparing project status to the approved baseline plan?
3. Describe two of the management *control knobs* that a PM might use to correct a project's plan variance.
4. How do scope change and risk affect project status measurement?

REFERENCES

Krawzcyhk, B. (2024). *What is the Tuckerman Model*, Log Rocket, https://blog.logrocket.com/product-management/tuckman-model-5-stages-team-development/ (Accessed May 12, 2024).

Low, G. (2024). *Management Courses to Skill up in 2024*, The Digital PM, https://thedigitalprojectmanager.com/personal/personal-growth/project-management-courses (Accessed May 12, 2024).

Mind Tools Ltd, (2008). *Coaching for Team Performance: Improving Productivity by Improving Relationships*. Mind Tools Ltd., http://www.mindtools.com/pages/article/newTMM_66.htm (accessed May 12, 2024).

Project Management Institute (PMI). (2021). *A Guide to the Project Management Body of Knowledge*, 7th edition, Project Management Institute, Newtown Square, PA.

Tuckman, B.W. (2001). Developmental sequence in small groups. *Group Facilitation: A Research and Applications Journal.*

Wong, Z. (2007). *Key Stages of Technical Development, Human Factors in Project Management: Concepts, Tools, and Techniques for Inspiring Teamwork and Mo*, New York, John Wiley.

CHAPTER 18

Project Model Reality Notes

This chapter overviews concepts related to handling variability and matching actual behavior. The material introduced here summarizes comments made throughout the text regarding reality gaps in model processes. Many of these comments will be redundant, but the focus here is broader in perspective. The theme throughout this chapter overview is focused on how the project plan can better reflect reality.

18.1 INTRODUCTION

Up to this point in the text, each chapter has focused on discussing a specific tool or management process. The chapters in Section V concentrate on various areas where the model compromises reality. The discussion now crosses over single-process boundaries to examine how a lack of reality match impacts the view of the plan's accuracy.

All of the chapters in Section V will look beyond the traditional waterfall model view, and each will highlight some specific management gaps related to reality or variability. None of the items in this section will present a management solution that will make all projects successful, but each will add a needed twist to improve the traditional view. Chapter 19 began to open the management model views to improve a reality view approach beyond what a typical organization uses. In other words, each chapter discusses a more contemporary view of reality that differs from the

DOI: 10.1201/9781003532989-22

traditional model. After this point in the text, each chapter describes a process or approach designed to better match the reality or variability of the project in some way.

The role of this advanced topic section is to set the background logic for showing a collection of little-used processes, leading to Chapter 23, which will outline a conceptual hybrid model that better deals with this goal and uses some of the earlier gap correction ideas. The key theme throughout this section is to highlight the need for a better model match regarding reality and variability. This chapter will summarize where the current model usage fails to reflect the entire project domain regarding matching reality. Said another way, the current practice of defining a project plan leaves significant gaps in the view shown versus how that result matches reality later. Continually seeing the same maligned result helps to explain why the traditional model needs to be upgraded.

Over the past 70 years, the evolution of project processes and usage culture has left the industry with many management model gap issues beyond what can be completely dealt with here. This section will illustrate some advanced tools and processes that better address the major reality issue. These are not new tools *per se* but tools or concepts not being utilized to the degree they should.

A review of tools and techniques used by project technicians yields an interesting perspective. In most models, the delivery management process consists of static events with simplistic and nonrealistic mechanics. Tools and techniques in areas such as Work Breakdown Structure (WBS), waterfall model, project plans, status tracking, and completion predictions are all essentially static models of a dynamic process. The internal processes related to these items are also more of a static view that ignores the underlying reality or variability. Does this mean the technician thinks a project consists of a family of static components (i.e., tasks, resources, scope, time, and cost)? Surely not! However, for some apparent cultural reason, project management refused to look critically at its processes realistically, which limits the profession's proper evolution.

18.2 GANTT CULTURE

One can trace the origin of current models back to the early 1900s when Henry Gantt's bar charts were first used to portray status. As project management began to evolve in the early 1950s, this crude presentation tool for a project became ingrained in the culture, and it remains so today.

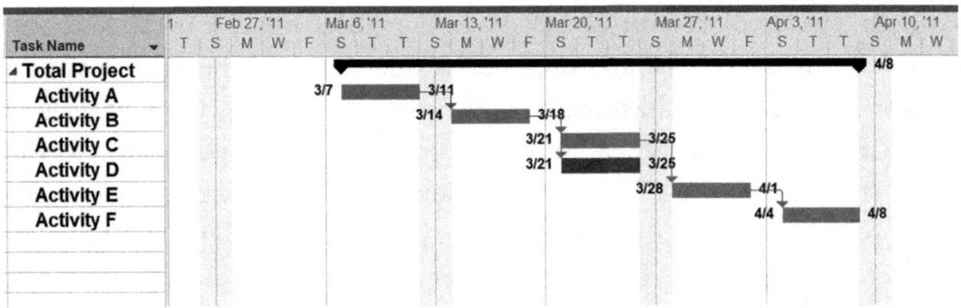

FIGURE 18.1 Classic Gantt chart plan.

Figure 18.1 illustrates a 21st-century computer-generated version of this early 1900s format.

Note that the graphic view of this project is defined by the start and finish dates for all tasks as though there was no variability to be expected. Also, note that none of the non-direct external variable components are shown in this view. This example is the essence of a static project plan. The beauty of the Gantt model is its clarity and simplicity. For example, a quick look shows the planned completion on 4/18. Given the underlying dynamics, one should not expect a deterministic value to always be accurate. The same is true for all other such values in the plan. The presence of a fixed calendar date is a stark reminder of the non-variability view that is embedded in the traditional project culture. Henry Gantt's clear pictures of status will be hard to remove from the current culture. The choice seems to be simplicity versus accuracy or total view. This type of view communicates none of the project complexity issues; therefore, one must question its operational or communication value. It is this oversimplistic cultural syndrome that is the focus of this chapter.

There is a somewhat interesting history story about the Gantt chart (for geeks anyway). This story started around the 1950s when networks were recognized as tools for computing schedules and critical paths. The use of networks did not catch on quickly for various reasons. One hurdle faced at this time was the lack of an existing computation technology to support reality-based concepts, and the manual mechanics to calculate the network schedule values were laborious. As a result of this technological time warp, the Gantt chart perpetuated and essentially became ingrained in the culture as a tool to show status. By the 1980s, small computer technology had invaded the work environment, and crude software calculation engines for planning networks became available. The early software vendors tried to display project plans as networks, but it was too late. The Gantt bar chart was the answer, but the question became how to turn a network into a Gantt chart. Microsoft Project (MSP) was the first vendor to satisfy both the technical and user side. This is the punchline to the story. If you examine the Gantt bar view of an MSP plan, you will see Gantt bars, but it is a flattened network. Notice that the bars have connectors between the tasks. This is a network link and not a classic Gantt bar that did not use or recognize task connectors. The result of the story showed that a Gantt bar had to be part of the visible schedule view, regardless of what was buried underneath. This story illustrates a lot about the project industry culture. It is also interesting to note that one of the display modes of MSP is a network display that is seldom used. Sometimes, reviewing history is a good teacher for introducing new technology. There are similar issues in the current world related to the introduction of improved reality models that fit the story of how Gantt bars have remained for almost 150 years.

18.3 TRADITIONAL MODELS

Traditional model structures in use today are processed through the coded logic embedded in MSP. This code structure strongly matches the original formulation defined by the Project Evaluation Review Technique (PERT) model design methodology from the 1950s, with the Gantt schedule presentation bar added to the view. Recently, there have been efforts to add the agile/SCRUM iterative development model view option to the software, and that model is still evolving.

The traditional model has two major foundation driving elements—the WBS to define the scope and the network task linkage to compute schedules. Both are valuable tools that will likely remain over the strategic time frame and will be the basis for any future evolution of new tools. Few of the tools in the project arena show a static nature more than Gantt charts, but the same logic prevails with all other less visible components.

As modern project management was initially formulated, task variability was recognized in the classic three-time estimate PERT model. This method was designed to identify and quantify the effect of variability. Probability and an advanced statistics theorem were used to convert this into a crude probabilistic forecast of completion (i.e., the Central Limit Theorem if you want to look it up). That view of task estimates was a good introductory example that recognized variability. However, the concept was dropped likely because of its calculation complexity, and it has never resurfaced in usage. The same fate awaited Earned Value as it marginally hung on in usage, supported mainly by the DoD's large project environment. The scheduling network logic described in Chapter 7 was so valuable that it found a way to remain and is the core of today's classic scheduling model. So, the variable time PERT model fell out of usage, and the resulting task model became deterministic. This left the current waterfall model less robust than the original 1950s view. Only the WBS and network scheduling concepts survived. EV and variable task time views were lost, even though both still have a better logical reality match than the surviving static views for task and performance now practiced. By the 1980s, the model view of projects began to focus on the WBS as a scope-defining tool that is essentially used to guide the definition of work tasks. Notice that the typical WBS does not show the external components beyond direct work units, so the work method is not specified other than "do something in this period." All task times are defined with fixed durations, and the completion values are also ultimately defined as fixed calendar dates; this logic cascades to a deterministic forecast completion date. Over time, additional techniques have been added to this static view to band-aid the recognition of variability; the most obvious example is the padding of task times to cover the apparent variability of the estimates. Unfortunately, this practice covers the actual underlying task variances and hides the root cause of the estimating variance.

18.3.1 Project Domain

If one were to ask a practitioner to define the components of a project, the answer would likely mention scope definition, tasks, dates, resource management, change control, various communication formats, etc. Also, the response might mention how difficult it is to define and manage each of these and to produce successful outcomes. Industry surveys point to various reasons for poor project outcomes, and some attempt to rank the reasons in order of frequency. Most responses would recognize that the users who specified requirements did not seem to know what deliverables were needed or how to identify those accurately. However, one might be surprised at the gap in this view compared to the need for a coherent level of general understanding regarding the total project domain. What that means is the total set of processes that make up the total composition of a project.

In reviewing hundreds, if not thousands, of project plans, observing the immature view of the total management picture is surprising. A typical view of the project plan is that its defined work tasks represent the total picture. Often, reality-based external items are not visible in most plans reviewed. We must confess that the sample programs used in the text also suffer from these external component shortcomings, but we have attempted to show these more as the chapters evolved. The following list summarizes the major observed missing plan components:

1. No indication of the support level of effort-type tasks (i.e., administration, PM, technical support).
2. No indication of contingency funds set aside for unplanned risk.
3. No indication of future scope changes that might occur.
4. No buffers in the plan to cover the schedule or cost overrun (maybe the task padding practice is assumed to cover this).
5. Status is measured by comparing current measured values to the static ones shown in the plan (i.e., plan versus actual). Recognition of the beeline concept is often missing.
6. Recognition of scope change was not well-tracked.

In fairness, some of the more mature organizations show references to variance for the static schedule and budget indicated, but this is not the norm. This would be indicated by the term "reserve" to highlight what that item represents.

Figure 18.2 shows an abstract schematic of the project domain outlining various "hidden" non-direct work elements external to the raw core work plan.

The diagram is trying to represent that risk and scope change are two external elements that will impact the work plan later as they emerge.

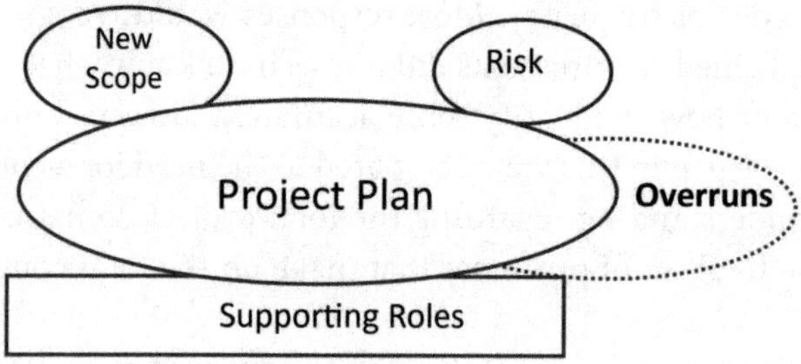

FIGURE 18.2 Project variability elements.

The administrative portion is challenging to show because these supporting elements that may be partially within the host organization represented in the plan budget are impacted by overhead allocations. Other administrative costs are more visible within the project and should be reflected in the plan. Finally, the fourth dotted area represents an expansion of the core project work plan. This area represents the expansion of planned elements, primarily task overruns. It would be dangerous to guess at the magnitude of these four elements as they vary with the type of project and local management culture; however, it is safe to say that they are not a minimal addition to the core. The topic of this section is reality and variability. These are working examples of those often given less attention and plan visibility in the management process.

Management models are great tools for conceptualizing the work required, helping with planning and control, and providing structure for communicating project status. For these to have full value, the overall view of the target components must be visible, including the extra domain objects shown in Figure 18.2. Even in the project core, where the specific direct work tasks are defined, the waterfall management model represents the tasks as fixed duration, with no indication that these values can vary. Showing these on the plan as padded tasks with fixed dates does not reflect their true character. Task padding may arithmetically cover a potential overrun characteristic but not in a coherent way. Every task does not necessarily overrun its estimate; the plan should represent that better. In many ways, the task representation error is not the shortcoming of the model but, more critically, a lack of understanding regarding how to portray the environment.

18.4 A MODEL ASSUMPTIONS VERSUS REALITY

For a management model to help guide the execution process, there needs to be a reality fit between it and the target project. Interesting perspectives emerge in mapping this notion to the current management models. Five core assumptions drive the traditional model. These relate to scope, time estimates, task sequence, risk events, and the definition of success. Comments for each of these defining points are summarized below:

1. ***Scope definition***. The concept of project scope is meant to define the work necessary to execute the desired output. If developed perfectly, the model guides a process to produce the required work and the calendar dates on which it will be performed. This area of the model

has been described as its Achilles' heel, meaning it is the weak spot. Failure of this initial step to accurately specify the project vision cascades through the plan. When performed poorly, the follow-up changes to the original scope bring chaos to the project team through excessive new work and changes to the plan. This statement is like Yogi Berra's (famous Yankee baseball catcher) saying, "If you don't know where you are going, you will end up someplace else" (Mitchell, 2015, para. 3). That well describes the story of poor scope definition.

When projects get into a high variance situation, one reaction is to freeze further changes to minimize further overruns. This action might allow the completion result to improve by meeting the plan forecast parameters, but that raises a critical question. Does the frozen defined work scope satisfy the now recognized and approved current requirement view, and does this curtailment make the resulting output unusable? If it does not, that project will be managed as a failure regardless of the improved outcome measures. Research evidence suggests that this type of action has been used with this result.

Another scope definition-related example occurs when the wrong project is selected for execution, leaving the better alternative without funding. Even if the approved project is perfectly executed, it is still not considered an organizational success because global resources were misapplied.

2. ***Scope specification technique***. The traditional method of work duration estimates comes with a work specification for the deliverable. For example, a physical part must produce 30 units of output according to the specification. If it fails, the item will be rejected, or more work will be applied until it meets that static constraint. However, from an engineering viewpoint, a value of 27 might be usable; 30 is still the target, and higher values would be even better. The subtlety of this idea lies in a more flexible specification scheme that is not fixed. There are many cases when a work unit has the potential for a more varied range of successful outputs that would be useful to know. Some types of work would fit this varied specification approach. The use of this could significantly impact the timing of work units. For example, if the time estimate was overrun, the choice could be made to stop at the 27-performance level. An overrun has been avoided with no harm to the final output. This three-level

specification technique represents a new way to manage work. Even though this might be considered somewhat radical, it is viable from an engineering viewpoint and undoubtedly beneficial from a management point of view.

The three-level specification method is defined in the literature as MoSCoW. This acronym translates to "Must have, should have, and could have." So, a work unit specification of 27, 30, 40 would be interpreted as indicated. This type of variable work execution process is a logical method to improve the schedule performance of a fixed-time plan. Once again, this is a contemporary concept that is not the traditional view but one to think about.

3. ***Task estimates.*** The process of estimating task durations is one of the fundamental components of the waterfall model. Unfortunately, using a deterministic value for this is one of the primary reasons why the model fails to represent reality. Common practice is recognizing that the value may be wrong, so extra time is added to avoid an overrun. At that point, the value has no real meaning. The poor practice of task padding has been referenced many times in the text, and we state it here again. This very logical decision destroys the value of a waterfall plan model and lies at the core of why this practice is not appropriate for proper management purposes. A padded task estimate hides a true plan because it is inserted to cover up overruns. The remaining chapters in this section focus on various aspects of improved methods for managing variability. Projects need to have motivators to execute faster, not relaxed practices that give the impression that more time is available. The resulting plan fails the reality test if a standard non-padded estimate is not the norm.

4. ***Task execution sequence.*** The waterfall model explicitly defines a fixed task execution sequence, and that specification is used to calculate the task network schedule. In some cases, there is a dynamic need to change this sequence later in execution. This is a legitimate management decision; however, the movement of the task sequence complicates the traditional status analysis that is theoretically based on a "frozen" baseline plan view. Changing the task sequence can make the conventional status comparison meaningless. The quandary here is how to choose the best strategy for project control. Is it a better goal to fix the plan for status comparisons or dynamically change it to fit the operational environment? The proper goal of task

management is to allocate the resources most effectively, given the current situation. However, there are very legitimate reasons to move tasks around during execution based on environmental dynamics. The classic example is lousy weather outside when other work can be moved inside. Another common cause is to move tasks based on resource availability. These changes to the defined task sequence affect the model process of status reporting. One way to modify the project status view might be to disconnect the work units from the plan sequence structure and view their status more independently as they are moved into an active state. This would provide a more accurate view of project performances but leave the overall status view undefined. Resolving this issue leaves a challenge for other techniques. Some of the Earned Value performance measurement concepts shown in Section V seem interesting to consider for this situation.

5. **Risk events**. No topic in project management is less well handled than this one. As described in Chapter 16, a theoretical risk management model might be helpful in pre-defining and dealing with potential events during the planning phase. If this process is aggressively used, the net result would be some risk avoidance plan task modification before execution. Also, an associated contingency reserve would be defined to handle those unknown events in the future. Review Figure 18.2. This is the unknown portion of the risk component shown there. Absent the early risk assessment effort, the team is left to wait for these items to occur. If the goal is to match the project plan to future outcomes, the future risk events must be recognized by a contingency fund that will cover the impact resulting from the yet undefined new work to be executed. Adding new work or newly uncovered risk events changes the original plan parameters. This also represents required work even if those task specifics cannot be specified at plan formulation time. Once again, task padding is not the answer to how this future work should be represented.

   Using a formal risk contingency fund is the recommended method to segregate this portion of the project from the original scope-defined work. Using historical data, mature organizations will develop a profile history that can help estimate this aspect of the plan based on the project's profile characteristics. The waterfall model can represent this by showing the risk contingency reserve as a "planning task," meaning an undefined work item with estimated cost and time. This would be attached to the end of the defined work task sequence to show an expanded schedule and a budget. Measuring a specific risk event impact and extracting that amount from the reserve fund may not be administratively viable. However, this approach does not help show an accurate work schedule because these events are random. The compromise could be to use the reserve fund as an estimated cushion and perform the new work without charging. This is not a great answer to the problem, but the plan shows at least the overall risk estimate.

6. **Project success**. Traditionally, project success has been defined by the match level of the three plan output variables: time, cost, and scope (deliverables). However, realistically, there are scenarios where other goals might be of equal or greater concern. The management process has techniques in place to manipulate each of these three variables to change outcomes. But, beyond this, other operational parameters have goals attached as well. Three other internal goals also fit into this question. Priorities related to resources, stakeholder satisfaction, and risk fit the definition of a project management goal, and each of these can be manipulated with various tradeoffs. The key reality question in this scenario is how to best satisfy this more global output goal list in the proper order. Management decisions often have the effect of adding or subtracting the status of these variables to improve focus on the desired one. Adding human resources may improve schedules, deliverables, or customer satisfaction and raise the budget beyond some target level. This might even increase (or decrease) risk. The traditional model looked at the topic of success once again as a static measure. Industry success surveys generally follow the traditional three-variable logic for their grading schemes. With this expanded view of success parameters, the question of success becomes much more complicated. The process would need to be more of a priority rank than an arithmetic one. The most important item to recognize in this overthrow of the traditional view is that the definition of project success is not a plan versus an actual comparison of three parameters. That might be fine in a static world, but this section has been working to show that static views do not accurately portray the management logic required. During the

execution process, every project will get into some level of variability that requires management intervention. The new parameters outlined above are more consistent with that decision process than the traditional three variables. When that situation occurs, a better goal priority definition is needed to guide the decision process for the best recovery. There should be agreement that each project is different in some meaningful way. This means that some formal recognition of this should be part of a priority decision structure that drives both the planning and execution phases of the project. The traditional model assumes that time, cost, and scope are equal goals defining success, but the argument here is that the decision options run deeper than the three traditional items.

Several years ago, a research project compared a slate of projects with variable success outcomes defined by traditional measures. The design purpose of this study was to see what measures counted most over time. Some projects were initially graded as successful, while others were much less. The longitudinal measure evaluated the user opinion of the project versus the success measures. This measurement process of user satisfaction continued for three years, and the expectation, according to traditional theory, was that the "better" performing projects would have higher user satisfaction levels. The surprising result from this was an anti-correlation with the traditional scorecard. In other words, traditional success measures did not agree with the user satisfaction level three years later. It is not a great surprise to note that this research study has not been advertised very much. Some believe it was flawed, while others seem to take a more analytical view. There is one personal thought that continues to emerge. Could fighting to achieve the stated success cut the project size concerning scope change and possibly lesser user involvement? Such a strategy might cause a drift away from the user's input and provide some explanation regarding this research finding. There is no further attempt to explain this unexpected result, but the art of project management must always look for new views that add insight into this complex undertaking. Notably, this research result has not surfaced much in the industry discussion forums, and one might guess that this news was not what the traditional model followers wanted to see. The hanging question related to the root cause of this result remains. Is it poor communication, poor scope definition, or poor change control? Maybe it is all those things, but it is a curiosity worth pursuing. Various survey results show that the management process does not focus enough on the user side of the equation during planning or later phases of the effort. This factor has been identified in project failure surveys as a common source.

18.5 STATUS TRACKING

Various stakeholders want periodic status of projects. Senior management typically asks, "When will it be finished, and at what cost?" Financial areas will ask questions about the budget forecast at project completion. Technical entities may want to know the status and projections for the widget's performance now or at completion. The perceived correct answer is a deterministic number in all these cases. Never mind that any deterministic forecast value of project status will be wrong by definition. This is a reality cultural issue that needs to be corrected. Why plan to produce wrong answers? As an example, Task X is deterministically scheduled to be completed on May 5, and the project completion is scheduled for July 4. All of the task values have a high probability of variability. Also, other operational-level systems contain similar forms of deterministic data (risk and change rates, for example). How can this result in a single deterministic value? Nevertheless, this is the current project culture. This situation is reminiscent of the local TV weatherman who forecasts, "We will have rain at 2:15 PM today." When related to a high variability topic, this type of statement is the essence of the variability problem. There may be a high probability of rain in this period. However, selecting a deterministic value for that event is inappropriate, just as it is inappropriate for forecasting project parameters. When describing such events, the cultural lesson that needs to be learned is how to explain the phenomenon in a suitable language. In the case of a future project completion forecast, the proper answer for this is to represent it in a probability-oriented view. All of the stakeholder questions above fall into that domain. As an example, a forecast for project completion should be formulated as something similar to the following:

> The deterministic estimated completion date is May 5th; however, there is a 70% chance this could extend to June 1st based on variability estimates. This range value is heavily influenced by uncertain supply chain variables.

This topic is a quandary for the contemporary PM. Do you provide the erroneous deterministic answer or try to educate your audience on the true variability of the subject? Perpetuating the current technique leads to distrust of the process since the value provided does not occur. This leads to the question of whether the project team knows what they are doing. Deterministic answers also feed a belief that this process is simple, with no great complexity. Giving an honest answer is the correct approach. If it is known that a deterministic answer is not accurate, why keep doing it and eroding our reputation? The receiver culture needs to be educated regarding the reasons for this new answer format.

18.6 UNCOVERING VARIABILITY

The fact that variability is not more visible in project plans has been traced to the early evolution of management methods and the need for improved models and computation tools. Evolutionary history has led to the current deterministic and static viewpoint regarding the process in all phases. Over the past few years, there has been some recognition that the existing processes need to match reality better, but these efforts have not been well orchestrated or aggressive. As indicated above, the current models have rejected improved variability techniques when they were known and replaced those views with simplified deterministic forms that did not reflect the accurate picture of the target topic. Previous chapter discussions have traced many of these points.

Around the 1970s, the highly successful Japanese quality movement migrated to the United States. One of the concepts that came with this movement was the idea of "lean," meaning minimizing waste. One of the target areas for this was an examination of inventory management practices. The traditional U.S. strategy for this was to have extra on hand to avoid stockouts (and that idea has not entirely gone away today). The quality program view was different for this. The new theme was to have essentially zero inventory and make the process run lean, a concept called Just-in-Time inventory. Taking the padding between stages away forced the process to be managed more efficiently. One key part of the Lean concept that has not made its way into project management is the formalized review of root cause issues. In other words, if something isn't working correctly, what is the source of the variation? One way of looking at this is to counter the variability issue, but in the quality view, the proper focus was finding the source of variability and fixing the underlying process. Variability always needs to be recognized and reported that way, while it needs to be identified and resolved in other cases. This latter view fits many areas that are relevant to project management. For example, if a new Lean-designed inventory system continually ran out of required items, a review step would try to find the cause. In the project world, when a padded task overruns, the solution might be to add more padding rather than seek out the underlying cause. The Critical Chain model theory outlined in Chapter 21 will highlight why this might happen, but there is little active data visibility in today's project world to understand or deal with such overruns.

The goal here is not to relive the old quality theory days but to highlight a classic idea that is part of that successful culture. The sections below will take a few contemporary project management problem areas and see what quality management cultural types of actions would help improve the process.

Many universal project-related questions need serious cause-and-effect analysis. Each of these has potential roots that link back to the theme of reality and variability. Some brief comments for each are offered here in the following four selected questions:

- *What are some of the primary sources of project failure?*
- *Why do projects take so long?*
- *Why can't users make up their minds about requirements?*
- *Why has Agile become so popular?*

Project failure rates. First, numerous textbooks have been written about this subject, and many causal lists exist. One thing that has been learned from this data collection process is that getting individual groups to react to these answers is an ongoing management gap issue. There is one somewhat surprising recurring class of failure-related causes, even though there are many hundreds of reasons mentioned. Surprisingly, four or five items repeat continually. These are related to ineffective support from senior management and stakeholders. Some have translated the consolidated core reason to be "communication." At least, that term is mixed up in many of the items described. For example, scope definition and user needs are collected via some communication channels. More specifically, if it is known that this segment of the project is not utilized correctly, why does it repeat so many times? Remember the learning organization idea mentioned earlier. The repetition of error sources is not characteristic of such a culture. If a common target has been identified, there would be active work to remove that from the list. However, that is not a visible industry movement—rather an excessively static culture. The culture needs to move toward identifying the problem and then taking action to fix it!

Project cycle time. Many factors can be traced to this; no short answer is satisfactory. If we assume the task list is valid, the question becomes how to execute that list quickly. The worker psychology theory described in Chapter 23 will offer more insights into why this is more than a mechanical or technical issue. Also, the reporting success of the agile, iterative school of execution described in Chapter 19 would be argued by some as an improved method of execution focused on enhanced customer satisfaction but possibly not a faster method.

If the chosen project planning and control model environment is the waterfall model, the question of execution speed is more focused on management skills and a better resource control process. Task padding has been mentioned many times as a poor estimating culture, and that comes back to the worker psychology theory again (Chapter 23). One of the major inhibitors of project completion cycle time is scope change, which is the Achilles' heel of the waterfall model. Continually adding new work to the original plan is very disruptive. The management problem with this statement is output accuracy versus completion speed. Scope change is the variable in the middle of this point.

Scope definition. When the question of users being able to define the deliverables accurately arises, it means the lack of success in a scope definition. Part of this issue is a fundamental flaw in the premise of the waterfall model that says project tasks can be clearly defined before execution. That has been proven to be a suspect problem for many projects.

If we look at the process as flawed, this would say that the collection process was ineffective. Some techniques do an excellent job of requirements definition and user review in a collaborative environment. Modern technology offers the capability to have joint sessions with key stakeholders, plus proven technology tools that facilitate voting and idea review. Processes such as this can improve the initial scope accuracy.

The U.S. DoD has long struggled with the question of the successful development of high-technology products. Realize that this environment pioneered most of the existing commercial project practices and should be considered a mature approach. However, only portions of what they do can be transferred to the commercial world. One of the scope definition processes pioneered by DoD to counter scope issues is to contact for a working prototype before pursuing a follow-on production effort based on the winning vendor's product. That dual prototyping approach has become a popular scoping method for high-unit-cost products and might be a productive niche for agile, iterative projects. In that case, the strategy often includes two competitors selected to build their similarly proposed models using modified management and control strategies. This might be characterized as using a performance-based scope definition and allowing vendors to follow a less rigorous control oversight. However, not every organization can afford to follow this double development cost mode, but it does raise the question of how feasible a non-viewable scope definition can be, particularly in high-technology environments. The DoD scope view also seems to view the first iteration as one designed to support multiple future versions with enhanced capability. This approach appears to be a strategy for faster delivery of a usable product with later planned upgrade phases. Remember, a project has no value until some finished component is placed in the hands of the user. This type of phased development logic would seem to have merit because scope change is one of the main negatives of project management. The focus would be more on getting the core capability finished quicker and following that execution cycle with enhancements on a priority basis. This doesn't change the problem with the initial poor scope definition issue but does avoid much of the impact of scope change during execution. The thought behind this is that it is better to have a working version early than to fight for a better scope-defined one later.

18.6.1 Task Padding

One of the most target-rich areas to examine for process errors based on variability can be found in the traditional waterfall model practice of task padding. This is an unusual example because it deals with what would be considered a legitimate variability correction factor. The previous description of the waterfall task view described its flaw in not recognizing that tasks were variable rather than deterministic. On one hand, the padding option seems logical to correct for this, and that is a common practice. This represents a rational attempt to make the project completion parameters come out correctly. On the negative side, the practice makes the execution process look more leisurely with the extra task times. For this reason, there is little internal project team disagreement about the padding practice. However, from a management viewpoint, nothing could be worse. Padding of each task overly expands the project schedule, thus contributing to excessive cycle times. That is, given extra time, the busy human worker will often procrastinate about starting the work. The net effect of this is that the task will still be overrun. As the quality model taught regarding handling excessive inventory, this is a similar issue with task times. And the underlying impact on performance is also similar. The ability to evaluate legitimate work times is compromised, and to compound this, the expanded estimates are still overrun because of the human procrastination aspect. This simple attempt to deal with variability erodes a key process in the management channel.

A second impact of the padding problem comes in status reporting. The traditional approach is to compare a planned value with actual results. In the task-padded culture, measured variations are viewed negatively, while positive variance values represent being ahead of schedule, which projects seldom seem to be. An earlier chapter described Earned Value (EV) as the proper method of quantifying project status. It is a superior technique to the plan versus actual method, but what if it compares padded work units? Is an EV measurement for completing a padded work unit equivalent to the exact measurement for a non-padded one? This is another subtle erosion of reality that seems to be ignored even by experts in the industry.

18.7 RESOURCE GAPS

One of the least tracked management variables is resource availability. This shortcoming affects not only the task execution process but also creates subtle schedule gaps between tasks. The latter is not visible in project management literature as a factor, yet it significantly impacts schedule performance. Multiple chapters have outlined the impact of resources not being available as needed to execute tasks, but the same result occurs when the resource is not available to start a task. The Critical Chain (see Chapter 21) model will best describe this logic.

Resource gaps are defined when the timely availability of required resources to start tasks on time is unmet. A review of the waterfall model structure theory shows tasks are directly linked in a Finish – Start relationship unless there is some special lag gap specified in the predecessor code (infrequent use noted). Related network calculations for tasks clearly show the intent is to finish one task and immediately start the next one, but the typical project environment does not focus on the gaps between tasks as a significant schedule impact from this source. This variability gap is one of the least recorded in the literature. Yet, it may be one of the essential sources explaining why the current model does a poor job of predicting outcomes. This is a strong statement and needs some further explanation. Using an EV calculation as an example, the computed parameters do a superior job of evaluating project cost performance but do somewhat less well with schedule evaluation. For instance, the Schedule Performance Index (SPI) formula has a known schedule-based idiosyncrasy as the project moves beyond the 65th percentile of its life cycle. That explains some schedule prediction errors, but we offer another unobserved one. Suppose SPI values were computed for each task and manually accumulated. Technically, this should define the schedule performance status of each task. The gap would be linked to missing resources if the overall schedule took longer.

The schedule EV formula flaw is well documented. This issue is covered in the literature under the title of *Earned Schedule* and is a well-known correction factor, although somewhat difficult to calculate. There is some logic to suggest that the EV measurement error is heavily influenced by resource gaps and task operational stability. At any rate, the inefficient allocation of resources to tasks is a major performance factor that causes schedule slippage beyond what the traditional status-tracking formulas indicate. One mental view of this slippage is envisioning a group of sequential task boxes showing idle work gaps between them compared to the plan, showing them directly connected with no gaps. Another more radical view of this is to envision shutting the project down after each task and doing nothing for a week. That affects the plan schedule in the same manner. Resource gaps are not represented in the models, but they occur and negatively impact the planned schedule. We will see some support for the idea of tight resource management in the upcoming chapter related to the Critical Chain model. The point of this is to highlight that the current toolset still struggles to catch up in its ability to accurately describe task performance. This also affects the related status-tracking logic because the model does not properly map how the variable task "floats" with no fixed schedule. EV is the most theoretically pure performance measurement tool available for this, but it also struggles in various status interpretation roles when executed using variable task estimates. Chapter 22 will describe a customized model to handle this situation.

18.8 INTERNAL SCOPE VARIABILITY

An earlier section described the impact of inaccurate scope specifications on the project plan. This factor represents one of the main reasons why Agile is so popular. An internal view of this same issue is noted in the model's fixed definition of scope elements. Like most other management processes, scope specifications in the waterfall model are defined with a single deterministic value. The use of a three-level specification called MoSCoW was discussed earlier. Rethinking how scope requirements are defined with variable values is an interesting method to improve task-planned completion values. Once again, the fixed static culture that projects are currently driven by does not support what is needed for a reality view. A more flexible view of requirements would allow a task to have a more flexible completion specification (and date). It will take some time to absorb this idea, but it has merit. Some engineers might frown at this loose view of the technical component specification, but it better supports the management process and helps smooth the task workflow.

All new ideas tend to run into the existing culture that doesn't like them. That is the case with increasing concerns about project variability and reality. Fixed measurement values make everything easy. Fixed task estimates gloss over complexities related to variability. Fixed dates are not confusing. The external user culture is trained to expect fixed answers to all questions. However, we must recognize that the culture of everything fixed dooms the traditional management models' ability to emulate reality. The waterfall model is not precisely correct for every project, nor is the agile methodology correct. Each project initiative should be driven by the specific goals of that project mapped against the delivery environment that will produce that outcome. In some cases, speed is the primary target, while in other cases, satisfying the customer is the most important, etc. Other situations can change the proper way to execute a specific project goal. Too often, the current model environment makes one feel like they are forcing a set of tasks into a pre-defined structure that does not fit the problem. If the existing project has variability, it is hard to simulate it accurately in the current models.

18.8.1 Schedule Culture

References were made earlier regarding the current cultural bias toward fixed dates for various items. Accepting task variability as a needed plan view will affect the scheduling process, but it is unclear how that change will be reflected in a modified management model. Current status processes are so ingrained in mapping the fixed dates that they will come unglued when no fixed control dates or values are defined. There will be no arithmetic comparison possible for a floating task schedule. In this view, a string of tasks "float" through time, and the schedule for that string is not the sum of some fixed value. It is more accurately defined as a probabilistic range. The question then becomes how to analyze status and communicate that to an audience looking for a deterministic value. This is a behavioral challenge to the future management process. That is one of converting it to a more accurate reality perspective. Project models will have to drop the usage of fixed values and resort to a more probabilistic view. This, in turn, wipes out simplistic plan versus actual comparison for status and introduces the need for equally probabilistic status measures. The concept of a single value overrun loses its meaning, as does a fixed date for any event. Although these points seem to be disastrous, these mean the model needs to reflect this variability and adjust how it is managed accordingly. The upcoming chapters offer some insights into this set of issues (but none provide a complete answer). Various models described in this Advanced Techniques section will take on different model perspectives for representing variability. Each tends to break from the current view, but none offers a single or complete, full solution to this problem.

18.9 CONCLUSION

Upcoming chapters will describe unique views on the topic of variability and reality. Each chapter will illustrate some factors that outline how traditional model views must be altered. Chapter 23 describes a hybrid model structure and a conceptual process that attempts to combine these fragmented elements into a single working view. Industry threads are starting to look in this direction to improve project results. Richardson and Jackson (2025) describe more specific details related to this approach in combining existing models into one structure. Regardless of one's view regarding what a project looks like or how it is best modeled, it is time to start thinking about better ways of packaging the requisite work into a more accurate structure and then using that to manage the following dynamics.

QUESTIONS

1. What are some reasons why the waterfall model has endured in the face of other alternatives?

2. Using the waterfall assumptions defined in the text, what are the management issues related to making these accurate?
3. Make a comparative list of advantages and disadvantages for both waterfall and iterative concepts.
4. What are some of the flaws in the traditional plan versus the actual approach to status tracking?
5. Why is the project industry not more proactive in finding better solutions?
6. Observe a local project (road, building, etc.) and comment on the management gaps you observe.

REFERENCES

Mitchell, H. (2015, May 12). *Yogi Berra dies at 90: Here are some of his greatest quotes.* Los Angeles Times. https://www.latimes.com/sports/sportsnow/la-sp-sn-yogi-berra-turns-90-quotes-20150512-story.html

Richardson, G.L. and Jackson B.M. (2025). *Optimizing Project Work, Management, and Delivery*, CRC Press. Boca Raton, FL.

V

Advanced Techniques

CHAPTER 19

Project Iterative Concepts

Project N Motion Videos for This Chapter

19AG	Agile Model	Agile Overview

19.1 INTRODUCTION TO ITERATIVE CONCEPTS

The goal of this chapter is to review Project Iterative Concepts. PMI (2021) states that "an iterative approach is useful for clarifying requirements and investigating various options" (p. 36). It will often produce acceptable, defined interim deliverables before the final deliverable. An iterative approach is "a development approach that focuses on an initial, simplified implementation then progressively elaborates adding to the feature set until the final deliverable is complete" (PMI, 2021, p. 241). Agile is the classic technique used to define the iterative approach to development.

There is evidence that this method has merit based on usage statistics and the explosion of many dialects spawning from the original agile view. Iteration is "a timeboxed cycle of development on a product or deliverable in which all of the work that is needed to deliver value is performed" (PMI, 2021, p. 241). Similar approaches are labeled as Scrum and Kanban, but there are many other slight modifications in practice, as later depicted in

DOI: 10.1201/9781003532989-24

Table 19.1. It is fair to encompass this group as the iterative approach to development. PMI defines the iterative approach as "a development approach that focuses on an initial, simplified implementation then progressively elaborates adding to the feature set until the final deliverable is complete" (p. 241). The fundamental concept fits into what one might call *lean thinking* because it shares classic lean concepts such as respecting people, focusing on delivering value, and adapting to change. Agile emerged quickly but was not entirely embraced by those following a more traditional method, where software developers followed the waterfall method of predefined requirements. In that prior approach, project tasks are worked sequentially, making it challenging to achieve flexibility in the design. Only time will answer how these two major schools of thought will converge or if a new process view will encompass both.

Agile methodology has continued to evolve, with significant updates and emerging trends reflecting changes in technology, workplace culture, and organizational needs. Summarized below are some key developments:

- Expansion beyond IT: While Agile began in software development, it is now widely adopted across sectors like healthcare, finance, education, and manufacturing. Non-IT departments, such as HR and marketing, leverage Agile to improve collaboration, flexibility, and speed in project delivery (StarAgile, 2024; OKR Institute, 2024; KnowledgeHut, 2024).

- Scaled Agile Frameworks (SAFe): As organizations aim to implement Agile across larger teams and departments, frameworks like SAFe and Large-Scale Scrum (LeSS) are gaining traction. These methods help maintain agile principles while addressing the complexities of scaling across large projects (StarAgile, 2024; OKR Institute, 2024).

- Integration with Artificial Intelligence (AI) and Machine Learning: AI is being used to enhance agile processes by automating repetitive tasks, predicting risks, and providing data-driven insights. These advancements optimize workflows, improve decision-making, and accelerate delivery cycles (OKR Institute, 2024; KnowledgeHut, 2024).

- Rise of Hybrid Models: Hybrid frameworks combining Agile with traditional project management approaches like Waterfall are emerging to manage complex projects requiring flexibility and predictability. This trend addresses the challenges of diverse project requirements (OKR Institute, 2024; KnowledgeHut, 2024).

- Remote and Distributed Agile Teams: The shift to remote work has driven the development of tools and practices supporting collaboration among distributed teams. These include enhanced real-time communication platforms, virtual whiteboards, and cloud-based project management tools (OKR Institute, 2024; KnowledgeHut, 2024).
- Emphasis on Customer-Centric Design: Agile is increasingly aligning with principles like Design Thinking to prioritize user needs, focusing on continuous feedback, and delivering business value beyond just product features (StarAgile, 2024; KnowledgeHut, 2024).
- Agile for Digital Transformation: Agile methodologies are being used to drive digital transformation across industries, enabling businesses to adapt rapidly to technological advancements and evolving market demands (KnowledgeHut, 2024).

These examples show that the agile concept has expanded beyond its initial starting place. The concepts have produced higher customer satisfaction in industries outside of software development, such as government and financial services. An iterative development focuses on small increments (chunks) of work at a time. Therefore, the project team can gain customer feedback and easily change what they work on next according to how well the previous work met the customer's needs. This helps to conquer the uncertainty surrounding project requirements. This approach has been found to reduce waste and rework because the project team better understands customer requirements, all while responding quicker and more accurately than using a complete project scope plan of written specifications. The concept of iteration can be viewed as a mindset of customer collaboration using the ultimate deliverable(s). These short work cycles focus on the teams' ability to respond quickly to the changing needs of the customer/end-user based on their iterative review of products to examine. In this mode, the team responds efficiently and effectively to scope changes.

Agile is "a term used to describe a mindset of values and principles outlined in the Agile Manifesto" (PMI, 2021, p. 235). The agile methodological approach was described in the Agile Manifesto. Four high-level value statements are included that set the tone for the interactions, and 12 principles lay out more specific operational steps. The sections below will elaborate on these classic design specifications, encompassing current approaches. Remember that "agility is a wide mindset broader than a development framework; agile approaches can be considered adaptive" (PMI, 2021, p. 38). Therefore, the term is sometimes used to describe similar iterative concepts where iterations in design are short, such as one to two weeks in duration, where the project teams are engaged with a demonstration of their accomplishments at the end of each iteration. The planned scope of what will be accomplished during an iteration is determined through a project team reviewing the prioritized backlog, estimating the work involved, and collaborating throughout the iteration. Beyond a high-level view of the underlying theory, a related goal of this chapter will be to provide a review of the breadth of approach and vocabulary regarding how various entities are embracing Agile.

19.2 AGILE MINDSET AND ENVIRONMENT

The agile design premise uses iterative and incremental approaches to produce deliverables. In 2001, a small group wanted to describe a method to change the project management environment by making software development easier (Hilbert, 2017). Software development has many challenges, from working with unknown technologies to tight deadlines. From this meeting, the Agile Manifesto was born, highlighting four values. The first value focuses on "individuals and interactions over processes and tools" (Productboard, n.d., para. 1). Given this focus, the team will be more responsive to changes, explicitly meeting customers' needs. Focusing on people also instantly increases customer communication instead of just at scheduled intervals throughout the project life cycle, as is more the case in a traditional model.

The second value is "working software over comprehensive documentation" (Productboard, n.d., para. 1). Traditional project management focuses on documentation, and this value statement is not to say it is unnecessary because it is. However, Agile focuses on working deliverables instead of predefined static documentation creation such as technical specifications, interface design documents, test plans, and approvals. These artifacts are not eliminated but instead are streamlined so that this activity does not delay the team in producing actual deliverables. The type of documents that Agile focuses on are *use cases*, *user stories*, etc., which are much less formal and needed to drive development, such as which new features to add. A use case describes situations regarding how the output will be used. These aim to help get developers into a correct and realistic mindset. These somewhat prioritized mini snippets of requirements provide a very high-level scope definition. Each user story gives developers enough information to identify the time estimate needed to implement it.

The third Manifesto value is "Customer collaboration over contract negotiation" (Productboard, n.d., para. 1). This value describes how customers are engaged throughout the development process, instead of initially and after development, with most negotiations occurring before development. With Agile, the customer actively collaborates during the entire project life cycle. In this mode, Agile focuses on flexibility and customer engagement throughout product development to meet customer needs more readily and easily. With Agile, customers are an active part of even daily meetings and interactions, so when questions arise, the team can dynamically incorporate the needs into the ongoing design instead of later when costly rework would be required.

The fourth Manifesto value is that the project team must be flexible and quick in "responding to change over following a plan" (Productboard, n.d., para. 1). Traditionally, the software development process developed detailed plans outlining what a deliverable would do and then viewed changes during the project life cycle as something to avoid since they were often expensive to incorporate. The Agile, iterative approach is the opposite, where changes are welcomed in each work segment. There are small periods between design iterations, and it is easy for a project team to dynamically make design changes, such as adding customer-required features. Agile views change as a way to better meet customers' needs.

Within the four core values are 12 more specific *Manifesto principles* that drive the process (Beck et al., 2001):

1. Our highest priority is to satisfy the customer through early and continuous delivery of valuable software.

2. Welcome changing requirements, even late in development. Agile processes harness change for the customer's competitive advantage.

3. Deliver working software frequently, from a couple of weeks to a couple of months, with a preference for the shorter timescale.

4. Business individuals and developers must work together daily throughout the project.

5. Build projects around motivated individuals. Give them the environment and support they need and trust them to do the job.
6. The most efficient and effective method of conveying information to and within a development team is a face-to-face conversation.
7. Working software is the primary measure of progress.
8. Agile processes promote sustainable development. The sponsors, developers, and users should be able to maintain a constant pace indefinitely.
9. Continuous attention to technical excellence and superior design enhances agility.
10. Simplicity—the art of maximizing the amount of work not done—is essential.
11. The best architecture, requirements, and designs emerge from self-organizing teams.
12. At regular intervals, the team reflects on becoming more effective, then tunes and adjusts its behavior accordingly. (para. 1)

The agile development environment does not have a predefined organizational structure with typical lines and boxes. It is more like multiple circles containing agile teams around an organization's leadership. With agile teams, servant leadership facilitates a team's activities (PMI, 2021). Servant leadership is "the practice of leading the team by focusing on understanding and addressing the needs and development of team members to enable the highest possible team performance" (p. 250). This involves communication and coordination with everyone, including listening, coaching, and helping team members grow. This approach allows teams to learn and become more capable. This type of leadership produces a collaborative environment for the team and leaders who serve as bridge-builders for everyone around them in solving problems and learning new skills. The approach to work is to establish a priority order of the purpose, people, and process. This leads to faster goal definition and work completion. This type of focus creates a team dynamic where everyone can successfully contribute to the project work. Also, this approach focuses on results-driven in terms of timely value delivered to the customer.

Design requirements for an agile life cycle are dynamic instead of fixed, changing as the customer drives. Agile involves repeating activities until

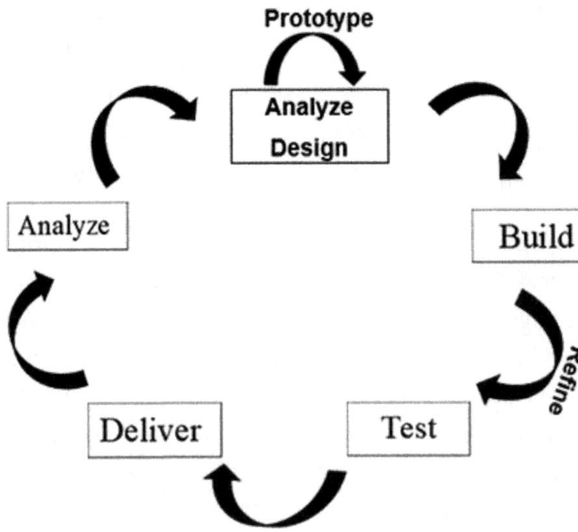

FIGURE 19.1 Iterative life cycle.

Adapted from ISE Tech, n.d.

correct with frequent small increments of deliverables (see Figure 19.1). Agile aims to achieve customer value through frequent deliveries and feedback. The life cycle includes repeating activities involving identifying requirements that the project team expects to change. This is executed by analyzing, designing, building prototypes, constructing the final objective, and incremental testing until the design is judged to satisfy the requirement.

In this development model, cross-functional team members with a combination of skills are needed to produce the product. These skills align with the iterative life cycle so that those on the team can carry out all the life cycle components. A product owner will also be responsible for working with the team daily to drive the team's direction regarding which product functionality to work on next. The product owner works with all stakeholders to provide transparency into the activities of a team and direct the team to focus on providing value for the customer. It is understood that there will be changes that the team will need to make to the product requirements as the features desired by the customer evolve. There will also be a team facilitator who is the team's servant leader. This role has different names depending on the dialect of the agile methodology used, but some of the common names are *agile coach*, *Scrum master*, etc. It is essential for the physical workspace for the agile team to be formally created. There will need to be a mixture of collaborative workspaces for the entire team and areas where the team members can work individually and uninterruptedly. Therefore, the desired workspace will combine private offices and group social areas to support collaborative work. Depending on the location of team members, support tools such as video conferencing and shared virtual space are also needed for storing project documentation.

Like other project teams, agile teams have a defined process to start, such as needing formal management approval for the effort. The minimum starting trigger an agile team needs is formal management approval for the effort and a project charter to provide the high-level project vision, purpose, release criteria, and intended workflow (PMI, 2021). The servant leader can facilitate the process of creating the project charter. The organizational structure for agile teams can be highly project-oriented, matrixed, or highly functionalized, depending on the organization.

19.3 AGILE METHODS AND PRACTICES

The Digital.ai (2023) 17th Annual State of Agile 2023-2024 Report shows that Agile adoption is prevalent across industries, with 71% of respondents using agile practices in their Software Development Life Cycle (SDLC). Small organizations report the highest satisfaction with Agile, while larger companies face challenges in scaling due to complexities related to hybrid work and digital transformations. Notably, 63% of teams favor the Scrum methodology, maintaining its status as the most popular agile framework. The top reasons for adopting Agile were (multiple reasons accepted):

- 60% of respondents cited enhanced collaboration as a critical benefit.
- 57% noted better alignment to business priorities, ensuring teams focus on delivering the most critical outcomes.
- Around 25% recognized improvements in software quality as a direct advantage of using Agile.

The top benefits of adopting Agile, as listed by Digital.ai (2023), are as follows:

- 60% Improved Collaboration and Communication: Agile fosters better teamwork and interaction across departments, breaking down silos.

- 57% Better Alignment with Business Priorities: Agile practices help ensure teams are focused on the most critical goals that drive business outcomes.
- 53% Increased Productivity: Agile enables teams to work more efficiently by adapting to changes and prioritizing tasks effectively.
- 47% Improved Transparency: Agile practices, such as regular standups and progress tracking, enhance visibility into projects for all stakeholders.
- 46% Increased Team Morale: Teams report higher engagement and satisfaction due to Agile's flexibility and autonomy.
- 25% Higher Quality Deliverables: Agile promotes continuous improvement and iterative testing, which leads to better software quality.

The top challenges that organizations face in adopting Agile, according to Digital.ai (2023), were as follows:

- 46% Resistance to Change: Many teams and leadership struggle with the cultural shift required to embrace agile principles and practices fully.
- 41% Insufficient Leadership Participation: A lack of executive support and involvement often hinders Agile adoption, as leadership buy-in is crucial for driving organizational change.
- 40% Inconsistent Processes and Practices: Variations in Agile implementation across teams lead to inefficiencies and misalignment.
- 35% Challenges in Scaling Agile: Larger organizations encounter difficulties scaling Agile across multiple teams due to complexity and silos.
- 34% Lack of Skills or Knowledge: Teams often lack the necessary training or expertise to implement Agile effectively.
- 26% Inadequate Tools for Collaboration and Tracking: Many organizations report that their current toolsets are insufficient for scaling agile practices.

A sample of several agile dialects is listed in Table 19.1, which briefly explains each of these methods and practices. When starting with an iterative approach, a product owner has a development team that works to produce value in small increments for the customer (Development Pays, 2017; Straughan, 2019). With each release, the customer provides feedback to the product owner, along with feedback from the stakeholders. This feedback is added back into the process by being prioritized in the product backlog, essentially a wish list of requirements for the product.

19.3.1 Scrum

The Digial.ai (2023) industry survey indicates that the Scrum dialect is the most popular approach. So, let us examine a typical Scrum team. This begins with a project team, including stakeholders and the customer, to create a prioritized wish list of requirements that a product such as software should have. This includes features, functionality, etc. (Scrum Alliance, 2015). Then, the process enters *sprint planning*, where the team outlines how to work on the top-prioritized wish list items; an example is shown in Figure 19.2. This planning meeting is facilitated by the Scrum master and attended by the customer and development team.

Task Name	Customer Priority
Sprint 1	
Feature 1	High
Feature 3	High
Feature 5	High
Sprint 1 Complete	
Sprint 2	
Feature 4	High
Feature 7	Medium
Feature 8	Medium
Sprint 2 Complete	
Sprint 3	
Feature 6	Medium
Feature 9	Low
Feature 10	Low
Sprint 3 Complete	

FIGURE 19.2 Sprint example.

Here, high-priority items are selected from the backlog and placed on the *sprint backlog* (Development Pays, 2017; Straughan, 2019). This planned-out work effort becomes the sprint backlog as in the prioritized wish list of work to perform and is the only item worked on during the sprint (Scrum Alliance, 2015; Development Pays, 2017). The team continuously assesses the value of the defined list of work on a project to determine whether these backlog items need to be adjusted (Francino, 2019). The focus of the discussion is on the prioritized backlog items. Items not at the top of the list are not discussed because other items may jump ahead, so there is no need to discuss items not at the top.

The next step with Scrum is to initiate the sprint, which is defined as a short period to produce the work defined on the top-prioritized wish list. Figure 19.3 provides a visual example of a sprint cycle. The *Scrum master* keeps the team on task through a daily brief Scrum meeting to ensure the work moves along. This meeting is typically limited to 15 minutes to identify progress and blockers (Development Pays, 2017; Straughan, 2019). A daily status board is used to communicate status. These are called a *Scrum board*, an *Agile board*, or a *Kanban board*. This board format is similar to Figure 19.4, except the content on the board only lasts as long as the sprint. PMI (2021) defines the Kanban board as "a visualization tool that shows work in progress to help identify bottlenecks and overcommitments, thereby allowing the team to optimize the workflow" (p. 241). At the end of the sprint, the deliverable should be something potentially usable or tangible. Completed work could be packaged for release, and any uncompleted work returns to the product backlog (Scrum Alliance, 2015; Development Pays, 2017). The sprint ends with a *sprint review* consisting of demonstrations to the customer of new features of the product

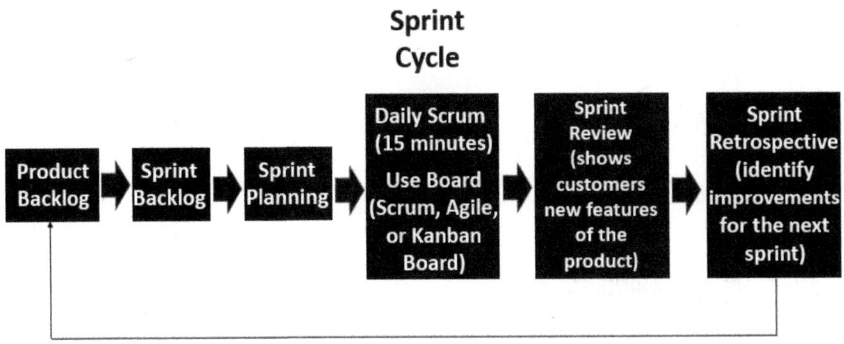

FIGURE 19.3 Example of a sprint cycle.

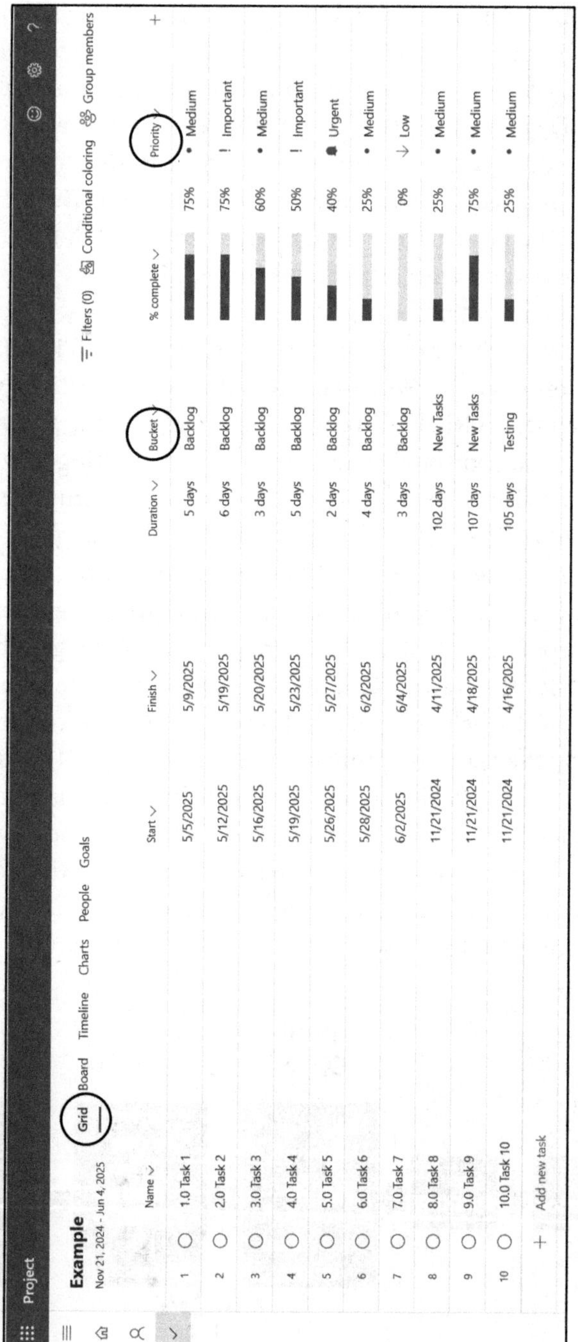

FIGURE 19.4 Grid view for sprint tasks.

(Development Pays, 2017; Straughan, 2019). After the sprint review, there is a *sprint retrospective* where the team discusses what went positive or negative during the sprint and identifies improvement targets. The sprint retrospective aims to help the next sprint be more efficient than the one before. The team then enters a review and retrospective phase (Scrum Alliance, 2015). From this, another *chunk* of the product backlog is selected, and the entire cycle repeats. With this active communication process, the team stays well informed on short-term items they are working on and is also assured that these are the most valuable aspects of their work.

Different agile-related tool features are offered within the spectrum of Microsoft Project (MSP) software versions. Figures 19.4 through 19.7 are from the MS Planner Plan 1 version. Figure 19.4 displays the grid view where project-defined task tools include placing tasks within various buckets that can be customized (i.e., Backlog, New Tasks, Testing). Furthermore, the user can identify the customer's priority of the different tasks (i.e., low, medium, important, and urgent). When a user clicks on a task, a pop-up window allows further description of task-specific details (see Figure 19.5). A sample board view is displayed in Figure 19.6 to show the category in which different tasks are assigned. Tasks can be moved with the cursor to change their work category. Also, new customized buckets can be created by clicking on the "Add bucket" menu item. Furthermore, the "Charts" view is another visual to assist the PM and team with the remaining tasks (i.e., not started, in progress, and late) and the tasks already completed. This is displayed in Figure 19.7, where two different formats are available to show status and category buckets.

19.3.2 Kanban

Kanban is another increasingly recognized dialect that utilizes similar aspects of the agile methodology (PMI, 2017). A Kanban team starts with *visual cards* containing a product backlog item, which then gets placed on the *Kanban board*, similar to Figures 19.6 and 19.8. These items are placed in a column on the board to represent where that item resides within the workflow. For example, it could be in the first column to represent new work or the last column to represent completed work. There will also be different columns in between to represent work in process. Figure 19.8 represents one example of columns on the Kanban board created in MSP Professional. This view is designed to communicate the work status in the

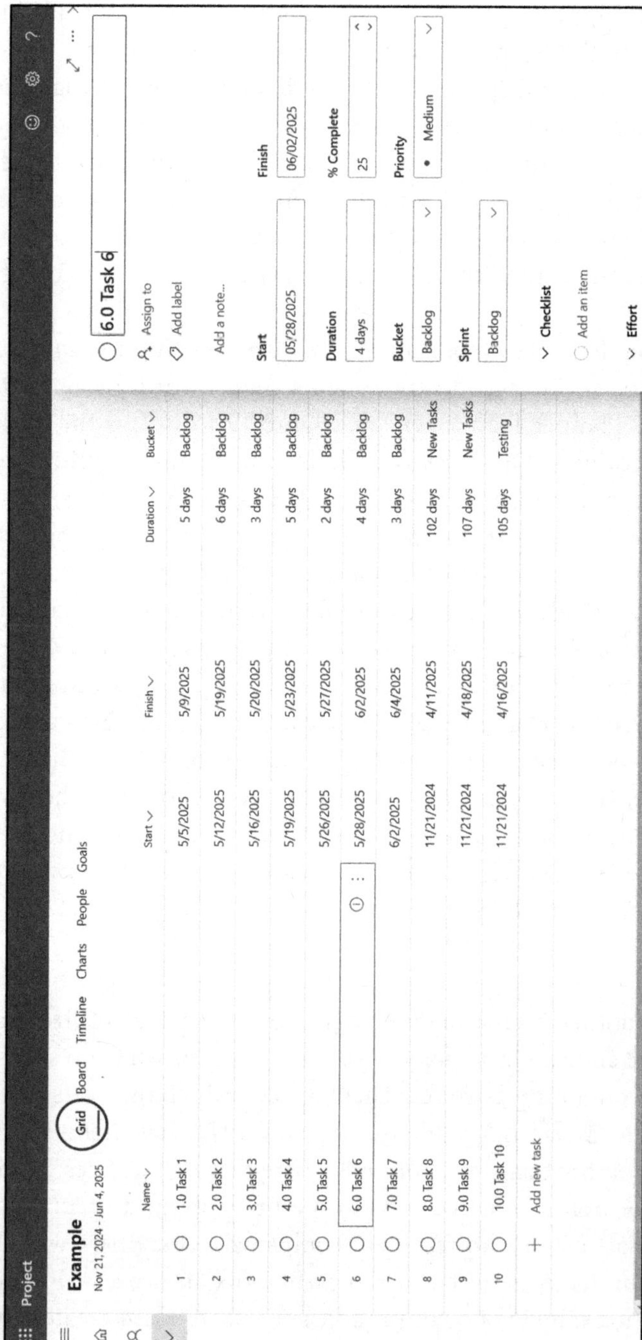

FIGURE 19.5 Grid view of sprint task details.

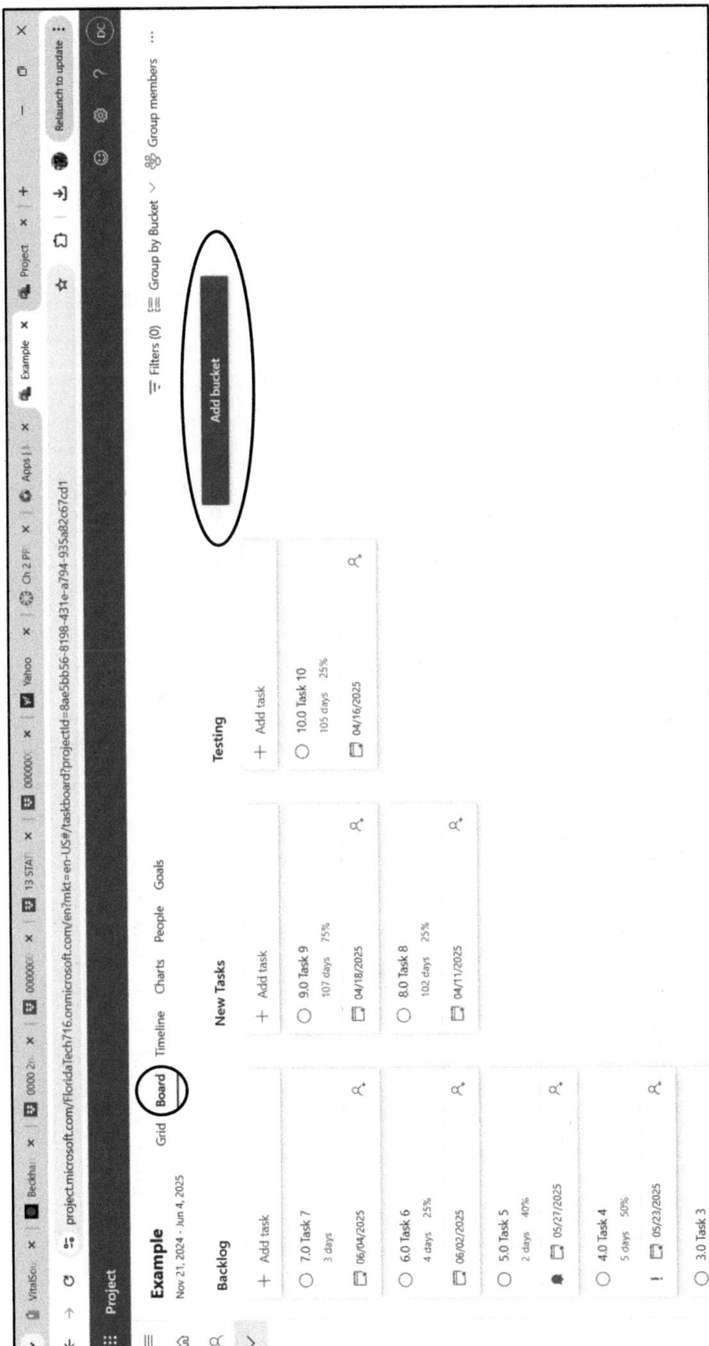

FIGURE 19.6 Board view of sprint tasks.

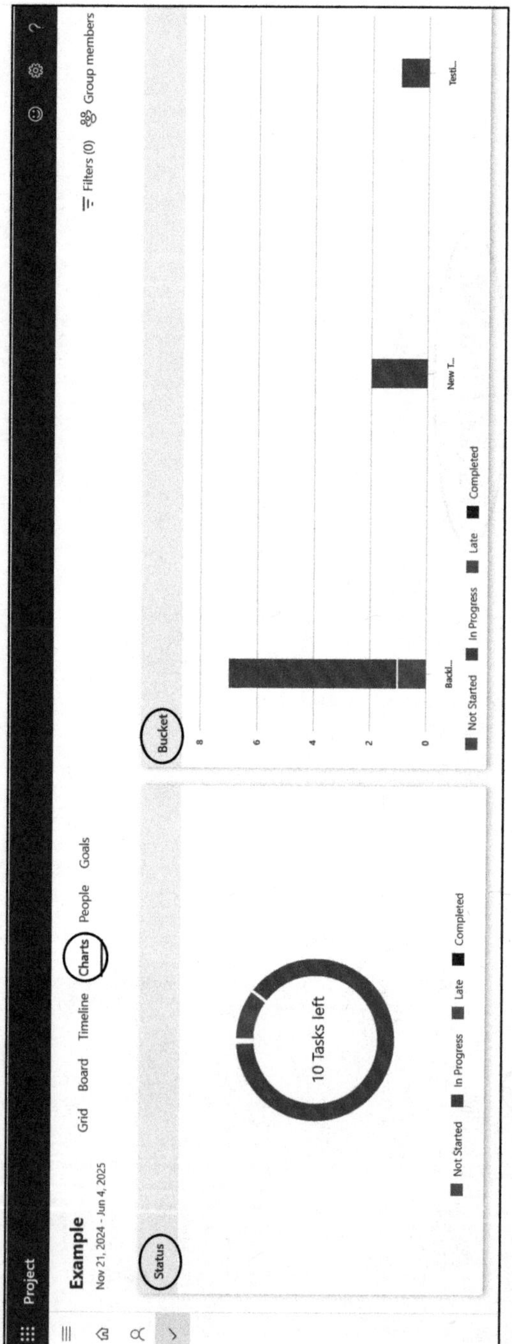

FIGURE 19.7 Charts view of sprint tasks.

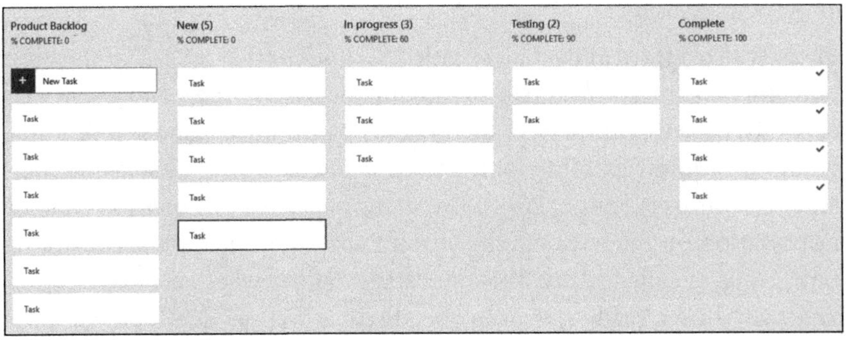

FIGURE 19.8 Kanban board example.

flow, whether it is new work, completed, etc. The cards move to the right to represent the completion of a work step until they reach the endpoint to show delivery to the customers. The Kanban method includes constraints for the number of cards residing within any status column. This number is called the work in progress (WIP) limit for a column. New cards cannot move into the column once a column has reached the WIP limit. Limiting the size of the WIP is designed to keep the overall work effort within manageable boundaries. Therefore, the team will focus only on the work in the column at the WIP limit. In this case, when a column is at the WIP limit, it serves as a bottleneck, perceived as a warning system to the team. This also helps to keep the team on task instead of constantly multitasking, switching, and introducing more work. Multitasking is one of the significant team productivity issues, and this is an excellent way to keep the total defined work under better control. This attribute is one of the primary reasons why the Kanban approach is also receiving increased attention from contemporary project teams.

Scrum and Kanban dialects are *pull systems* that ensure work moves from the product backlog to the customer in the shortest possible time (Development Pays, 2017; Straughan, 2019). However, these methods differ in what happens between the product backlog and the customer. The Scrum master role is similar to a traditional PM, while in Kanban, this is more of a coaching role. Scrum typically uses sprints of two weeks in length, and Kanban is more of a continuous backlog workflow with WIP limits at each stage related to the team's capacity to execute that phase.

19.3.3 ScrumBan

There is also a third agile dialect called *ScrumBan* that takes features from both the Scrum and Kanban methodologies. *ScrumBan* approaches the effort as a fixed requirement for work to be produced under a deadline, but it also allows flexibility for the team to decide how they schedule work and WIP limits (Clayton, 2018). The status board for this method contains information on the backlog: the current sprint is called *on deck*, and the final group is called *ready*. This final stage defines what needs to be specified and what already has been specified, as well as development tasks, pending testing tasks, testing tasks, and deployment tasks. This process defines work being organized into sprints with daily *Scrum meetings* to ensure that work is continuously focused on the dynamic product requirements and reduces team members' idle time. The team collaborates daily through meetings, so people work together to resolve any impediments. The team has established triggers to decide which work will take place next.

19.3.4 Summary of Agile Dialects

As shown in Table 19.1, a fundamentally similar approach is used in each of the defined agile dialects.

TABLE 19.1 Agile Methods and Practices

Type of Methodology	Description
Scrum	Scrum is an agile framework using fixed-length iterations (sprints) to deliver incremental, iterative improvements, emphasizing roles, ceremonies, and artifacts to manage projects and optimize team performance (Schwaber & Sutherland, 2020; Cohn, 2010).
Hybrid	Hybrid Agile refers to the blending of multiple agile frameworks, such as Scrum, Kanban, and Lean, to suit the specific needs of an organization or project, integrating elements that best fit its workflows and goals (Schwaber & Sutherland, 2020; Cohn, 2010).
Scrumban	Scrumban is a hybrid methodology that combines Scrum's structured sprint planning with Kanban's continuous flow and visual task management to allow teams more flexibility and focus on continuous improvement (Keup, 2021; ProjectManager, 2021).
Kanban	A visual management system that optimizes workflow by limiting WIP and focusing on continuous delivery. The system is flexible, without fixed timeframes like Scrum, and aims to reduce bottlenecks and improve efficiency by visualizing tasks on a Kanban board (Cohn, 2010).

(*Continued*)

TABLE 19.1 (Continued) Agile Methods and Practices

Type of Methodology	Description
Iterative Approach	"A development approach that focuses on an initial, simplified implementation then progressively elaborates adding to the feature set until the final deliverable is complete" (PMI, 2021, p. 241).
Spotify Model	Spotify is a music, video, and podcast streaming company that organizes its company into agile teams called squads that have successfully maintained an agile mindset without sacrificing accountability (Mankins & Garton, 2017). Following their model in scaling Agile beyond the team to a larger organization is called the Spotify Model.
Lean Startup	A method that focuses on reduced product development cycles using iterative product releases.
Extreme Programming (XP)	This method delivers high-quality software quickly by practicing continuous feedback, frequent releases, and techniques like pair programming and test-driven development (Beck et al., 2001). It emphasizes values like simplicity, communication, and customer involvement in decision-making. There is also a hybrid version that combines both Scrum and XP methodologies.

19.4 AGILE TECHNIQUES

Various agile techniques use different management and technical approaches to structure work techniques. This section will address some of the more common ones. There is the *daily Scrum* stand-up, where the purpose of this activity is not status management but rather to help the team self-organize work (Francino, 2019). This involves a very short meeting, not to exceed 15 minutes, held simultaneously daily. Team members share what they did the day before and what they are working on for the current day. Challenges that might be encountered are discussed. Another common term is *retrospective*, which occurs when the team reflects on how to be more efficient and effective. This is designed to support a continuous improvement goal. Table 19.2 summarizes other techniques found across the agile landscape.

Various vocabulary terms are a standard part of the general agile vernacular. One of the main terms is backlog, which is "an ordered list of work to be done" (PMI, 2021, p. 235). This is a prioritized wish list for the software or product. Another term is the *user story*, defined as "a brief description of an outcome for a specific user, which is a promise for a conversation to clarify details" (PMI, 2021, p. 252). User stories tend to be brief and simplified descriptions of the desired feature from a user perspective.

TABLE 19.2 Agile Techniques

Type of Technique	Description
Daily Stand-up	Short daily meetings for teams to organize themselves (Francino, 2019).
Sprint/Iteration Planning	"A meeting to clarify the details of the backlog items, acceptance criteria, and work effort required to meet an upcoming iteration commitment" (PMI, 2021, p. 241).
Retrospective	"A regularly occurring workshop in which participants explore their work and results to improve both the process and product" (PMI, 2021, p. 247).
Sprint/Iteration Review	"A meeting held at the end of an iteration to demonstrate the work accomplished during the iteration" (PMI, 2021, p. 241).
Short Iterations	A short period, typically two weeks or less, to develop a deliverable.
Release Planning	A very high-level plan contains a guideline for multiple Sprints with expectations about the timeline for features that will be implemented (International Scrum Institute, 2019).
Planning Poker/Team Estimation	Identifies numerical values for points estimation of a user story that involves all team members holding playing cards with their estimates on them (Cohn, 2010)
Kanban	A visual management system that optimizes workflow by limiting WIP and focusing on continuous delivery. The system is flexible, without fixed timeframes like Scrum, and aims to reduce bottlenecks and improve efficiency by visualizing tasks on a Kanban board (Cohn, 2010).
Dedicated Customer/ Product Owner	This person is deemed responsible for the final product.
Single Team (Integrated Development and Test)	Integration testing is a vital testing technique to ensure that the parts of all the software and hardware components fit together, resulting in more value being delivered to the customer.
Frequent Releases	Agile teams frequently release a product into the hands of end users and obtain feedback.
Common Work Area	Agile teams tend to have common areas where an entire team can fit in the same space, with smaller rooms surrounding the shared workspace for when team members need to work in smaller groups.
Product Road-mapping	"A high-level timeline that depicts milestones, significant events, reviews, and decision points" (PMI, 2021, p. 248).
Story Map	"A visual model of all the features and functionality desired for a given product, created to give the team a holistic view of what they are building and why" (PMI, 2021, p. 250).
Agile Portfolio Planning	The planning activity is to identify which products should be worked on and in which order and period.
Agile/Lean UX	"A term used to describe a mindset of values and principles as outlined in the Agile Manifesto" (PMI, 2021, p. 235).

Figure 19.9 displays a sample template that helps to structure and draft user stories. In practice, these are often written on index cards or sticky notes stored on a shared display board where the team can use the data for planning and discussion purposes. *Personas* is another tool that helps a team understand more about typical users for a designed deliverable. A persona shows a picture of a person along with demographic information to describe a typical day for this person. Figure 19.10 is a sample persona template.

There are various metrics used to define success. These metrics are somewhat different from those used for traditional projects. Because the process of defining work differs from the traditional model, different status views are observed with the agile approach. Table 19.3 shows sample metrics that agile teams may use to measure status. The measurement of

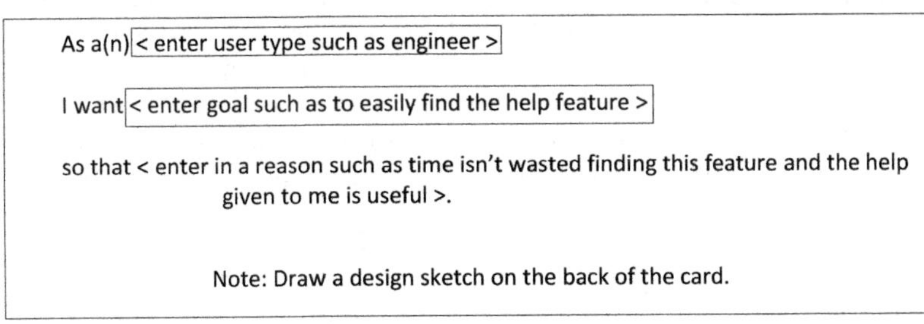

FIGURE 19.9 Template for user stories.

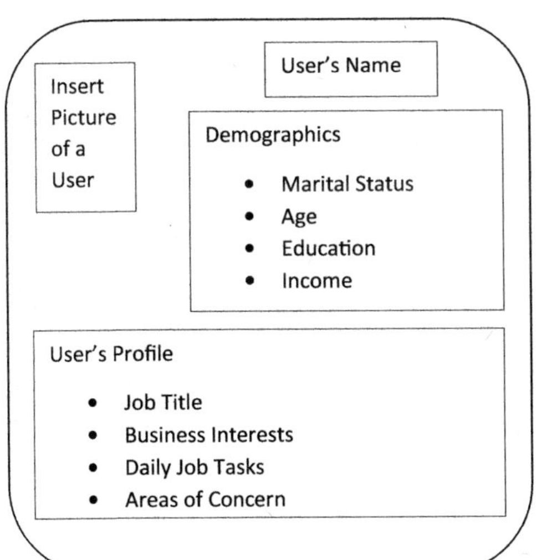

FIGURE 19.10 Template for personas

TABLE 19.3 Agile Metrics

Metrics	Description
Customer/User Satisfaction	Metric to measure whether project deliverables meet customers' expectations, such as sales figures, usage statistics, etc. (Hashemi-Pour & Chai, 2023).
On-time Delivery	Metric to measure whether project deliverables were delivered on time with the expectations about what will be delivered (Badgley, 2019).
Business Value	"The new quantifiable benefit derived from a business endeavor that may be tangible, intangible, or both" (PMI, 2021, p. 236).
Quality	Metrics include measuring maintainability, portability, functionality, performance, compatibility, usability, reliability, and security, but can also measure delivery models for software, such as delivering what was committed at the start of a sprint (Ghahrai, 2018).
Productivity	A measure can be the count of stories or features over time (Badgley, 2019). The number of story points committed versus those completed in a sprint can indicate a scope issue (Ghahrai, 2018).
Predictability	Metrics that measure how likely a team will be able to complete work and consistently perform work. A metric could be how much work has been completed sustainably on average (Badgley, 2019).
Project Visibility	"A visualization tool that shows work in progress to help identify bottlenecks and overcommitments, thereby allowing the team to optimize the workflow" (PMI, 2021, p. 241).
Process Improvement	The cumulative flow chart signifies how well work flows and shows bottlenecks or slowdowns, letting a team know if they are improving (Badgley, 2019).
Product Scope	"The features and functions that characterize a product, service, or result" (PMI, 2021, p. 245).

quality is one of the unique management aspects of Agile. As an example, one of the key quality metrics includes having user stories that are clear and concise with acceptable criteria (Ghahrai, 2018). A second quality objective involves continuous checking to ensure the deliverables at the end of a sprint represent what the team committed to producing. Third, there is a focus on measuring defects. Finally, rollbacks are a standard metric because they could indicate something is wrong with the current version. This characteristic tracks defects that should have been better handled through testing. Table 19.3 offers a more general overview of some agile status metrics. These represent general flavors of the traditional status view and some unique perspectives.

Ghahrai (2018) identified the following eight additional status metrics to consider that are not listed in the discussion above.

1. *Maintainability* metrics measure how easy a product will be to maintain, such as tracking defects that would symbolize software bugs or tracking the number of amendments needed in maintaining the code.
2. *Portability* metrics measure how easy product installation will be for the customers.
3. F*unctionality* metrics measure the difference between planned and actual functionality regarding how the product contains the intended function.
4. *Performance* metrics measure the product responsiveness under user loading.
5. *Compatibility* metrics measure how well the software works with other technologies.
6. *Usability* metrics have several aspects, such as the time it takes to use the product or the number of errors users make with other similar products. It can also refer to how intuitive it is to use without instructions.
7. *Reliability* metrics measure how well the product can be trusted to work.
8. *Security* metrics measure how safely users use the product, such as safety from hackers.

19.5 AGILE TOOLS

Different tools are used in agile projects (see Table 19.4). Many different commercial vendors provide these tools. Some of the terms shown here have not been previously described but will be in later sections of this chapter. One term is *swimlanes*, which is a theme found in Kanban. This is how status data is displayed as they resemble swimlanes in a pool (see Figure 19.11).

TABLE 19.4 General Tools

Tool Name	Description
Kanban Board	"A visualization tool that shows work in progress to help identify bottlenecks and overcommitments, thereby allowing the team to optimize the workflow" (PMI, 2021, p. 241).
Bug Tracker	This is a tool that tracks reported software bugs. Open-source software projects may provide a means for end users to report bugs directly.
Task Board	"A visualization representation of the progress of the planned work that allows everyone to see the status of the tasks" (PMI, 2021, p. 251).
Agile Project Management Tool	Companies have tools that assist them in following agile methods efficiently. These tools provide the team with different features depending on the tool, such as project planning, monitoring, response plans, stand-up meeting assistance, collaborative tools, release management, retrospective analysis, and reporting.
Spreadsheet	Provides a tabular form to store information and analyze data.
Wiki	Website for information to be stored and edited by those with access.
Automated Build Tool	These tools automate the process of building software. These tools compile computer source code into binary code, package binary code, and run automated tests.
Unit Test Tool	Type of software test where individual units or a software component is tested.
Continuous Integration Tool	This is a tool used to help integrate each team member's work.
Requirements Management Tool	A tool used to manage the current definition and requirements of deliverables, including documenting, analyzing, tracing, agreeing upon, monitoring, prioritizing, and performing version control.
Traditional Project Management Tools	Software that assists with task management so that projects get implemented and executed within a required time frame.
Project & Portfolio Management (PPM) Tool	PPM tools allow companies to manage project risks, resources, budgets, timelines, etc.
Automated Acceptance Tool	This tool automatically drives output products through acceptance tests to quantify acceptance.
Index Cards	Index cards are a fundamental principle of Agile because they help write user stories and backlog items.

(Continued)

TABLE 19.4 (Continued) General Tools

Tool Name	Description
Story Mapping Tool	This is a tool that assists an agile team in creating its backlog. It displays customers' activities and tasks associated with the product or software being developed. The story map created ensures that team members are on the same page from the beginning of the product or software development to the ongoing delivery of new releases. This tool helps teams decide what to build first, how to scope the project, and what to prioritize, all while visualizing the project's progress.
Refactoring Tool	Tools to assist in modifying existing source code so it can be reused in another context for another purpose.
Customer Idea Management Tool	Tools that allow for collaboration between employees, customers, communities, etc., in creating shared solutions to complex challenges so the best ideas can be used and reused.

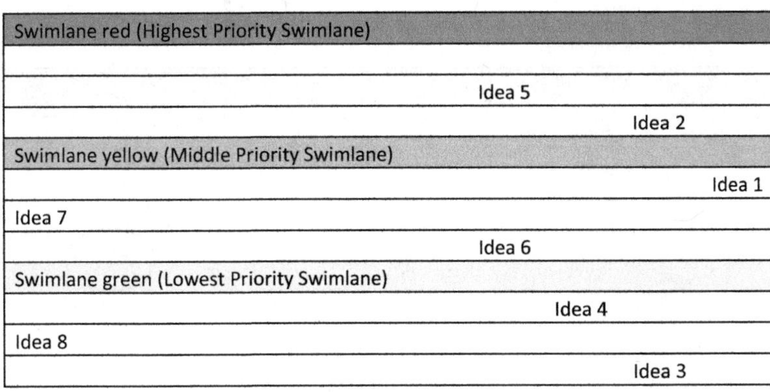

FIGURE 19.11 Example of swimlanes.

19.5.1 Task Board

MSP has capabilities that assist in supporting agile teams. This part of the chapter demonstrates some primary features related to the Task Board. Figure 19.12 displays basic methods for utilizing the *Task Board* feature. The areas linked by the black circles outline the basic navigation steps. To initiate this process, go to the "View" tab, then click on "Other Views," and then select the menu option of "More Views…" This will open a "More Views" window to appear. Then, select this, and a new "More Views" Window will display the "Task Board" option. Select the "Apply" button to start using the option.

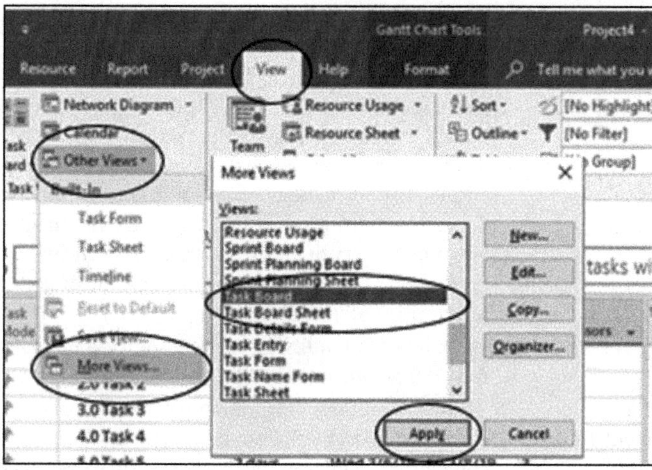

FIGURE 19.12 Navigating to the task board.

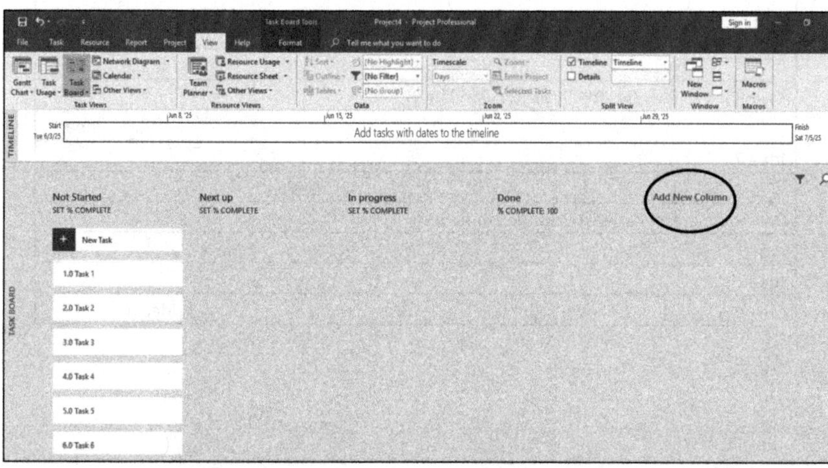

FIGURE 19.13 Adding a new column in a task board.

Figure 19.13 displays the blank Task Board view. Adding new columns to this view, as outlined by the black circle, is also possible. To do this, click on the "Add New Column." Once clicked, Figure 19.14 displays how the "Add New Column" button changes to a text-entry box. The black circle outlines how to name a column "Testing."

Figure 19.15 displays the mechanics of moving a column to the left or right. The user can click this option to move a column to the appropriate place.

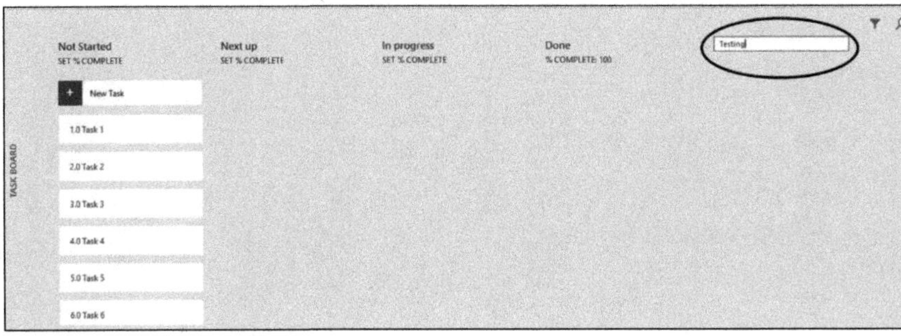

FIGURE 19.14 Naming a new column in a task board.

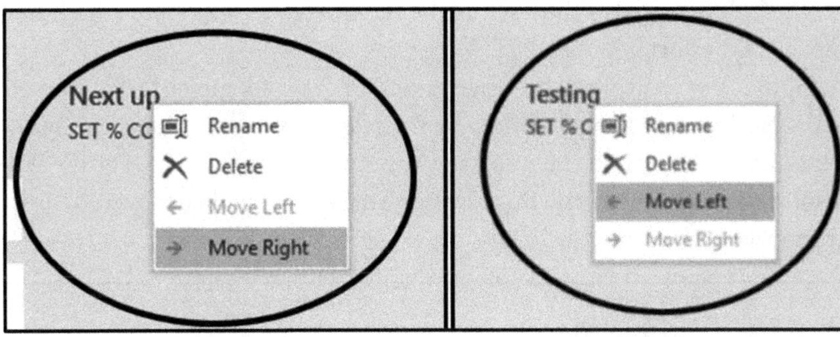

FIGURE 19.15 Moving a column.

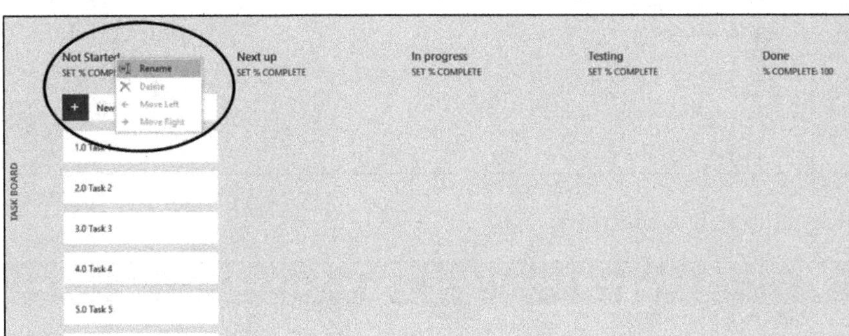

FIGURE 19.16 Renaming an existing column.

Figure 19.16 displays how a column can be renamed. The black circle shows how to perform this by right-clicking from the column heading to bring up an option to "Rename" a column. Figure 19.17 displays the resulting Task Board with all columns moved to the appropriate place within

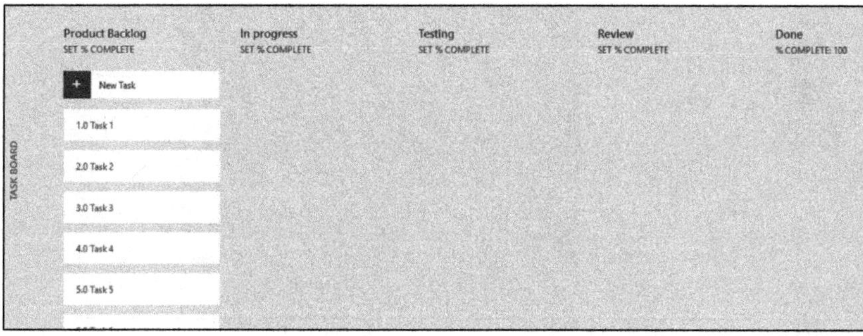

FIGURE 19.17 Task board with columns moved and renamed.

the Task Board. It also shows how the renamed columns were edited to fit the current effort.

Figure 19.18 displays the mechanics of how to move tasks into the appropriate column. As outlined by the black circle, step one is to click on a task and drag it with the mouse to the appropriate column. Figure 19.19 displays the revised Task Board after all tasks have been moved to the appropriate column.

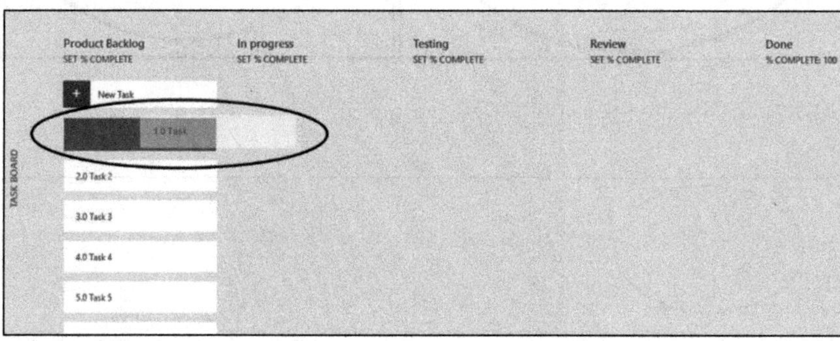

FIGURE 19.18 Moving tasks.

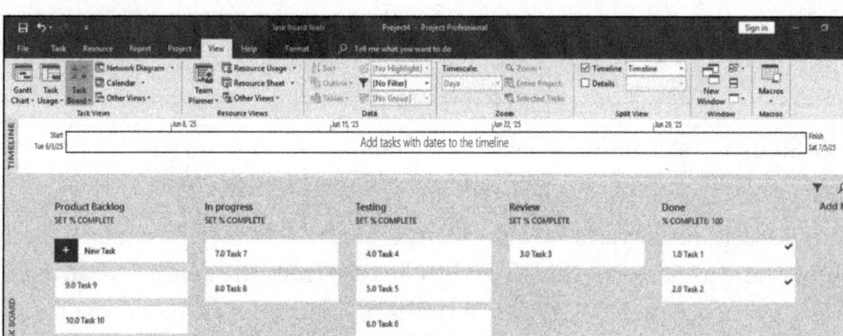

FIGURE 19.19 Display showing tasks moved.

Figure 19.20 displays how a new task can be added to the Task Board directly from the Task Board view. The MSP user must click on the "New Task," which will bring up the text-entry box outlined by the black circle. Once the user has entered the task name, the "Add" button, also in the black circle, needs to be clicked. This will then add a new task that can be moved to the appropriate column.

Figure 19.21 displays a shortcut for updating the percent complete data for a specific task. This is accomplished by clicking on a task such as Task 3, outlined with a black circle, going to the Task tab, also outlined with a black circle, and then clicking one of the percent complete options outlined by a black circle. In this example, you will see that the 50% complete option has a gray shading to indicate that this option was selected. Figure 19.22 displays how to invoke the task information box, another way to enter the percent complete data for a specific task. If a task has a percent complete that is not among one of the shortcut options, then the task information box will be the only way to reflect the percent complete accurately. Figure 19.23 displays how to enter the percent complete data for the column. This is another tool to help the team view the work remaining in

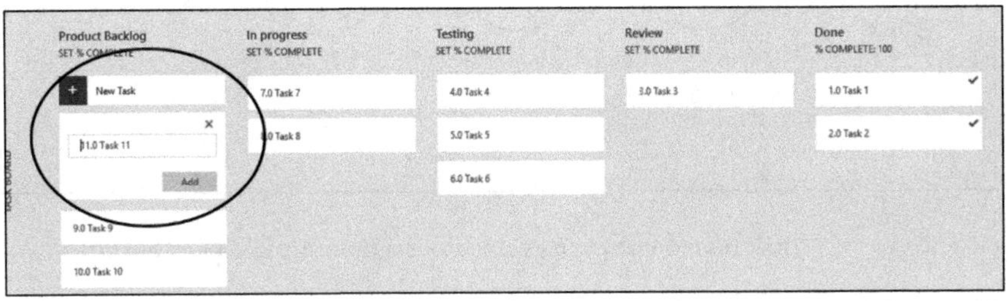

FIGURE 19.20 Adding a new task.

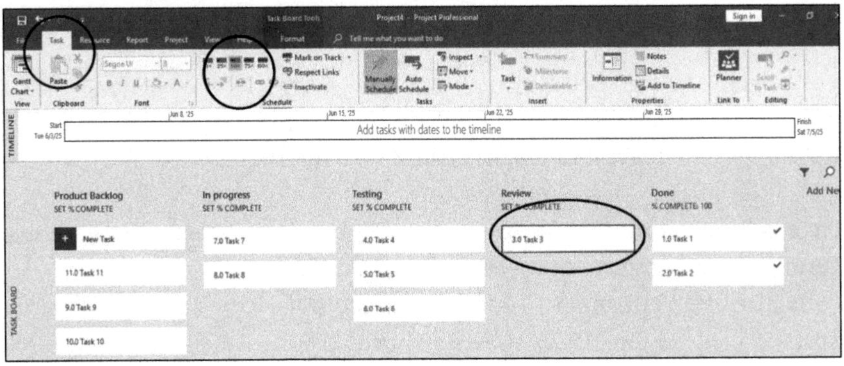

FIGURE 19.21 Shortcut to updating the percent complete on a specific task.

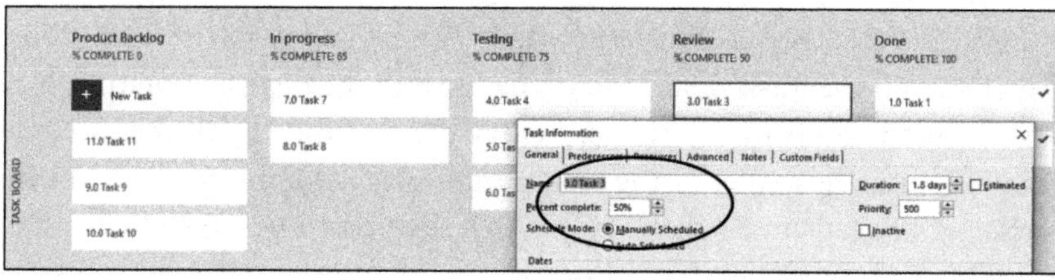

FIGURE 19.22 Using the task information box to update the percent complete on a specific task.

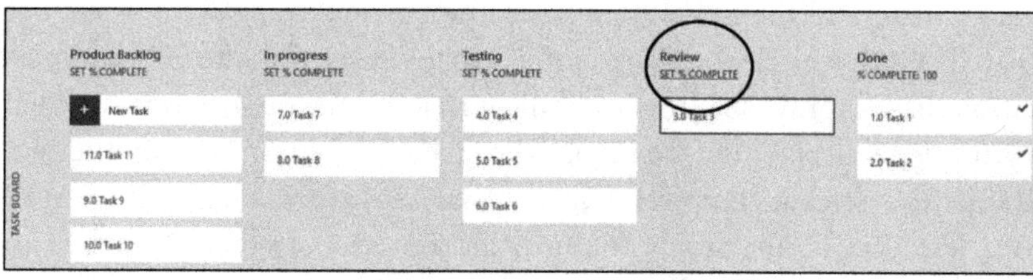

FIGURE 19.23 Entering the percent complete per column.

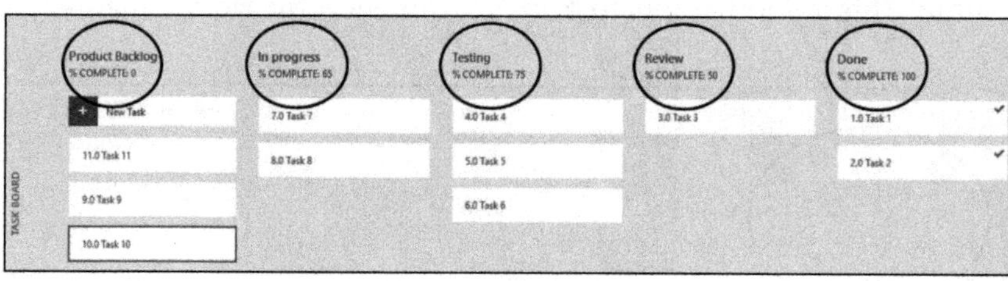

FIGURE 19.24 Task board with the entered percent complete per column.

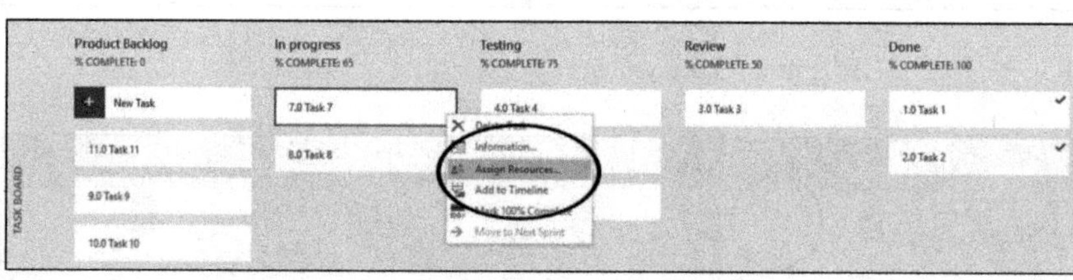

FIGURE 19.25 Shortcut to assign resources to a specific task.

each column group. Figure 19.24 displays the Task Board after each column percent complete number is recorded.

Figures 19.25 through 19.27 illustrate the steps to assign resources to a specific task. Figure 19.25 shows a shortcut outlined in the black circle where the MSP user can correctly click to assign resources for a specific

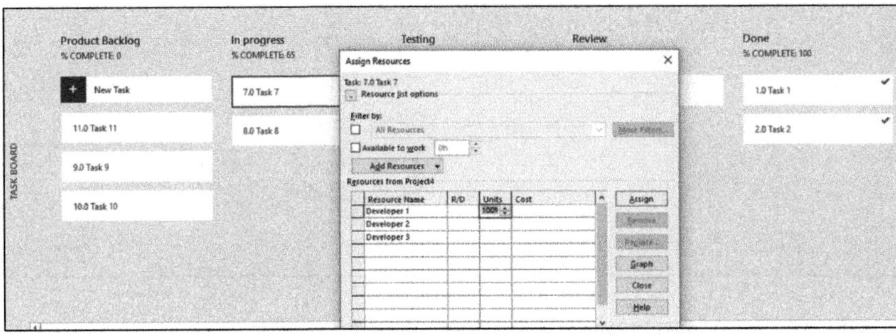

FIGURE 19.26 Assign resources window.

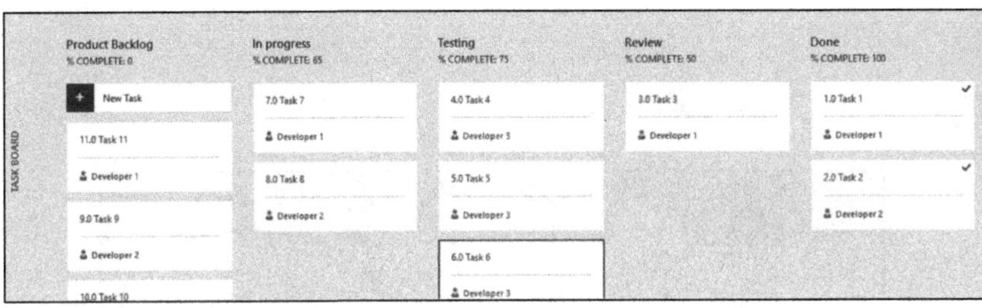

FIGURE 19.27 Display with task resources.

task. Clicking on assign resources causes an assign resources window to appear (see Figure 19.26). The assign resources window shows all available resources used in a Task Board. Once entered, it will store this information to assign other resources. Specific task assignments can be entered. Another way to add the resources is to double-click one of the tasks on the Task Board. This process will display the task information box, as shown in Figure 19.22. Within the task information window, different task information, such as the resources, can be entered because it is the same task information box displayed when double-clicking a task from the Gantt chart view. Figure 19.27 displays the Task Board with all the resources for each task.

MSP can also display Task Board task assignments on the timeline to help get a big-picture view regarding task status. To access this feature to add tasks to the timeline, view Figure 19.28. Within this figure, the user performs a right-click on a specific task as outlined in the black circle, then select the "Add to Timeline" menu option. This will result in the tasks appearing on the Timeline (see Figure 19.29).

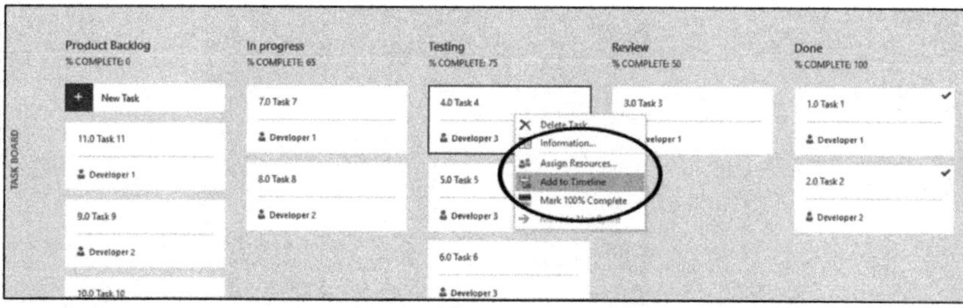

FIGURE 19.28 Adding tasks to the timeline.

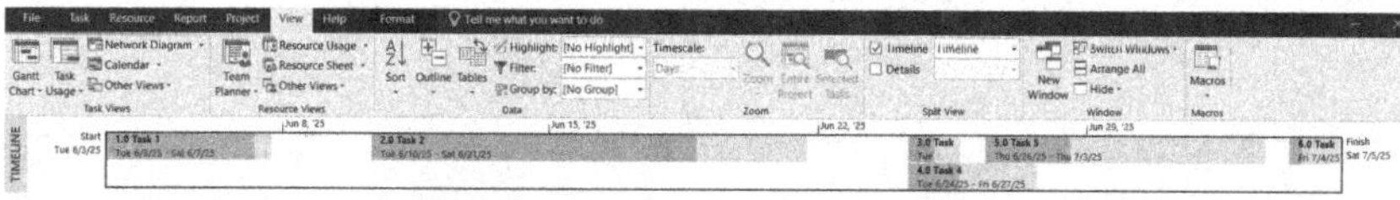

FIGURE 19.29 Display of the timeline.

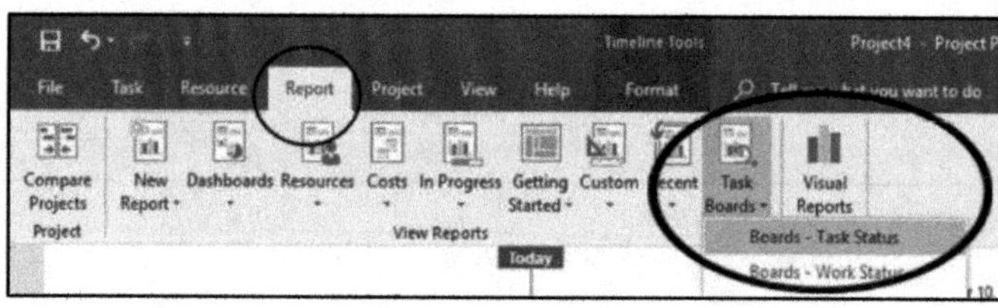

FIGURE 19.30 Navigating to the task status report.

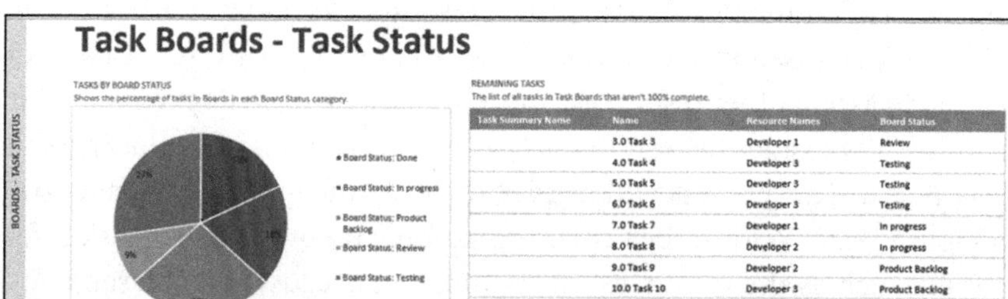

FIGURE 19.31 Display of the task status report.

Figure 19.30 displays the steps to take data from the Task Board to produce the task status report. The black circles show this process. Figure 19.31 outlines the percentage of tasks in each board status category. It also provides a table of the remaining tasks that are less than 100% complete. Figure 19.32 displays the steps required to take data from the Task Board to access the work status report (see black circles). Figure 19.33 provides the

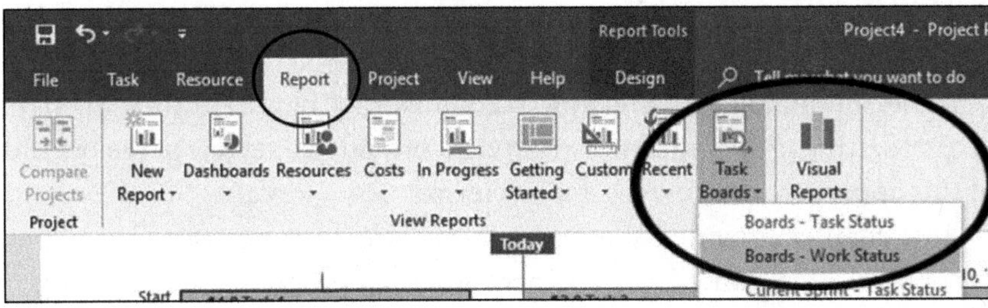

FIGURE 19.32 Navigating to the work status report.

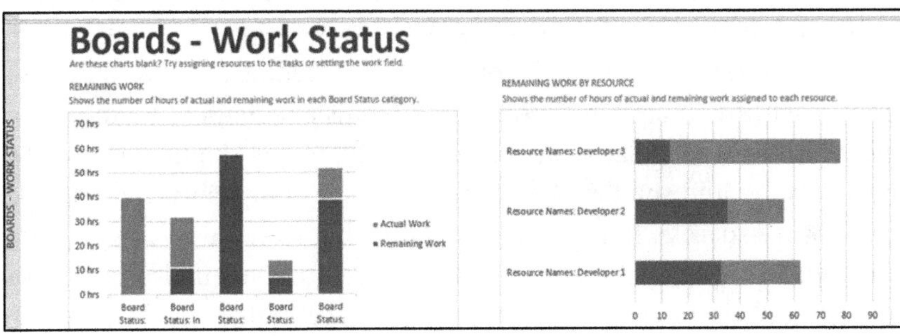

FIGURE 19.33 Display of the work status report.

resulting report displaying a graph of the remaining work in terms of the number of actual and remaining work hours in each board status category. It also shows a graph of the remaining work by resource regarding the number of hours of actual and remaining work assigned to each resource. Both reports in Figures 19.31 and 19.33 are created from the information found in the Task Board. These reports can be both edited and printed.

19.6 CONCLUSION

This chapter provided a broad outline of Agile, iterative methodologies, practices, tools, and techniques. This discussion traced the evolution of agile dialects from the initial Agile Manifesto, which defined the base conceptual model. Scrum, Kanban, Scrumban, and other dialects were described from this origination point. Even though there is no singular agile definition across the various dialects, there are core-level similarities in philosophy across the spectrum. There is also a general focus across the board regarding the Manifesto principles. Sample embedded agile features offered in MSP were demonstrated. This utility option allows teams to use the Task Board feature within MSP and generate status reports from that data. Also, many vendors are moving into the tool area to support this development view.

Agile methodologies are now recognized as having utility in many project settings. The industry survey data provided a review of the breadth of this approach and key vocabulary terms.

This chapter provided some general recommendations regarding how an organization can best implement agile methodologies, according to Schiel (2017). A key example of this is to recognize the importance of increased focus on customer input throughout the life cycle instead of just focusing on completing the initially defined deliverables as specified in the traditional model. Organizations also need to recognize that Agile is truly a mindset and not a framework of mechanics. Teams need the autonomy to work within an agile mindset. Small teams also tend to be more productive compared to larger team sizes. Agile teams should work within shorter cycles and smaller batches to minimize complexity and optimize the project's success. Finally, organizations that value soft skills also tend to be able to implement Agile effectively. These fundamental agile-focused approaches help a development process be successful.

QUESTIONS

Student Review Questions

1. Describe the four values of the Agile Manifesto.
2. Describe the 12 Manifesto principles.
3. What are some of the characteristics of the agile environment that differ from the traditional waterfall project model?
4. Describe and differentiate between the Scrum, Kanban, and Scrumban methods.
5. Describe five agile metrics to measure success.
6. Describe five agile tools.

In-Class Questions

1. Research how software-specific methods relate to general agile methods.
2. Find three examples online of different status-tracking techniques used in various agile dialects.

3. Find software online that supports each agile tool listed in the chapter.
4. Research best practices in implementing Agile within organizations and teams that this chapter has not covered.

EXERCISES

1. If you were going to organize a project in your field, which agile method would you use, and why?
2. Use MSP to create a Task Board and run the task and work status reports from the Task Board.

REFERENCES

Beck, K., Beedle, M., Bennekum, A. V., Cockburn, A., Cunningham, W., Fowler, M., Grenning, J., Highsmith, J., Hunt, A., Jeffries, R., Kern, J., Marick, B., Martin, R. C., Mellor, S., Schwaber, K., Sutherland, J., & Thomas, D. (2001). Manifesto for Agile Software Development. Retrieved from https://www.agilemanifesto.org/principles.html

Clayton, M. (2018). Project Management in Under 5: What is Scrumban? https://www.youtube.com/watch?v=kiI3IweyAeQ

Cohn, M. (2010). *Succeeding with AGILE: Software Development Using Scrum.* Addison-Wesley.

Development Pays. (2017). Scrum vs. Kanban – What's the Difference? https://www.youtube.com/watch?v=rIaz-l1Kf8w

Digital.ai. (2023). 17th Annual State of Agile Report. Retrieved November 20, 2024, from https://stateofagile.com

Francino, Y. (2019). The 5 Most Common Agile Techniques. *TechBeacon.* https://learn.techbeacon.com/units/5-most-common-agile-techniques

Ghahrai, A. (2018). How do we Measure Software Quality in Agile Projects? https://www.testingexcellence.com/how-do-we-measure-software-quality-in-agile-projects/

Hashemi-Pour, C., & Chai, W. (2023, August 29). *What is customer satisfaction (CSAT) and how to measure it.* TechTarget. https://www.techtarget.com/whatis/definition/customer-satisfaction-CSAT

Hilbert, M. (2017). The Real Origins of the Agile Manifesto. https://www.red-gate.com/blog/database-devops/real-origins-agile-manifesto

International Scrum Institute. (2019). Scrum Release Planning. https://www.Scrum-institute.org/Release_Planning.php

ISE Tech. (n.d.). Iterative Development. Retrieved November 27, 2024, from https://isetech.co/iterative-development/

Keup, M. (2021). What is Scrumban? How it Differs from Scrum & Kanban. *ProjectManager.* Retrieved from https://www.projectmanager.com

KnowledgeHut. (2024). Agile's Evolution and Its Future Role in Digital Transformation. Retrieved from https://www.knowledgehut.com/blog/agile/agile-methodology-trends

Mankins, M., & Garton, E. (2017). How Spotify Balances Employee Autonomy and Accountability. *Harvard Business Review*. https://hbr.org/2017/02/how-spotify-balances-employee-autonomy-and-accountability

OKR Institute. (2024). The Future of Agile: 5 Game-changing Trends to Look Out for in 2024. Retrieved from https://okrinstitute.org/blog/agile-future-trends-2024

Productboard. (n.d.). Agile Values. Retrieved November 20, 2024, from https://www.productboard.com/glossary/agile-values/

Project Management Institute (PMI). (2021). *A Guide to the Project Management Body of Knowledge*, 7th edition. Newtown Square, PA Project Management Institute.

Project Management Institute (PMI). (2017). *Agile Practice Guide*. Newtown Square, PA: Project Management Institute.

Schiel, J. (2017). 10 Things You Must Do to Become Truly Agile. Agile Implementation eGuide. *TechWell*. https://www.agileconnection.com/sites/default/files/webform/file/2017/eGuide-Agile.pdf

Schwaber, K., & Sutherland, J. (2020). The Scrum Guide. Retrieved from https://www.scrumguides.org

Scrum Alliance. (2015). What is Scrum? https://www.youtube.com/watch?v=TRcReyRYIMg

StarAgile. (2024). Top 7 Agile Trends for Agile Transformation in 2024. Retrieved from https://www.staragile.com/blog/agile-trends

Straughan, G. (2019). *Scrum vs. Kanban Cheat Sheet*. https://www.developmentthatpays.com/files/DevelopmentThatPays-ScrumVsKanban-CheatSheet-1_6.pdf

CHAPTER 20

Project Simulation

Project N Motion Videos for This Chapter

20S1	Simulation Modeling	Simulation Mechanics

It is first important to recognize that the topic of simulation fills many books and classrooms with a wide variety of methods and target areas. It is not realistic to try to describe or cover this level of breadth, so the target of this chapter is to describe a process for forecasting the future behavior of a project as modeled using the Monte Carlo technique. In this case, the target is project performance and quantifying the internal variability anticipated in the project. Various status views collectively reveal new insights into the project behavior.

20.1 INTRODUCTION

Many events can lead to project failure, but one visible fact that cannot be argued is the significance and visibility of task overruns beyond the planned estimates. Regardless of the reasons for this, it is sometimes easier to estimate time variance than to guess at all the possible reasons it might occur. Recall from previous discussions that this does not mean one should pad task estimates but instead find an appropriate method to deal with their variability. The waterfall model is the most frequently used structure for time management, yet many previous comments have been offered regarding why this model does not accurately predict actual

DOI: 10.1201/9781003532989-25

outcomes. This chapter provides examples designed to demonstrate an improved modeling technique for mapping project performance and forecasting completion dates, as well as supporting improved internal analysis of the project; this technique is simulation modeling.

20.2 BRIEF HISTORY OF SIMULATION MODELING

The use of simulation modeling techniques is appropriate for evaluating complex situations where no algorithmic solution is available. Crude early attempts to deal with this class of problem can be found in the literature, but these were of limited complexity due to inadequate computational support tools to handle the iterative nature of this process. Compared to all the techniques described in this text, the mechanics of simulation modeling require the most robust automated computational support.

One of the most interesting early documentation of a simulation technique is "Buffon's needle," which involved a manual process designed by the French naturalist Buffon in 1777. Random events were created by throwing baguettes (French bread loaves) onto a tile floor. It was considered a hit if the bread landed on a tile intersection. The simulation attempted to prove the relationship $\pi = 2N/X$, where N is the number of attempts, and X is the number of times the bread π intersects a tile line. Each "hit" is recorded as being successful.

The modern beginning of simulation modeling can be traced to the early 1940s when brilliant engineers and scientists were working on the Manhattan atomic bomb project that eventually ended World War II. Stanislaw Ulam and John von Neumann were searching for an analysis technique to help them understand neutron patterns moving through radiation shielding (Harrison, 2011). They collaborated to devise an analytical technique using random data points to simulate defined events. Their technique had probability characteristics like the famous roulette wheels found at the Monte Carlo Casino. As a result of this comparison, the technique was named Monte Carlo, and that label remains today. Once this new form of problem analysis process was recognized, it became popular for examining various complex problem types that exhibited random behavior too complex to convert to traditional algorithmic results. The most notable examples can be found in finance, physics, and engineering, although this class of problem exists in a multitude of areas.

The maturation and availability of the modern electronic computer around the 1950s are credited with making the simulation process a realistic problem-solving technique. One of the characteristics of these models is their ability to predict realistic outcomes based on probabilistic interactions of events. This characteristic brought this topic to project management, where outcomes are dictated by complex interactions between various variables that affect task outcomes.

20.3 SIMULATING RANDOM EVENTS

Given the need for an arithmetic-oriented computer to behave randomly, it seems appropriate to digress here to demonstrate one of the core logic components of simulation modeling, that being the creation of random numbers. A short answer to this more complex internal arithmetic mechanic is to envision the multiplication of a selective seed number with an algorithm that causes the calculated number to overflow. Segments of these calculations are then taken as pseudo-random numbers. Mathematicians have tested various seed numbers to find those that have the best random character. The second half of this explanation comes from the definition of the event to be simulated. For this introductory example, let's assume simple coin flipping with the two outcomes of Heads and Tails. Excel has a random number function called *rand()*, which we can use to illustrate the process. For step one, we produce ten random numbers as shown in Table 20.1.

This table data demonstrates that an automated computational device can produce a set of random numbers, but the question that remains is how do we equate these to a Head or Tail? These numbers are all in the range of 0 to 99, so if all values less than 50 could represent a HEAD and all others greater than or equal to 50, then would we represent it as a TAIL? So, to simulate the coin toss answer, one must test the randomly generated

TABLE 20.1 Random Number Generation

Trial	RN	Event
1	96	
2	7	
3	100	
4	95	
5	32	
6	14	
7	17	
8	83	
9	71	
10	73	

TABLE 20.2 Random Logic Generation

Trial	RN	Event
1	96	TAILS
2	7	Heads
3	100	TAILS
4	95	TAILS
5	32	Heads
6	14	Heads
7	17	Heads
8	83	TAILS
9	71	TAILS
10	73	TAILS

cell value with the conditional logic. Performing that conditional test yields the comparative results displayed in Table 20.2.

It is important to note that the result of this execution shows six TAILS and only four HEADS. As a result of this unexpected outcome, the simulation would suggest that this coin is not fair because the results should have been 50/50. From this result, one must recognize an important lesson regarding the simulation process. A small number of iterations does not adequately represent the process. Simulation models must be executed hundreds or even thousands of times to match the underlying characteristics of the target. From this computational characteristic, we see the need for automated support tools, which were not widely available until the 1980s. Even though this example logic is very simplistic with only a two-state result, the basic mechanics described reveal the core process.

20.4 DESIGNING THE SIMULATION MODEL

The mechanics of simulation modeling require the following five steps:

1. Defining the random profile of the event (i.e., coin 50/50 logic)
2. Generating random occurrences
3. Defining the link between the random number and the associated event
4. Iterating the model thousands of times and collect results
5. Displaying the results (typically in graphical or statistical form)

The following sections of this chapter will describe these steps in more detail.

20.5 PROBABILITY DENSITY FUNCTIONS

To understand how random events are defined, one must first understand how to model a random event. This is done through statistical functions called Probability Density Functions (PDFs). Essentially, these statistical formats map the probabilistic event behavior. In the coin-flipping example, the two results were assumed to be equally likely but that is not the normal event complexity. From a general technical view, various PDF formats are used to model different behaviors. Each of these distributions is defined by complex mathematical expressions such that a random number can be mapped to more complex outcomes based on the PDF structure. This concept may be confusing at this point but should be clearer with upcoming examples.

The most familiar PDF is the bell-shaped, "Normal" or Gaussian distribution curve that teachers often use to define class grades from raw test data. Figure 20.1 shows a graphical format of this shape.

In this exam profile, the average grade is the mean value and is also the median value that divides the class grades into two parts. Various segments of this curve could be given further arbitrary grade grouping such as who receives an A, B, C, D, etc. Doesn't everyone remember the arbitrary line between major grades?

The logic behind this curve shape involves the idea that this exam population is homogeneous, and the grade values represent preparation. That

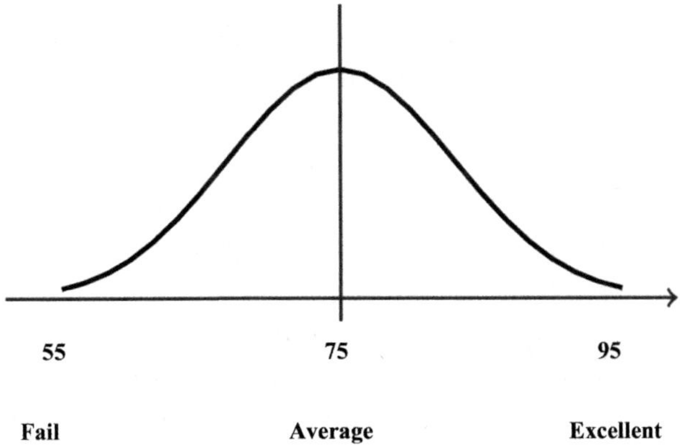

FIGURE 20.1 Bell-shaped (Gaussian) PDF.

point might be arguable but, amazingly, this distinction is so well-known and used that it is often accepted without disagreement. However, if the exam population class was filled with Nobel Prize candidates one might quibble with this assumption, but that is not our role here. Regardless of the underlying assumptions, the role of a specific PDF is to allocate values to the event model. Let's not get lost in this more personal exam grading example and get back to the broader question.

View the PDF as a probability distribution that defines the way "event" variables are assumed to occur whether that is class grades, task execution times, or some other even behavior.

There are numerous mathematical probability distribution options available for this role. In the project environment, the list of appropriate event mapping forms is smaller. A sample of the most used PDF options is displayed in Table 20.3.

The sample list of PDF options has the following general event mapping characteristics:

- A Gaussian or Normal distribution fits situations with randomness about a central value with defined upper and lower range values.
- The beta distribution fits a classic model of work performance for a task that has the potential for significant overrun (non-symmetrical with three variable shape values); event values are defined as optimistic, most likely, and pessimistic.
- The uniform distribution assumes equally likely values across a defined range. Coin flipping is a two-state uniform event.
- A triangular distribution is a pseudo-statistical view that uses three shape values and is often mapped as a beta distribution.

20.5.1 Normal Distribution

As illustrated earlier, the Normal distribution offers the easiest example to show shape parameters. The classic shape of this distribution is defined by two Greek symbols, called shape parameters. These are the central tendency value μ and a measure of distribution labeled by the Greek 6, which defines the range characteristic. In more familiar terms, the signal (6) symbol is known in statistics terms as the Standard Deviation, and it essentially serves to quantify the probabilistic area under the range curve.

TABLE 20.3 Common Probability Distribution Function Types

Type of Probability Distribution	Description	Visual
Normal	This well-known symmetrical bell-shaped curve is concentrated in the center and decreases on either side with the values in the middle near the mean are most likely to occur. It has less of a tendency to produce unusually extreme values, compared to some other distributions. It describes natural phenomena such as inflation or individual's weights.	
Lognormal	This curve has positively skewed values. It is used to represent values that don't go below zero but have unlimited positive potential. Examples of variables described by lognormal distributions include real estate property values, stock prices, and oil reserves.	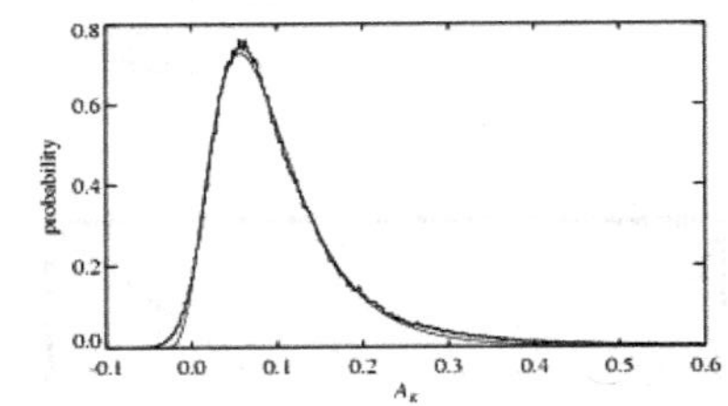
Uniform	All values have an equal chance of occurring, such as tossing a coin with equal chances of getting heads or tails. The minimum and maximum values need to be identified. An example of a variable is sales revenues for a new product.	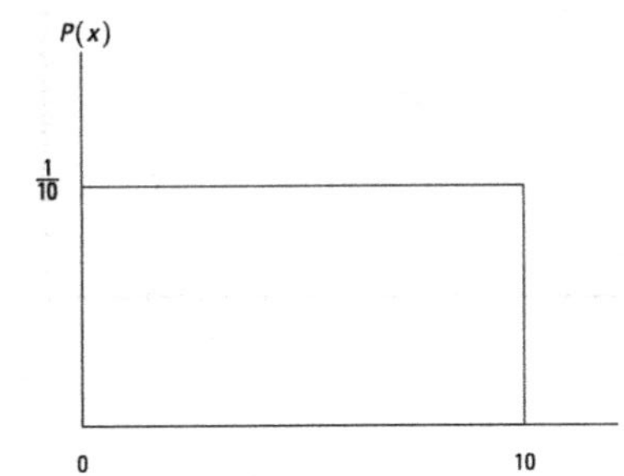

(*Continued*)

TABLE 20.3 (Continued) Common Probability Distribution Function Types

Type of Probability Distribution	Description	Visual
Triangular	It favors the most likely value and gets its name because the most likely value which is referred to as the mode creates the top point of the triangle. There is no requirement that the distribution is symmetrical about the mean and therefore depending on estimates for the minimum, maximum and most likely, the shape of the triangle may be skewed to the left, minimum, or right, maximum, values. Although the triangular distribution can model a variety of different circumstances, it can produce too great a value for the mean of the risk analysis where the maximum for the distribution is very large.	
PERT/Beta	The distribution is used to simulate skewed estimates, either positive or negative. Three shape parameters are required. This is the founding distribution of the PERT model and is used frequently in simulation modeling.	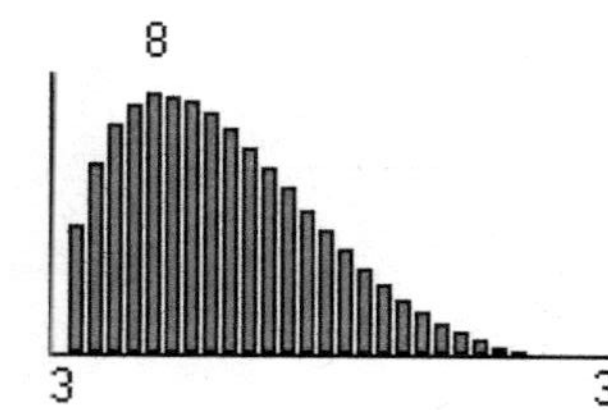
Discrete	This distribution is used when there is a specific probabilistic value that occurs with each event. For example, flipping a coin 20 times will result in a specific integer o 0 to 20 tails occurring. An example might be the number of accidents on a college campus in a given month.	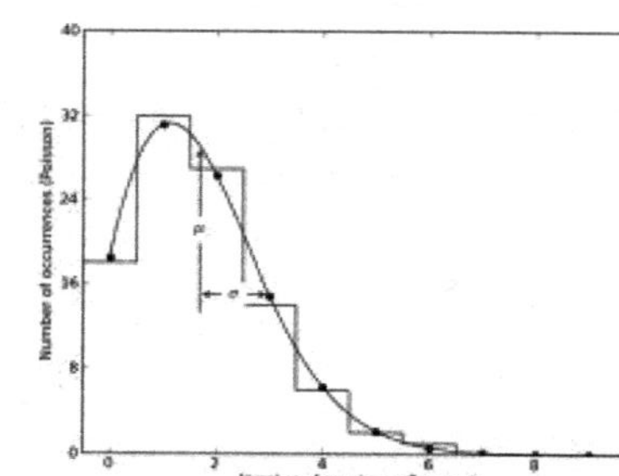

Figure 20.2 demonstrates how the shape parameters define the distribution of values across the range specified in terms of Standard Deviation (б) values.

As shown in the standard normal curve, one SD represents approximately 34% of the curve value on each side of the central value or 68% of the central portion of the total distribution. Quantitative values for the total area under the curve are as follows:

1. SD = 68%
2. SD = 95%
3. SD = 00.7%

If one were modeling a normal distortion, the random numbers generated for this PDF would select 68% of the events to occur within one SD. More granular values for the area under the curve interpretation can be explored through Excel functions such as NORMINV and NORMDIST. For standard SD values, if one wishes to know the probability of an event exceeding 2 SD, that would be the 2.5% shown in the right-tail section of the distribution. If the average time for a normally distributed task was 10 days and the SD was 2, one interpretation of that is "There is a 2.5% chance the task time will exceed 14 days." At the extreme point (3 SD), there is only a minimal probability (0.3%) that the event will require more than 16 days. The important idea related to this topic is that the defined events are not deterministic, and their behavior is based on their underlying PDF profile. This is the essence of the simulation process regardless of the specific PDF.

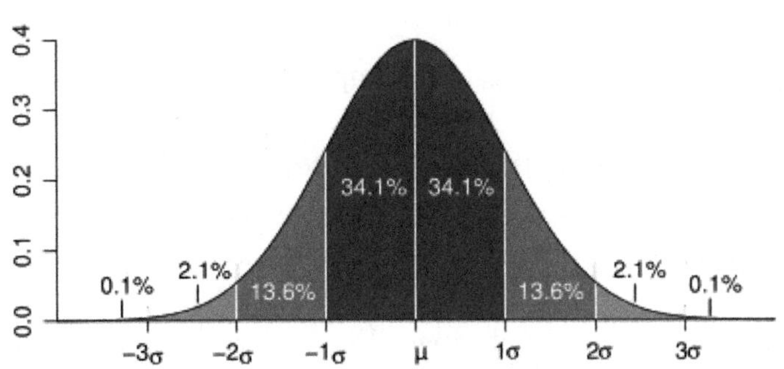

FIGURE 20.2 Normal distribution shape characteristics.

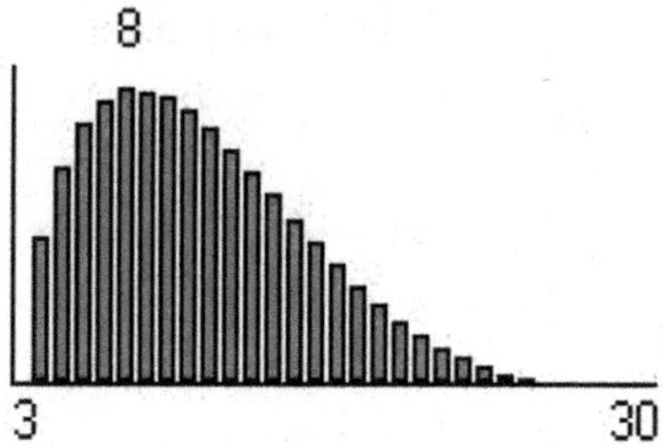

FIGURE 20.3 Beta PDF.

20.5.2 Beta Distribution

The beta distribution is useful for defining work performance variability in which the profile does not fit the Normal symmetrical view. The use of a Beta format for task estimating was initially described in the 1950s classic Polaris project where task estimates were very uncertain based on their high technology nature with no historical background experience. Note the shape of this distribution has a long tail (skew) to the right indicating that the range is not symmetrical in that direction. Figure 20.3 shows a sample beta PDF format that is skewed to the right.

If the specifications for this PDF were 80–100–150, that would indicate an optimistic value of 80% but a pessimistic value of 150%. This would drive the simulation events to produce time estimates based on this profile.

20.6 RANDOM EVENT GENERATION

One of the more difficult mechanical aspects of simulation is how to randomly produce events from a PDF profile. The coin-flipping example only provided a peek into that complex topic. There are two components to that effort:

- Generating a random value
- Matching the random value to a PDF profile

20.6.1 Automating Random Value Generation

The coin-flipping or dice-throwing examples may work well to illustrate simple model events but are not good enough to illustrate more complex modeling. The first requirement is to have a mechanism to generate random values across a defined range. The modern method for doing this is to generate such a number through defined computer utilities, which essentially overflow computer number registers to produce adequate random values, which are technically pseudo-random but sufficient for the modeling process. The previous example illustrated the availability of random number utilities in Excel, but this capability also exists in most modern computer devices. Although these are easy sources to find the capability, they do not represent the easiest-to-use sources. The most likely production option is to find a packaged commercial vendor solution that fits the problem area and that is the approach that will be demonstrated here.

The logic of how PDFs can be used to fit event formats has been demonstrated; however, using this in the automation model requires an additional step. To match an array of possible answers to a random number value, the best way to model this is using a companion function called a Cumulative Distribution Function (CDF). There are various techniques to accomplish this type of match, but the use of a CDF seems easiest to understand.

20.6.2 Role of the CDF

A CDF is an inverted format for a PDF, meaning the values are accumulated through the range rather than shown as discrete for each value. The general shape of a CDF is shown in Figure 20.4.

The Y-axis is labeled with values between 1 and 100, while the X-axis contains the associated event height values. The sample X–Y value indicates that approximately 85% of the individuals in this population are less than 72 inches tall or less. Technically, this is a "continuous" distribution rather than a "discrete" one since there can be partial values. In the project

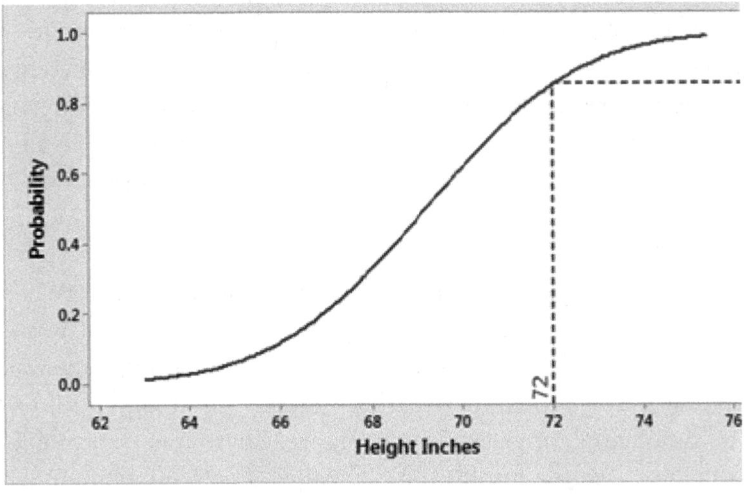

FIGURE 20.4 CDF schematic format.

modeling case, only integer values will be defined but the concept is the same. Note that the shape of the CDF causes the central tendency values to be mapped more frequently than the outer range values. Looking at this logically, it says there is a higher probability of the value being around the middle range and a lesser probability that it is in the outer portion. *It is this process by which the linear list of random numbers generates non-linear results.* In other words, this relationship supports the idea that a random number can represent a complex relationship. For each random number(y) generated, a simulated event X is returned. Through hundreds of iterations that process produces the probabilistic frequency defined by its defined PDF. Internally, the complex mathematics related to this process will generate the CDF structure shown. Calculation utilities for this process exist and are encapsulated in the commercial models for easy access.

20.7 MODEL CONSTRUCTION

Complex simulation models have more than one probabilistic event to simulate, and these must be combined into higher-level logic to map the entire view. In the case of simulating a project, there is task variation defined, but each task is only one component of the overall project schedule. So, the modeling process requires a random duration generation of each task, then running the project plan to evaluate results. The data for each iteration is captured, and the process continues for 500 or more iterations. The result of this is a collection of results that emulate what would statistically occur if the tasks did vary as the assumed PDF specified. It is important to point out that this is called a descriptive simulation, meaning that it produces an output based on assumptions that could be wrong. One of the tests that needs to be executed for all simulations is to test different event assumptions to see the variability of output. Simulation is a powerful tool if used properly by a skilled operator but should not be considered magic without proper model understanding and testing the assumptions.

20.7.1 1980s Simulation

Let's digress for a moment and look back into the recent history of simulation. During the 1980s, the "simulation" view of a project was coded into MS Project using various discrete design assumptions of the waterfall model that were outlined in an earlier chapter. The result of this exercise was a "simulation" of project performance. There was no variability or PDFs involved, so the result might be called a deterministic simulation. Nevertheless, the goal was to develop a view that simulated the results.

Even though this might look crude, just examine the change in perspective from that point and the current probabilistic view that will be shown here. We often lose sight of capability advances such as this. If this demo makes the current model look crude, imagine what another 50 years may bring in tool enhancements.

One might argue that MS Project does not represent a simulation but given its base assumptions, it did indeed simulate (describe) the project behavior as modeled. A key vocabulary term here is *verisimilitude*—does the model behave the same as the environment it is defined for? If all the assumptions coded into that original waterfall model structure were true, it would accurately simulate the outcome. Over this intervening period, all delivery methodology models have attempted to simulate project planning and execution. The Monte Carlo computer automation addition is just another enhancement to these early attempts to emulate project outcomes. Adding a role for PDFs and an iteration process uplifted the idea to a new generation of project simulation understanding. Full Monte is a project simulation utility that is part of a growing collection of commercial vendors who have recognized the va of this tool class (Barbecana, 2020).

A point to recognize from this historical perspective is that the 1980s version of MS Project was designed to "simulate" project outcomes, and it does that if all the underlying assumptions are true, just as the example here is true if the event PDF profiles represent actual behavior. It seems logical to conclude that using PDFs expands the reality of this process over the deterministic approach. But in both cases, the resulting output is only as good as the model parameters matching the underlying events.

20.7.2 Deterministic Base Plan

The sample project plan used for the simulation demonstration is titled Car Problem First Cut. This plan consists of 50 tasks that create sufficient complexity to help show the value of simulation. The traditional version of this plan is shown in Figure 20.5.

Only 26 of the 50 tasks are shown here, as space is limited. This plan will be simulated using a beta distribution to define task variability. The profile duration variation for all tasks ranges from 75% of the central value to 125%. This represents the optimistic, most likely, and pessimistic task values. Even though this is called a beta distribution, the profile shown is essentially a symmetrical Gaussian bell-shaped curve. So, it will behave like an event randomly moving around a central value. Changing the

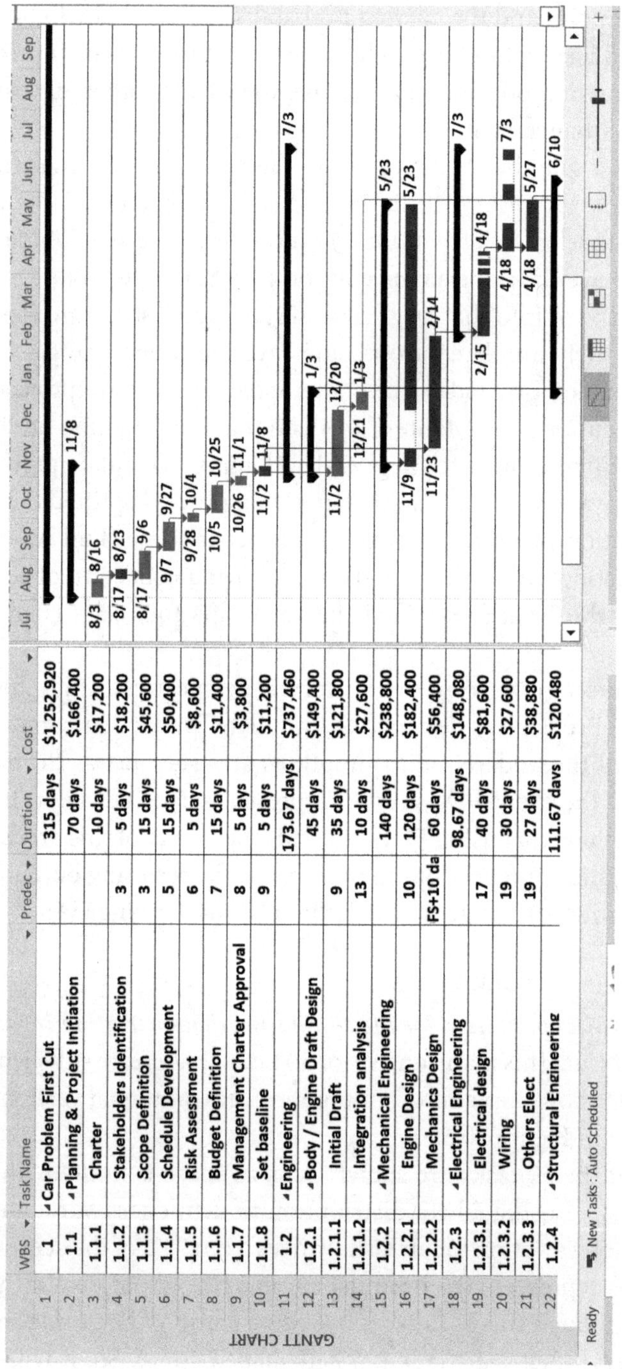

	WBS	Task Name	Predec	Duration	Cost
1	1	Car Problem First Cut		315 days	$1,252,920
2	1.1	Planning & Project Initiation		70 days	$166,400
3	1.1.1	Charter		10 days	$17,200
4	1.1.2	Stakeholders Identification	3	5 days	$18,200
5	1.1.3	Scope Definition	3	15 days	$45,600
6	1.1.4	Schedule Development	5	15 days	$50,400
7	1.1.5	Risk Assessment	6	5 days	$8,600
8	1.1.6	Budget Definition	7	15 days	$11,400
9	1.1.7	Management Charter Approval	8	5 days	$3,800
10	1.1.8	Set baseline	9	5 days	$11,200
11	1.2	Engineering		173.67 days	$737,460
12	1.2.1	Body / Engine Draft Design		45 days	$149,400
13	1.2.1.1	Initial Draft	9	35 days	$121,800
14	1.2.1.2	Integration analysis	13	10 days	$27,600
15	1.2.2	Mechanical Engineering		140 days	$238,800
16	1.2.2.1	Engine Design	10	120 days	$182,400
17	1.2.2.2	Mechanics Design	FS+10 da	60 days	$56,400
18	1.2.3	Electrical Engineering		98.67 days	$148,080
19	1.2.3.1	Electrical design	17	40 days	$81,600
20	1.2.3.2	Wiring	19	30 days	$27,600
21	1.2.3.3	Others Elect	19	27 days	$38,880
22	1.2.4	Structural Engineering		111.67 days	$120,480

FIGURE 20.5 Sample program (partial task list shown).

pessimistic value to 150% will shift the simulation results significantly, so making this change with the downloaded plan data would make a good results sensitivity learning exercise.

20.7.3 Executing the Model with Full Monte

To add some reality to this example, imagine you are presenting a project completion forecast to your sponsor based on the derived plan (execution has not started at this point). At this point, you can only offer deterministic values provided by the earlier traditional plan, even though history suggests something might happen to change these values. You may feel uncomfortable regarding the accuracy of the completion value indicated in the plan. Let's see what a more modern simulation model might add to this perspective.

Given the complexity of constructing a complex model, the most likely strategy for performing simulation is through third-party vendor "canned" models with specific capabilities. PDF profiles can be easily set up, and the existing plan structure will be used for all except the variable task specification. For this example, a traditional model is used as the base plan, so remember that the result only reflects that portion.

"Canned" models offer ease of use and a quick opportunity to evaluate what this class of tools brings to operational understanding. [20S1] In many ways, simulation models can serve as educational tools to understand why projects are so hard to control and predict their outcomes. Profile assumptions can be changed for comparisons to allow a deeper analysis of outcome results.

The first step in using Full Monte is the selection of the event PDF options for each task (or using the same profile for all tasks to obtain a quick look). The examples described here will utilize a single beta function specification with the same three-time estimate profile for all tasks. This is valid for a quick first-cut look at general variability, but a more serious review would require more individual task specifications. A beta format fits situations where the task time has the potential to overrun beyond what a symmetrical (Gaussian) range would show. A Normal or Gaussian distribution is commonly used for tasks with a more symmetrical range around a central value. Depending upon the specific probability distribution shape chosen for a task, the required data entry to describe that shape is 2 or 3 data values, assuming the same profile is used for all (Barbecana, 2020).

Once the PDF parameters are defined, the model (plan) is iterated for many cycles (1,000 iterations, in this case). Each iteration captures the plan status for that set of assumptions.

The mechanics for dealing with the individual iteration results likely involve the most mystery for a new viewer. What does a single iteration reveal? Nothing! The key to the process is measuring the randomness quantified by the iterations. As seen in the coin-flipping example, it takes many iterations for a variable process to reach a steady state. The simple two-state coin example might take 50 iterations, but 500 or more would be more commonly required for a larger mode with more events.

The mechanical advantage of simulation is its ability to show model behavior close to the real environment. The more that is known about the statistical behavior of the model, the closer the results will match actual behavior. One attribute of simulation that should be considered is its ability to find root causes. It may be able to find strong correlations for causal factors by changing various assumptions, but that is more of a secondary view than a primary one. Results obtained through simulation also do not mean that this represents an optimal state. It is a descriptive view only. And, of course, if the host model does not contain valid profiles and linkages, its results will not replicate reality. Here is a little ditty to highlight this point:

A fool with a tool is still a fool if he doesn't understand the tool.

20.7.4 Setting the Run Specifications

Figure 20.6 shows the main data entry portal that describes the project events.

It is important to note that the entire project will be simulated using a beta profile specification of 80–100–150 for all tasks. Changing these values would also affect the characteristics of the output. A more detailed profile requires that each task have a customized profile based on its characteristics. The interested reader can research from the Full Monte User Guide (2017) A free 14-day trial of Full Monte can be downloaded from the vendor's website. (see https://www.barbecana.com/full-monte/free-trial-MSP) to explore the mechanics of this tool.

Full Monte is accessed from the "Add-In" menu bar. When invoked, it can pass its values directly to MSP for the execution process using Full Monte's randomly generated task values.

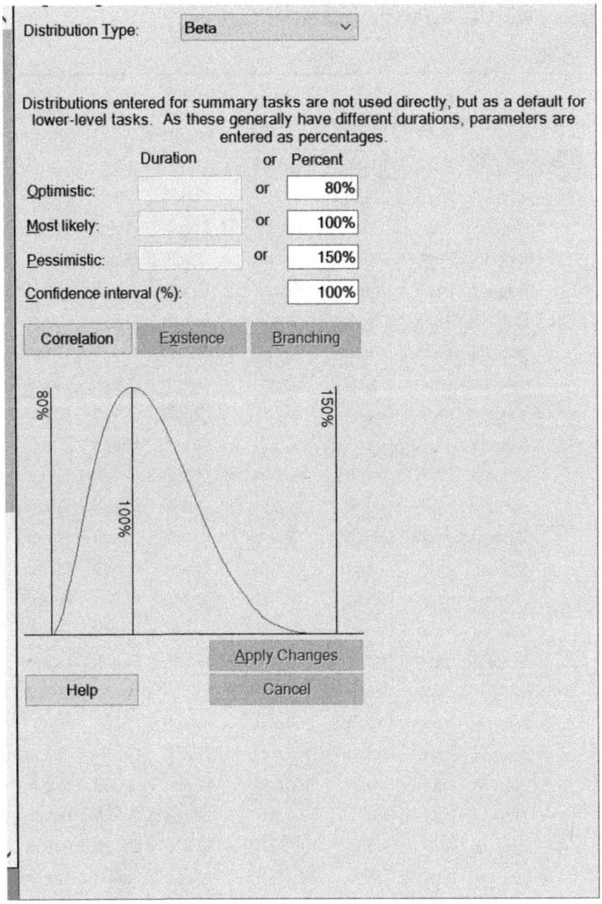

FIGURE 20.6 Defining the full Monte™ run options.

After the profile parameters are defined and saved, the "Risk Analysis" process executes to produce the performance histograms. Figure 20.7 shows the list of schedule and cost performance graphs produced.

The column headings indicate the type of results, and the row values represent the task. Any row and column cells can be selected to produce a histogram showing the behavior for that intersection situation (i.e., project cost for ID 0, late finish for ID 23, etc.). Two of the most common displays are for the final project duration and total costs (ID 0). The example output described below will show samples of these. Another common interest target is the distribution of late finishes. This library contains significant value for understanding the internal workings of the project and a general understanding of how it is created. The histograms are an easy

PERT Car Problem Variable - Full Monte View: Graphs (custom)

File View Help Risk Analysis

ID	Task Name	Cost listogran	Duration listogran	Early Start Histogram	Early Finish listogran	Late Start listogran	Late Finish listogran	Free Slack listogran	Total Slack listograr
0	⊟ PERT Car Pr	Graph	Graph	8/3/21	Graph	8/3/21	Graph	0	0
1	⊟ PERT Car F	Graph	Graph	8/3/21	Graph	8/3/21	Graph	0	0
49	Close Pro	$0	0	Graph	Graph	Graph	Graph	0	0
44	⊟ Formal Cl	Graph	Graph	Graph	Graph	Graph	Graph	0	0
47	Formal S	Graph	Graph	Graph	Graph	Graph	Graph	0	0
46	Close R	Graph	Graph	Graph	Graph	Graph	Graph	0	0
45	Docume	Graph	Graph	Graph	Graph	Graph	Graph	0	0
48	Project Bu	Graph	Graph	Graph	Graph	Graph	Graph	0	0
2	⊟ Planning &	Graph	Graph	8/3/21	Graph	8/3/21	Graph	0	0
8	Budget I	Graph	Graph	Graph	Graph	Graph	Graph	0	0
7	Risk As	Graph	Graph	Graph	Graph	Graph	Graph	0	0
10	Set base	Graph	Graph	Graph	Graph	Graph	Graph	0	0
9	Manage	Graph	Graph	Graph	Graph	Graph	Graph	0	0
4	Stakeho	Graph	Graph	Graph	Graph	Graph	Graph	Graph	Graph
3	Charter	Graph	Graph	8/3/21	Graph	8/3/21	Graph	0	0
6	Schedul	Graph	Graph	Graph	Graph	Graph	Graph	0	0
5	Scope D	Graph	Graph	Graph	Graph	Graph	Graph	0	0
39	⊟ Testing	Graph	Graph	Graph	Graph	Graph	Graph	0	0
43	Safety t	Graph	Graph	Graph	Graph	Graph	Graph	0	0
40	⊟ Compon	Graph	Graph	Graph	Graph	Graph	Graph	0	0
42	Mecha	Graph	Graph	Graph	Graph	Graph	Graph	0	0
41	Electr	Graph	Graph	Graph	Graph	Graph	Graph	Graph	Graph
11	⊟ Engineeri	Graph	Graph	Graph	Graph	Graph	Graph	0	0
22	⊟ Structur	Graph	Graph	Graph	Graph	Graph	Graph	0	0
25	Design	Graph	Graph	Graph	Graph	Graph	Graph	Graph	Graph
24	Other I	Graph	Graph	Graph	Graph	Graph	Graph	Graph	Graph
23	Body	Graph	Graph	Graph	Graph	Graph	Graph	0	0
26	⊟ Body / E	Graph	Graph	Graph	Graph	Graph	Graph	0	0
29	Mecha	Graph	Graph	Graph	Graph	Graph	Graph	Graph	Graph
30	Electri	Graph	Graph	Graph	Graph	Graph	Graph	0	0
27	Chass	Graph	Graph	Graph	Graph	Graph	Graph	Graph	Graph
28	Other	Graph	Graph	Graph	Graph	Graph	Graph	0	0
31	Enginee	Graph	Graph	Graph	Graph	Graph	Graph	0	0
12	⊟ Body / E	Graph	Graph	Graph	Graph	Graph	Graph	0	0
14	Integra	Graph	Graph	Graph	Graph	Graph	Graph	0	0
13	Initial I	Graph	Graph	Graph	Graph	Graph	Graph	0	0
15	⊟ Mechan	Graph	Graph	Graph	Graph	Graph	Graph	Graph	Graph

FIGURE 20.7 Histogram graphs available for review.

format to explain the randomness shown and how that value is calculated.

20.7.5 Displaying Results

A total project (ID 0) schedule histogram can be extracted by clicking on the "Duration Histogram" cell in Figure 20.7. Various cost histograms are produced by clicking on the Cost column. Examples of these are shown below.

20.7.6 Schedule Histogram

Various options for schedule-related histogram results can be seen on multiple cell labels of Figure 20.7. Each histogram produced is accompanied by statistical measures such as mean and standard deviation. Figure 20.8 shows a simulated project schedule histogram.

Displays of this form provide a clear example of simulation modeling value; note that this reporting format shows both most likely and range probability values. In addition, the individual task characteristics can be reviewed based on their defined profiles. This type of forecast view represents a more robust approach to determining future outcomes than the deterministic traditional view that offers none of this. Rather than stating a single value that is wrong, this view shows that the outcome is uncertain with some defined range characteristics. From an educational viewpoint,

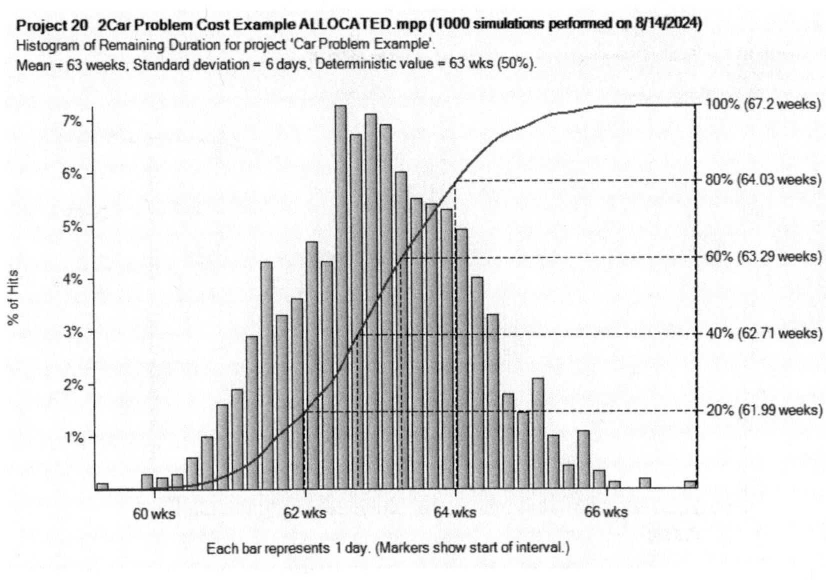

FIGURE 20.8 Schedule histogram display format.

this should help stakeholders understand that Gantt charts do not best represent project plans.

This quick, simple setup evaluation option may suffice, but comparing this output with other values to judge the sensitivity of the results is recommended. If justified, a full task specification might be required.

20.7.6.1 Cost Histogram

The cost simulation display output format is like the schedule view. Figure 20.9 shows a summary cost display.

The mean value for the project cost is $1.422 million, with a standard deviation of $22,000.

20.7.7 Comparison Results

The following list shows a comparison of the deterministic and simulated output:

Deterministic plan results

Duration—315 days (63 weeks)

Cost—1.252 million

Simulation results

Median simulation duration—315 days 6 (3 weeks)

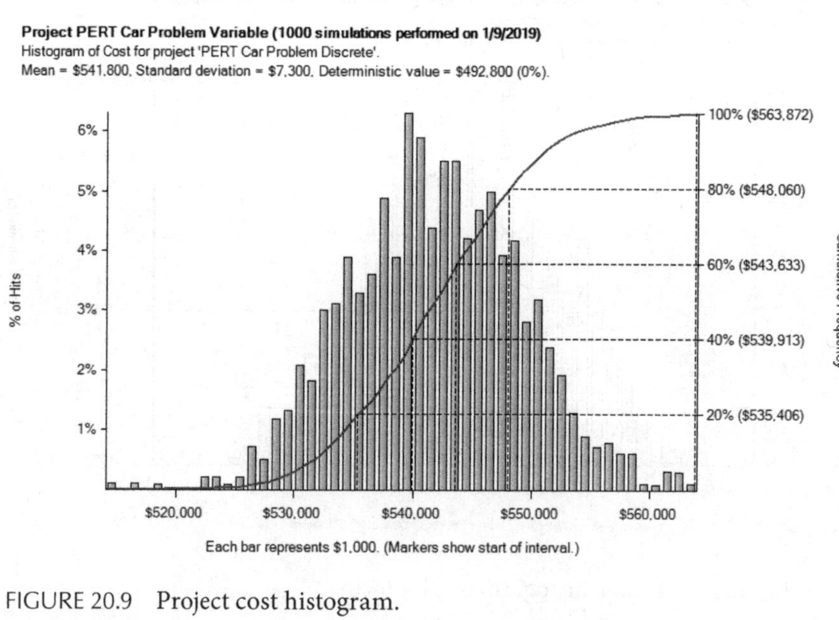

FIGURE 20.9 Project cost histogram.

Cost—\$1.422 million

SD = \$22,000

Upper-schedule range—3335 days (67 weeks)

Upper cost range—\$1.500 million

The reason for less variation in the simulated schedule is based on using a symmetrical profile (75–100–125). If a larger profile factor had been used, the simulation variation from the deterministic plan would have been higher. Testing other range values would be a good test case for sensitivity analysis. Note that the simulated cost mean value is approximately \$20,000 higher than the traditional view, and the upper range variation is even higher. Assuming the initial profile value fits the project characteristics, these output values show more of the project's true expectation hidden by the deterministic view. Explaining this output format to non-mathematical types is easier than justifying why the value is not variable.

Generating these results is relatively easy for performing "what if" type evaluations. One additional stimulus this class of exercise might bring is an increased interest in collecting more historical data to better profile the tasks. There is a clear linkage between the profile used and computed output, which is also educational. If asked today, little is known about the profile character of various project tasks. A good profile guess is a skewed beta structure, but that may not be valid in all situations. Beta profiles can represent a range variation that can skew the output to anticipate unexpected overruns.

20.7.8 Evaluating Cause and Effect Factors

Assume there is a recorded history of tasks overrunning estimates along the lines of an example profile of 80–100–150. An internal review indicates that poor resource management might be the most prevalent cause of task overruns. Simulation mechanics can test such events to measure the result. Even with such quantitative evidence, one must be careful in assuming this is the causal factor. Simulation mechanics ignore that question, but this process allows evaluating such points.

Because of the various internal factors that affect schedule and cost variation, one must be careful in assuming the project is affected only by task variation. Some cost components in the task may or may not be correlated to task variance times. For example, if a worker takes longer to do a defined task, does that mean he uses more material? Material may be fixed regardless of the execution time, so one must understand the general mechanics of the task to know how to treat the simulated output. Likewise, overhead and other non-direct costs may or may not be related to time overruns and certainly are not the cause. It is to understand what the variance trigger event is regarding both duration and cost. In this demonstration case, it is assumed that all component cost expenditures are based on execution time, which seems to be a weak assumption. Splitting labor and other cost types may be necessary to obtain a more accurate view of cost expansion. This type of effort will require additional data manipulation outside of the MSP data umbrella.

Beyond seeing the value in this form of more realistic output, the counter warning is to look at simulation correctly. It is not a clear cause-and-effect view. It is an analytical approach to producing a "what if" examination based on complex assumptions. Hopefully, both sides of this message are now clear. The model has provided a signal of potential variation based on a simple assumption. The output histograms are compelling and look very realistic. The operational rule is not to blindly accept computer output until the underlying logic is reviewed. The core question here is, "What causes a cost variance?"

As indicated earlier, simulation modeling has two major functions. First, it forecasts future schedule and cost status. Second, it provides a more robust tool for analyzing the internal process. Two advanced analysis processes are described below: tornado charts and risk-related plan view.

20.8 SENSITIVITY ANALYSIS

Sensitivity analysis is yet another form of plan analysis. It is designed to highlight which task areas of the project have the greatest impact on variability. The classic schedule and cost histograms displayed are useful for describing task date and cost variation. However, they do not help identify which tasks are the major drivers of this output variation. Moving to the task level of analysis opens the category of sensitivity analysis. This focus area involves a family of techniques to help understand each task's relative impact on the outcome. A common method of describing sensitivity is a tornado diagram, which has the shape of a famous storm. In the case of project management, the variation drivers are project work tasks.

20.8.1 Tornado Charts

Tornado or *butterfly* charts display sensitivity task ranks for the project. These diagrams rank the most sensitive task impact on the plan (schedule or cost), with the highest item shown at the top and declining values below that. In the project domain, this diagram describes the sensitivity of tasks regarding cost or schedule variability.

20.8.2 Tornado Schedule View

A slight modification to the basic sensitivity chart is shown in Figure 20.10. This view focuses on the schedule impact created by individual task variations. The columnar data is straightforward in terms of meaning. A simplistic interpretation of the graphical view says that the higher bars in the stack represent tasks with the most significant influence on outcome variability. In this example, tasks 48, 13, and 23 would be the ones most sensitive to affecting schedule outcomes, which is the same conclusion derived from the earlier chart. Simulated range values are also shown here, and as an example, task 13 shows that it could finish as early as 9/14

Task or esourc	Task or Resource Name	Sensitivity Bar Basis	ensitivit ptimisti Cost of	ensitivit essimisti Cost of
Task	Engine Design	Estimated	$1,4...	$1,53...
Task	Mechanics Design	Estimated	$1,4...	$1,50...
Task	Initial Draft	Estimated	$1,4...	$1,49...
Task	Electrical design	Estimated	$1,4...	$1,49...
Task	Body	Estimated	$1,4...	$1,47...
Task	Wiring	Estimated	$1,4...	$1,47...
Task	Electronic Compo...	Estimated	$1,4...	$1,46...
Task	Others Elect	Estimated	$1,4...	$1,46...
Task	Scope Definition	Estimated	$1,4...	$1,46...
Task	Other Body Parts	Estimated	$1,4...	$1,46...
Task	Chassis	Estimated	$1,4...	$1,46...
Task	Integration analysis	Estimated	$1,4...	$1,46...
Task	Part Ordered	Estimated	$1,4...	$1,46...
Task	Schedule Develop...	Estimated	$1,4...	$1,46...
Task	Safety test	Estimated	$1,4...	$1,46...
Task	Finishing	Estimated	$1,4...	$1,46...
Task	Electric - Electroni...	Estimated	$1,4...	$1,46...
Task	Test drive	Estimated	$1,4...	$1,46...
Task	Mechanic compon...	Estimated	$1,4...	$1,45...
Task	Charter	Estimated	$1,4...	$1,45...
Task	Station Electric	Estimated	$1,4...	$1,45...
Task	Documentation	Estimated	$1,4...	$1,45...

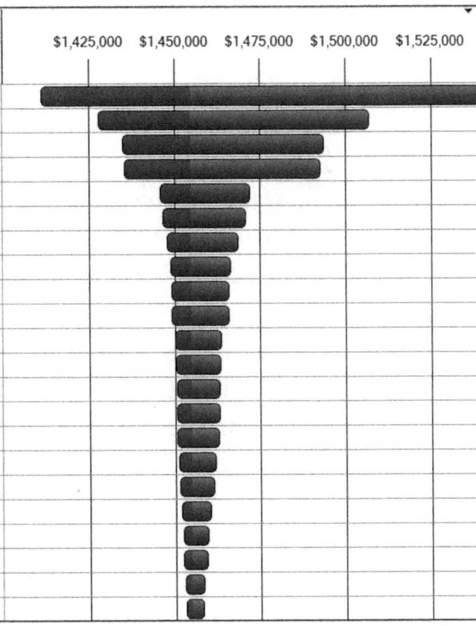

FIGURE 20.10 Cost tornado diagram.

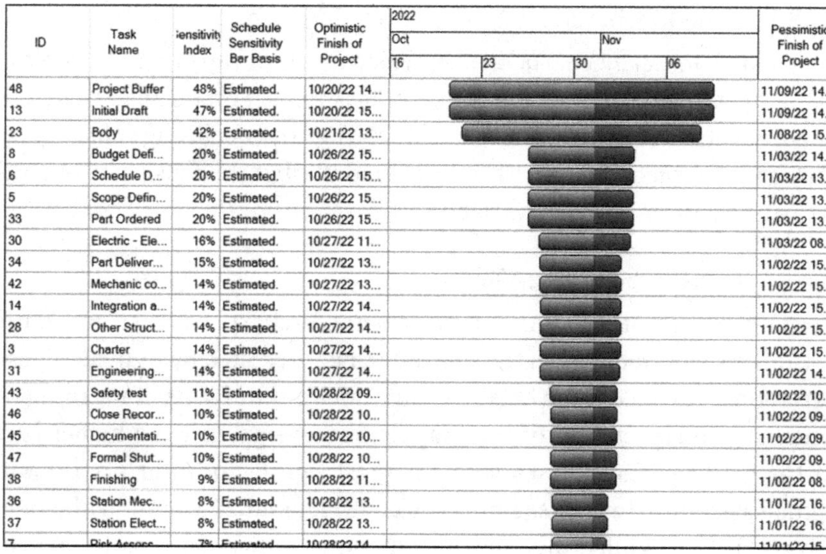

ID	Task Name	Sensitivity Index	Schedule Sensitivity Bar Basis	Optimistic Finish of Project	Pessimistic Finish of Project
48	Project Buffer	48%	Estimated.	10/20/22 14...	11/09/22 14...
13	Initial Draft	47%	Estimated.	10/20/22 15...	11/09/22 14...
23	Body	42%	Estimated.	10/21/22 13...	11/08/22 15...
8	Budget Defi...	20%	Estimated.	10/26/22 15...	11/03/22 14...
6	Schedule D...	20%	Estimated.	10/26/22 15...	11/03/22 13...
5	Scope Defin...	20%	Estimated.	10/26/22 15...	11/03/22 13...
33	Part Ordered	20%	Estimated.	10/26/22 15...	11/03/22 13...
30	Electric - Ele...	16%	Estimated.	10/27/22 11...	11/03/22 08...
34	Part Deliver...	15%	Estimated.	10/27/22 13...	11/02/22 15...
42	Mechanic co...	14%	Estimated.	10/27/22 13...	11/02/22 15...
14	Integration a...	14%	Estimated.	10/27/22 14...	11/02/22 15...
28	Other Struct...	14%	Estimated.	10/27/22 14...	11/02/22 15...
3	Charter	14%	Estimated.	10/27/22 14...	11/02/22 15...
31	Engineering...	14%	Estimated.	10/27/22 14...	11/02/22 14...
43	Safety test	11%	Estimated.	10/28/22 09...	11/02/22 10...
46	Close Recor...	10%	Estimated.	10/28/22 10...	11/02/22 09...
45	Documentati...	10%	Estimated.	10/28/22 10...	11/02/22 09...
47	Formal Shut...	10%	Estimated.	10/28/22 10...	11/02/22 09...
38	Finishing	9%	Estimated.	10/28/22 11...	11/02/22 08...
36	Station Mec...	8%	Estimated.	10/28/22 13...	11/01/22 16...
37	Station Elect...	8%	Estimated.	10/28/22 13...	11/01/22 16...

FIGURE 20.11 Schedule tornado diagram.

or as late as 9/23. Since this task has a high Sensitivity Index value, the project goal should be to carefully manage it and strive to achieve the earlier time target (Figure 20.11).

20.8.3 Tornado Cost View

The cost tornado diagram has the same basic interpretation for cost sensitivity, which is outlined above. Figure 20.10 shows the output for the sample program. In this case, the most sensitive cost variability is linked to four design-oriented tasks.

After the computed variation of the schedule and cost is examined, two additional views add insight into the inner workings of the plan. Tornado diagrams point out areas in the plan that the model shows to be the tasks that most affect the variation. This class of diagram falls into the category of *sensitivity analysis*. For example, which tasks have the most significant impact on project variability, and how might the critical path change as tasks vary in duration? In a deterministic task view, the critical path is also deterministic. That may not be the case where tasks vary. A companion to that view is to see which tasks are causing the most variability. Both points are new management ideas beyond what the traditional process mentions.

20.9 RISK-ADJUSTED SCHEDULE

Another valuable analysis of control-oriented tool is *risk mapping.* Full Monte has a method to produce such a view. Figure 20.12 shows the *Risk-Adjusted Schedule* with Gantt bars. This view compares the traditional project plan with parallel bars showing each task's variation. In many ways, this view offers the same general capabilities as individual views described above, with the added perspective of familiar Gantt graphical bars. The total figure is too large to display here, but the data format should be clear from the example plan segment.

As a comparative analysis example of simulation results versus the deterministic value, note the top task values in Figure 20.13:

Deterministic finish—10/18

Simulated Early finish—11/4

The comparative Gantt graph view of the two plan types highlights the range of simulated values versus deterministic outcomes. Analysis data of this type gives the project team added insights into potential task variances in the plan and where they are forecast to occur. Also, this helps to warn management about schedule variances with easy-to-explain

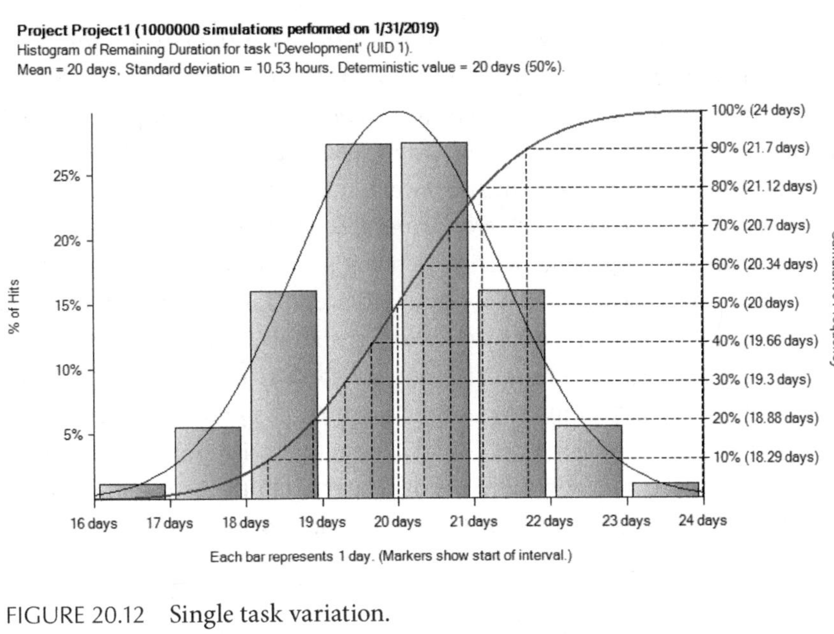

FIGURE 20.12 Single task variation.

ID	Task Name	Percen Critical	Start (MSP)	Finish (MSP)	Early Start Percentile (80%)	Early Finish Percentile (80%)	Early Finish Histogram
0	⊟ PERT Car Pro	100%	8/3/21	10/18/22	8/3/21	11/4/22	Graph
1	⊟ PERT Car F	100%	8/3/21	10/18/22	8/3/21	11/4/22	Graph
2	⊟ Planning &	100%	8/3/21	11/8/21	8/3/21	11/18/21	Graph
3	Charter	100%	8/3/21	8/17/21	8/3/21	8/18/21	Graph
4	Stakeho	0%	8/17/21	8/24/21	8/18/21	8/26/21	Graph
5	Scope D	100%	8/17/21	9/6/21	8/18/21	9/10/21	Graph
6	Schedul	100%	9/6/21	9/27/21	9/10/21	10/4/21	Graph
7	Risk As	100%	9/27/21	10/4/21	10/4/21	10/12/21	Graph
8	Budget I	100%	10/4/21	10/25/21	10/12/21	11/3/21	Graph
9	Manage	100%	10/25/21	11/1/21	11/3/21	11/11/21	Graph
10	Set base	100%	11/1/21	11/8/21	11/11/21	11/18/21	Graph
11	⊟ Engineeri	100%	11/8/21	4/25/22	11/18/21	5/11/22	Graph
12	⊟ Body / E	100%	11/8/21	1/10/22	11/18/21	1/28/22	Graph
13	Initial I	100%	11/8/21	12/24/21	11/18/21	1/12/22	Graph
14	Integra	100%	12/27/21	1/10/22	1/12/22	1/28/22	Graph
15	⊟ Mechan	0%	11/8/21	3/1/22	11/18/21	3/29/22	Graph
16	Engine	0%	11/8/21	3/1/22	11/18/21	3/29/22	Graph
17	Meche	0%	11/22/21	1/4/22	11/18/21	1/6/22	Graph
18	⊟ Electrica	0%	1/4/22	4/25/22	1/6/22	5/11/22	Graph
19	Electri	0%	1/4/22	2/15/22	1/6/22	2/24/22	Graph
20	Wiring	0%	2/15/22	4/25/22	2/24/22	5/11/22	Graph
21	Others	0%	2/15/22	3/25/22	2/24/22	4/7/22	Graph

FIGURE 20.13 Risk-adjusted schedule.

background reasons, should one wish to debate the logic of this variance. Finally, project variance data offers an exciting perspective on buffer sizing that should help explain the need for adding resources to the plan. Simulation helps make the need more salable and with some reasonable quantification as to size.

The top Gantt bar shows the deterministic task range, while the companion bar below shows the simulated value. Also, the project plan is sorted into periods. Analysis data in this format gives the project team added insights into task variance impact on the plan and where this is forecast to occur. Also, this helps to warn management about the general magnitude of schedule and cost variances with easy-to-explain background reasons should one wish to debate the logic of this variance. Note that these variances do not represent different data but rather the way variations in the original data can manifest themselves in the plan.

Suppose one were trying to implement the Critical Chain logic with buffers. In that case, simulation modeling offers insights into sizing if one were trying to implement the Critical Chain logic with buffers. If the simulation were accurate, one could see the expected buffer requirement from the histogram range. Experience with simulation improves this understanding over time. John Owen, CEO of Barbecana, suggests using *Schedule Margin* versus the term buffer as a less controversial term for defining plan variability (Owen, 2019).

20.10 ADVANCED THEORETICAL CONCEPTS

The Full Monte modeling logic includes several other perspectives that can offer additional insights into the project outcome. Each of these examples in this section highlights some of the shortcomings of the traditional waterfall model design encapsulated in the traditional MS Project plan that is essentially based on a 70-year-old network model view. The various logic examples show why the simulation approach adds a more advanced understanding regarding project outcomes. Each section below describes a conditional example that needs to be better covered in the classic deterministic project model.

20.10.1 Branching

One of the more serious shortcomings of the deterministic model is its limited capability to reflect task sequence (predecessors). Simulation offers the ability to handle both conditional and probabilistic task branching logic. One common usage for probabilistic branching is to estimate the impact a probabilistic test will have. For example, a test will fail some percentage of the time and must be redone, or it may have multiple different behaviors. There is no accurate way to model this situation in the deterministic structure, but it is very doable with the simulation option. This situation creates calculation errors in the traditional model.

20.10.2 Task Criticality

Another phenomenon that becomes more visible in probabilistic analysis is dynamics related to the critical path, as we have previously described here in the Tornado diagrams. That is because plan tasks are now viewed as probabilistic rather than deterministic, the process of simulation uncovers the potential for recognizing multiple critical paths as individual tasks are randomly changed during the analysis. A variable titled *Criticality* is used in this chapter to measure this situation. This metric reveals the "percentage of overall time that a task/activity is critical, including the times when it was not" (Full Monte User Guide, p. 72). This dynamic critical path perspective is yet another way in which simulation adds insight into the operational characteristics of the plan.

20.10.3 Correlation

"Correlation is the name given to the phenomenon of interdependence between random variables, such that the frequency for one of them depends on the actual values of the others" (Full Monte User Guide,

p. 66). Task correlation values of zero indicate that the task performance is independent, while values of positive or negative means that it is perfectly (positively or negatively) correlated with the performance of another task. In simple terms, this means that one variable has a strong behavioral linkage to the value of the other. This can only partially be recognized in the traditional network logic.

20.10.4 Merge Bias

> Merge bias is probably the single biggest reason why deterministic schedules are optimistic and, therefore, why simulation is so important. This situation results when two or more parallel paths merge either at the end of the project or serve as predecessors to some intermediate task or milestone and when two or more of these paths have a chance of being critical.
>
> (Full Monte User Guide, p. 77)

This phenomenon can build up throughout a project, creating greater variances between a deterministic and probabilistic plan view, as we saw in the sample program used for this chapter. One should realize that this essentially makes the deterministic plan inherently more optimistic than one might intuitively think.[1]

20.10.5 Merge Bias Example

All task durations are subject to uncertainty. In the best case, this uncertainty is symmetrical around the estimated duration. Consider a single task with an estimated duration of 20 days. If we model the effect of a ±20% symmetrical task uncertainty, the best-case duration would be 16 days and the worst-case duration 24 days. The resultant probability histogram is shown in Figure 20.12.

As expected, given the symmetrical uncertainty, there is a 50% chance of delivering by the end of the 20th day and a 90% chance of delivering by the 23rd day. Now consider two identical parallel 20-day tasks, with ±20% uncertainty, leading to a common successor. Intuitively, what are the chances of the successor starting on day 21?

Figure 20.14 shows the probability of the two parallel task successors starting on time (4/21) is approximately 25%. Intuitively, given the symmetrical uncertainty of the predecessors, we might expect the successor

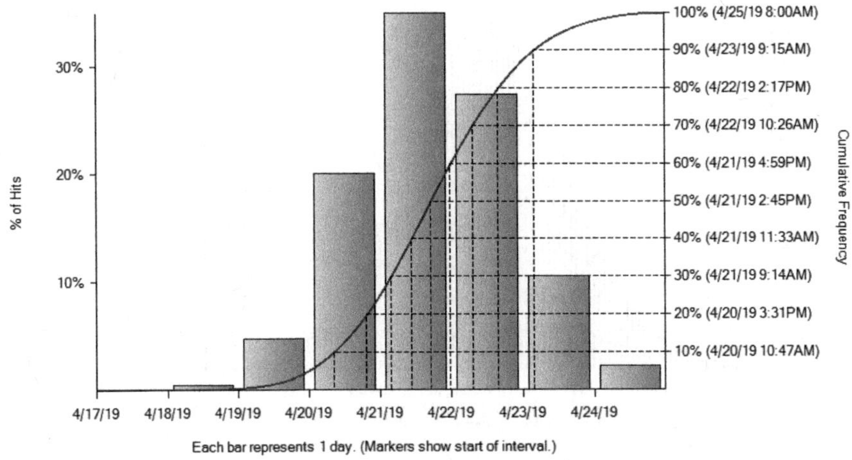

FIGURE 20.14 Start date for two parallel tasks with a common successor.

to have a 50% chance of starting on time, but because the two predecessors are in parallel, and both must be complete before the successor can start, the successor only has a 25% chance of starting on time. This effect is called *Merge Bias* and is one of the reasons that models created using the traditional deterministic network models are inherently optimistic.

For a more complex example, let's apply symmetrical uncertainty (± 0%) to all the tasks in our demonstration project and see the effect of merge bias. Task 90 is the final task and represents project completion, so it is used for the analysis (see Figure 20.15).

Because multiple parallel activities are being performed in the project, and all must be completed before successors can proceed, merge bias causes cumulative delays regarding the probability of completion by February 2 (the traditional network-defined date) as only 42% probable in the simulation view.

Note that the simulation model indicates a 50% chance of completion by February 4. However, most projects require a higher level of confidence than a 50% chance of delivering by a specific date. Many organizations require an 80% level of confidence in forecast dates. In this case, the 80% confidence date on the right-hand Y-axis is February 12. Do realize that this example is based on unrealistic symmetrical uncertainty, equally applied to all tasks. In practice, some tasks are more variable than others.

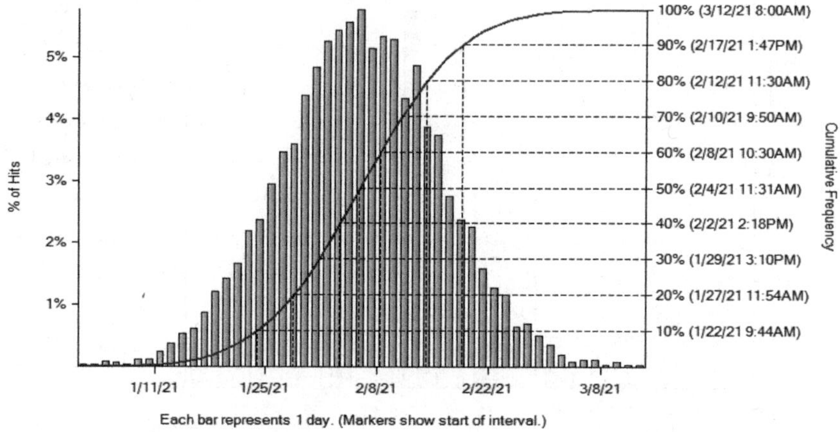

FIGURE 20.15 Task 90 completion histogram.

20.11 RISK HEURISTICS

Duration uncertainty, AKA duration "risk" or duration "confidence" can be quantified in two ways:

1. Capture discrete three-point duration (i.e., Best, Worst, Most Likely) estimates.
2. Use generic percentages for Best, Worst, and Most Likely durations based on experience.

For this example, the task list is broken into four risk groups (i.e., high, medium, low, and none). Table 20.4 outlines the risk grade definition for each grade.

The next step in the analysis is to define the technical specifications for these groups further (generally based on experience). Table 20.5 shows the technical task settings, assuming no historical data is available.

Using manual task data entry, the selected PDF can be applied.

20.11.1 Risk Path Mapping

Early discussion of plan execution management described the importance of monitoring the critical path as it dictates the shortest completion date. We have now entered a new domain of understanding. That is, there may be more than one critical path, assuming the project tasks are variable.

TABLE 20.4 Risk Grade Definition

None (Blank)	No Concerns with the Estimated Duration in the Schedule
Low Risk	Some uncertainty regarding the estimated duration.
Medium Risk	More concern with the reliability of the estimated duration.
High Risk	The estimated duration is considered provisional and may change based on the results of work earlier in the project logic, or the work is known to be technically challenging, or the work is dependent on unknown external suppliers.

TABLE 20.5 Technical Risk Grade Definition

Type	PDF	Optimistic (Best Case) (%)	Most Likely (%)	Pessimistic (Worst Case) (%)
Low Risk	Beta	90	100	115
Medium Risk	Beta	95	100	125
High Risk	Beta	95	110	150

Some describe this situation as *near critical paths*. Simulation modeling identifies which tasks exhibit this behavior by tracking which ones became critical during one of the simulation iterations. Cataloging this situation is *Risk Path Mapping*. This view focuses on the frequency of a task to exist on the critical path as the simulation iterates through hundreds of cycles. This view focuses on more than one critical path tracking concern, rather than the single deterministic path.

20.12 CONCLUSION

Simulation modeling is becoming an increasingly popular tool to evaluate the variability of project outcomes based on probability distribution estimates for various tasks. This output is a generational improvement in examining project behavior. This method has the advantage of not requiring a complex statistical theory background or heavy manual data manipulation to produce the results. Also, the growing number of commercial simulation utilities, such as Full Monte, makes this analytical approach to project plan analysis much more reasonable and flexible. Simulation modeling is a discipline that allows one to develop an increased level of understanding regarding the project variation and interaction of the tasks within a project. Simulation modeling is content-rich. The examples shown in this chapter have attempted to demonstrate the basic specification data requirements and core outputs of this process regardless of the brand of utility used. For all the reasons outlined, this chapter's material represents a core project plan analysis skill area that needs to be in the project technician's tool kit. The next chapter will offer another view of task variation handling with the Critical Chain theory.

The reader needs to understand that the topic of simulation is much more complex than shown in this chapter of introductory examples, but the analytical value is unprecedented in comparison to other analysis tools. Hopefully, the sample capabilities described here have provided sufficient motivation to place this subject higher in the personal growth proficiency category. The variety of improved project status insights alone should make this a worthy analysis and status presentation tool. Realize that the value shown in the traditional MS Project plan is, at best, a guess and usually a poor one. Simulation modeling offers a technique to counter that by showing the variability of the future output defines an erroneous project completion date. This modified view of project schedule and cost would be a more informative approach for management and stakeholders.

The following list summarizes improved analytical capabilities that can be improved using simulation modeling:

1. Ability to display scheduling characteristics of projects to include time, cost, and risk variability.
2. Ability to model complex relationships between activities.
3. Ability to model newly recognized complex characteristics of an activity (i.e., conditional occurrence, multiple critical paths).
4. Aid in analyzing resource allocation issues resulting from variable time results.
5. Aid in identifying the real constraints of the project—activity variability and resource profiles.
6. Allow evaluation of alternative scenarios with complex assumptions.
7. Show multi-project interactions.

To obtain the value from this class of modeling, more recognition of the value is needed. Properly utilizing this new process will take some time to incorporate into the organization's management process since the new views will be quite different from the static traditional models. These new project analytics are a key planning and control skill set that needs to be developed.

REVIEW QUESTIONS

1. What is the primary goal for pursuing simulation in project planning?
2. What basic capabilities does simulation offer over the traditional waterfall plan?
3. What are the basic steps in the Monte Carlo modeling building process?
4. What are the benefits of simulation modeling over traditional plan views?
5. What does task criticality mean?
6. What new perspectives are shown in a tornado diagram that help manage a project better?
7. Use the Internet to identify and describe the functionality of two different project management software tools for performing the Monte Carlo simulation.

CASE EXERCISES

Use the sample *Car Problem* from this chapter and reproduce the simulation results using a different task profile specification (Try a beta with 70–100–150). Compare the schedule and cost histograms and assess differences with the traditional output values. From the simulation output, answer the following questions:

a. What is the estimated range for the project duration with a 90% confidence level?
b. What is the latest project finish time estimate you would anticipate?
c. What is the estimated range for the project cost with a 90% confidence level?
d. Identify the top two main drivers for schedule and cost variability from the tornado diagrams.

NOTE

1 Special appreciation to Mr. John Owen, CEO, of Barbecana for his help in writing this section and for supplying access to Full Monte software.

REFERENCES

Barbecana. (2020). *Full Monte™ Software*, Barbecana, Inc., https://www.barbecana.com/

Full Monte™ User Guide, (2017). https://www.barbecana.com/?s=user+guide

Harrison, R. (2011). Library of Medicine (LM), https://www.ncbi.nlm.nih.gov/pmc/articles/PMC2924739/ (Accessed April 25, 2024).

Owen, John. (2019). *A personal interview, January 17, 2019*, Houston, TX.

CHAPTER 21

Critical Chain Model

Project N Motion Videos for This Chapter

21CC2	Critical Chain Model	CC Overview
21CC3	Critical Chain Model	CC PowerPoint Slides

This chapter offers a quick peek into an interesting process touted to improve project execution cycle times, called the Critical Chain (CC) model. This model contains techniques that have stood the test of time in enhancing project execution but are still underutilized in terms of acceptance. This model is based on the application of Dr. Eliyahu Goldratt's (1997) *Theory of Constraints (TOC).* [21C2]

21.1 INTRODUCTION

The CC approach to project planning and execution offers many positive attributes for improving the speed of completion. However, PMs must abandon traditional task estimating and project control practices. The mechanics described here will not include the complete set of options outlined in the original Goldratt (1997) model, as that seems to be one of the acceptance resistance issues. Instead, the view shown here will focus on the process components that add the most value for the least complexity. The core material described in this chapter will focus on task estimating and simplified buffers.

DOI: 10.1201/9781003532989-26

Somewhat surprisingly, the root of this model comes from the psychological side of the work issue, rather than some modified technical engineering process. At the core, the theory says that workers procrastinate about doing work. There is always something else to do, and the tendency is to wait until the last minute and often find insufficient time to complete the task as scheduled. This is called the student syndrome based on noted behavior of students' approach to school homework over the years and a learned trait that follows later. Mapping this logic to the project work, there is a known tendency to pad time estimates to protect overruns. A worker then knows that a task is padded, so they have time to defer starting it. Later, they find that the task is more complex than planned, so the padded task still overruns. This is one of the great revelations from the CC theory. The waterfall planning structure supports this behavior and has much to say about why traditional task-padded projects are still overrun. Also, the conventional process needs to be motivated to execute faster. Quite the contrary. So, how does the CC model "magically" speed up the execution cycle? It is not magic; instead, it deals with worker psychology and a disciplined approach to resource management.

21.2 CC MODEL PROCESS

Goldratt (1997) describes the expedited process as removing excess time by challenging the team with estimates based on a 50/50 basis, meaning they have a 50% chance of being either under the estimated time or an equal chance to be over. This concept falls against the current project control culture that forces task overruns, which does not seem logical and may be a primary reason the model has been rejected in practice. However, there are compelling reasons to think about this approach. As a starting point to examine the work process, Figure 21.1 illustrates three sequential tasks padded with estimates of ten days each.

The traditional scheduling approach would plan this string of tasks for 30 days. Now, we lay the student syndrome logic on this view. Knowing that there is padding, when does the worker tend to start working? The typical response is as late as possible. There are competing demands on

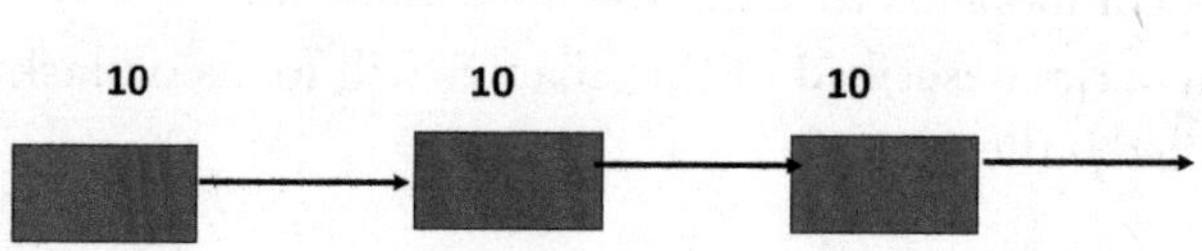

FIGURE 21.1 Sequential padded tasks.

one's time, and the reaction is similar to the school homework scenario. What happens here is the same as observed in real projects: padded tasks still have overruns despite extra time. So, this was a bad practice based on execution cycle time, but this is only a part of the answer.

21.2.1 Buffer Types

Goldratt (1997) proposed using buffers to protect schedules. So, a buffer would absorb that overrun when a task chain overran the time allocated. It is here that we need to deviate from the formal Goldratt theory. The following four buffer types are defined in the full model description:

1. Project—project the project
2. Feeder—project sub-chains within the project
3. Resource—project against nonavailability of critical resources
4. Capacity—protect against insufficient capacity between tasks

These are all valid overrun protection mechanisms to maintain a smooth workflow. However, adding all of these to the project plan does not add equal value, and they add increased complexity to the structure (authors' opinion). However, the functions related to project and feeder buffers are mandatory in the modified process. Each of these has an essential status-tracking role. Functions related to the other model buffer types will be transferred to an internal resource management process. The first step in accepting the basic logic of this model is understanding the role of buffers in protecting schedules.

21.2.2 Model Design Steps

The project, by definition, has a critical path, so that is one of the key task chains to attach a buffer. There can also be noncritical sequences of tasks that require similar treatment. Any chain of tasks forms a starting place to attach protection buffers. Figure 21.2 shows the same task chain outlined

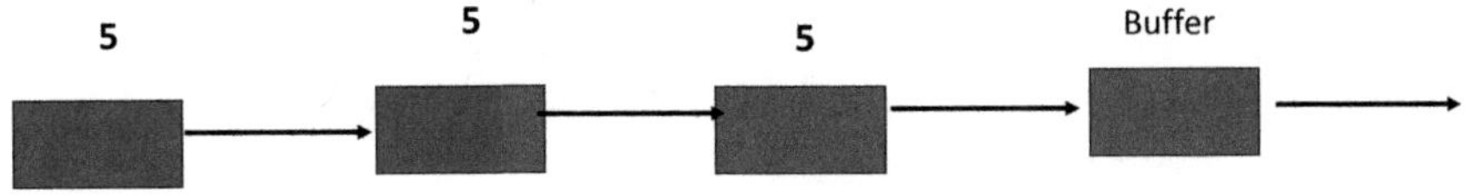

FIGURE 21.2 Sequential compressed chain tasks.

earlier but with the duration modified using the 50/50 logic and a buffer to protect the final date. This is the model of a task chain.

Simply cutting the task times and adding a buffer cannot be the answer, although it essentially is. The answer lies in the worker motivation layer and the associated buffer. These key ones need to be understood as the core mechanisms that drive the process.

The next most frequent question is, How big are these buffers? That answer requires usage history, but Goldratt (1997) recommended starting with buffer lengths of 50% of the chain. Subsequent experience with various project types will provide the background to refine this value. The chain length is still less than the original value, even at this large number.

The third area that will need to be explored later is the removal of all slack paths in the plan. This modification would signal that there is no idle time to waste. As a strong point, we will focus on buffer placement and sizing.

There are various reasons why projects do not operate this way, but none of those reasons include the fact that the alternatives achieve faster completion times. In this modified view of work, there is more pressure to work faster—the most likely reason this approach is not favored. It lies in the political view of what management wants regarding project status. That being fixed calendar dates. The 50/50 duration logic causes all scheduled dates to be meaningless, and the associated overrun culture cannot seem to absorb this approach. That is not a logical reason but a politically based one. Remember that traditional planned task dates are seldom met, even when padded. To absorb this new planning culture, various management aspects will require retooling the status-tracking mechanics, and buffer status will be the core of that new view.

Let us take a philosophical sidebar for a moment. If one were to ask whether the goal of a project was to supply management with specific dates or use a method that would finish the project quicker, which would be the better answer? This answer may look obvious, but traditional project management has deep roots in its fixed task culture, and the industry has rejected a change of this type. The question for this chapter is to evaluate whether this is the best response.

21.2.3 CC Definition

A short definition of the CC model provides insight into its logic.

> The CC method accomplishes improved execution speed through a network of linked task work chains with restricted task duration

and formally defined buffers to protect chain overruns. Also, a disciplined resource management process minimizes task execution delays.

On the surface, this definition looks like a motherhood statement for a project.

Compared to the CC model, the traditional view of the project schedule is like a padded city bus schedule with the workers standing on the corner waiting for the bus (next task) to arrive. Various portions of the traditional plan are designed to minimize overruns as these are viewed as adverse events. To avoid this, tasks are padded to accomplish this goal. The net effect of the practice is to elongate the plan duration. As Goldratt (1997) points out, the padding process does not achieve the desired goal and thus causes the overall cycle time to be longer than needed. We need to look deeper into worker psychology to understand this point.

21.3 WORKER PSYCHOLOGY

The CC model design logic is firmly based on worker psychology. Precisely, the following three cultural ideas lie at the core of this model:

- Student syndrome—execution procrastination
- Parkinson's law—overly conservative estimating
- Multitasking—multiple concurrent work tasks are inefficient

Each of these background ideas requires explanation.

21.3.1 Student Syndrome

A typical student will quickly understand this concept. Homework is assigned, but it is not due for a few days. Should I start right now or wait? Waiting is the typical answer to such questions, and that same logic works in the project world. If there is perceived extra time to get a task done, the typical reaction is to wait. However, then, a second issue emerges. The task estimate may have been underestimated, just like that previous homework assignment. For a padded task, the observed approach is to start late and then overrun the estimate because of the underestimation of the time required. From the project planning viewpoint, this is a significant work management point. Padding was created to protect overruns, but because of worker psychology, it does not accomplish that goal and produces longer cycle times than required.

21.3.2 Parkinson's Law

The central premise of the 1955 Parkinson's law essentially says that "work expands to fit the time available" (Cherry, 2021). In many ways, this is a behavioral companion to the student syndrome. One translation of this says that no matter how long a defined task is scheduled, the worker will overrun it. Like the student syndrome, both are categorized as human behavior rather than a technical fact, but both are valid descriptors of project work.

21.3.3 Multitasking

This is a third worker productivity factor that impacts the outcome. Multitasking involves assigning multiple concurrent tasks to an individual and, somehow, expecting more output than if assigned in sequential priority order. Every project participant has experienced this phenomenon and probably tried to make it work. Moralis and Dinan (2022) concluded that multitasking makes workers less efficient and fatigued. Some have described this dilemma as inefficient based on the extra overhead of starting and stopping work. Another view is goal shifting, where attention is moved from one task to another, requiring different "brain rules."

The three behavioral traits summarized above are the core drivers for the CC model. The internal model work rules minimize the time allocated to a task and focus on a single prioritized task. This part of the model is straightforward, but mechanically constructing that view into a project environment involves new processes. The one obvious problem is the visible decrease in estimated task times. Let us look at a simple project plan with decreased times to see what one of the model fixes might look like.

21.4 SAMPLE CC PROJECT

Figure 21.3 shows a simple traditional padded project plan consisting of four sequential tasks.

Task Name	Duration	Actual Duration	Predecessors
Schedule B	50 days	0 days	
A	10 days	0 days	
B	10 days	0 days	2
C	10 days	0 days	3
D	20 days	0 days	4

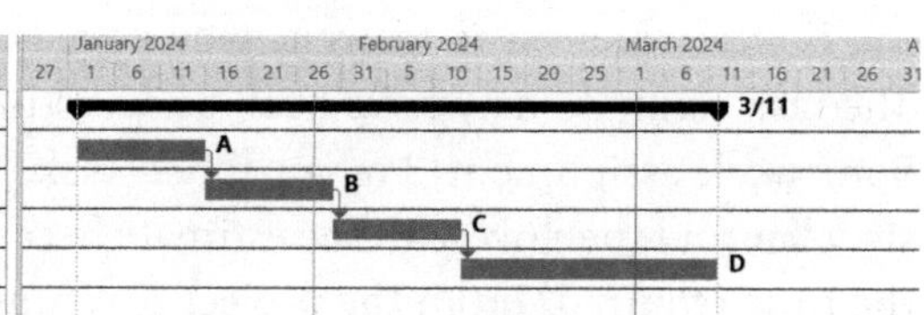

FIGURE 21.3 Simple project plan.

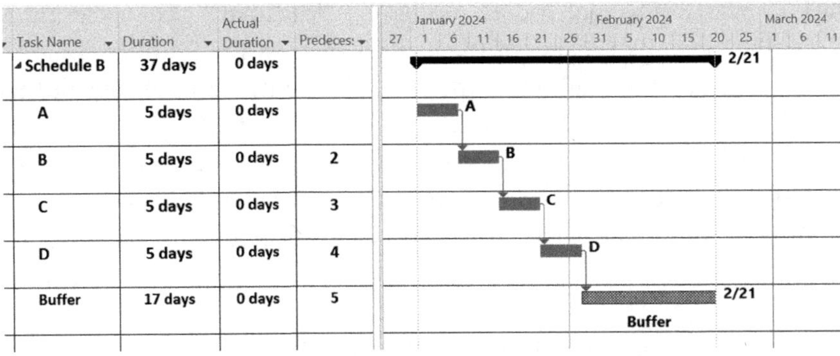

FIGURE 21.4 A 50/50 plan with buffer.

One of the basic concepts of plan estimating is to size project tasks at 50/50, which means there is a 50% chance of meeting that time and an equal probability of exceeding it. Assume the original plan was padded at some expanded amount to cover overruns. Converting the original plan to a CC format requires the following two steps:

1. Convert all tasks to 50/50—halving the original value in this case.
2. Add a project completion protection buffer at the end of the task chain to protect the completion date—this buffer is sized at 50% of the project chain tasks.

The resulting modified plan is shown in Figure 21.4.

Please note that the CC plan duration is 37 days, compared to 50 days for the traditional equivalent. Planned completion dates are 3/11 for the traditional plan versus 2/21 for the CC version. In terms of a modified planning approach, this is a simple conversion of plan task philosophies with a single buffer; however, that is overly simplistic for an accurate project view with multiple work chains.

Modifying this concept into the environment of multiple work chains requires introducing another buffer type. In each case, a buffer is introduced to protect that work chain.

21.4.1 Changing Plan Task Geometry

As stated numerous times throughout the text, every good idea brings "challenges" to the status quo, and this is certainly no exception here. If we admit that this compressed task time is aggressive, we must expect overruns in the chain. So, what is its scheduled finish date if we expect a task to overrun? Bingo! It does not have a "legitimate" scheduled finish date. This one event tears out the control roots of the traditional mode. Tasks are now "floating" along the chain. The only status statement that could be made is a measured task overrun, but the buffer is theoretically in place to protect against the overrun. Even though the traditional fixed task times have no true status meaning, the loss of plan-defined scheduled times seems to be one of the primary reasons this model was not accepted.

21.4.2 Buffers

Historically, buffers were frowned upon as padding, so the related process went underground to pad multiple tasks and hide the issue. On the surface, CC buffers look like sophistical padding (and they are). However, in this case, they are visible padding rather than the hidden version from the traditional view. Management has historically viewed visible buffers as schedule fluff that must be removed. This historical and cultural view is a more complex topic but recognizing what 50/50 logic means should help justify the concept of protection. The only legitimate technical issue is how large the buffer should be, not whether one is justified. In addition to this, the CC buffers are visible and outside of the individual task level.

21.4.3 Workflow Management

Managing the CC workflow also has significant changes to be recognized. Suppose the new project environment will function like a track relay race, with the tasks representing race segments and task resources representing the baton. In that case, the concept of resource management is elevated in priority. The CC resource management system must have the following characteristics:

- When resources work on an active task, they are dedicated to it (no multitasking).
- Because the chain has no fixed schedule, a system must alert the next resource group to get ready so they can start work immediately (no gaps between tasks).
- Aggregate management of overall resources must address capacity issues in advance (minimize resource shortages).

Allow a personal project observation regarding these points. Some evidence suggests that none of the three items above are well performed in most projects. Schedule overruns are heavily based on a lack of timely resource management. What is represented by the CC operational requirements falls into the definition of management discipline. There is nothing unusual or magical in the CC model. It completes the project faster by focusing on each task and supplying the required resources.

21.4.4 Critical Path Perspective

In purist form, the goal of a CC plan is to shrink all tasks in the plan so that they are all critical. (i.e., no extra time anywhere). This mechanically complicates the plan structure, as the example below will show, but in theory, this fits the task expediting goal. Sometimes, a valid theory segment must be compromised to gain acceptance, which is assessing this goal.

21.4.5 Status Tracking

Now, to the final plan geometry change requirement. The traditional project has defined start and finish times for each task (padded as they may be). During execution, status is defined by how well these tasks have been performed (i.e., plan versus actual). Review Chapter 13 for details on the traditional approach.

The loss of fixed task plan dates significantly changes the project status culture, which is heavily influenced by fixed Gantt bars to show the plan. The CC design does not fit this project view in many ways, even though the padded bar values do not accurately show performance. Ignoring that the traditional model does not accurately show the unpadded cycle time, this view is heavily ingrained into the current status-tracking culture. It will take more discussion later to fully describe what the variable time chain logic does to the status-tracking process.

21.4.6 Model Assessment

Reviewing the CC model, one finds insightful management ideas on improving task throughput, representing management gems for all projects. Breaking away from fixed time estimates needs to be a valid strategic goal because it better matches reality. Also, buffering to protect work chains is an area to pursue beyond the CC formal view.

Since its introduction some 35 years ago, the CC model has been dramatically under-recognized in the project industry in mechanical and behavioral forms. Descriptions in this chapter are judged to be the model components that add the most value to the result. Industry evidence is that using this model can cut project cycle times by 25%. If senior management believed in this capability, would they demand a change in CC processes? It was rejected earlier because it required more management discipline than the project culture wanted to deal with. We leave the reader to deal with an answer to this.

21.5 TWO-CHAIN MODEL

We now have sufficient theory to move into a more complex view of the model structure. This version will show a primary chain and a secondary chain. More importantly, this example will better illustrate some of the new geometry concepts outlined above.

The sample program shown in Figure 21.5 contains a traditional plan structure with a critical path and a slack path.

The slack path can be viewed as a feeder chain in this case. The two work chains are easy to see. Also, let us assume that these tasks are based on 50/50. At this point, no buffers have been defined. This version is essentially a traditional project plan, except we now anticipate task overruns, so showing fixed completion data is erroneous.

Three more steps exist to modify this plan to fit the CC design rules. These are as follows:

1. A primary chain buffer needs to be inserted.
2. A feeder buffer must be attached to the feeding chain task chain (CDE) to protect its completion, impacting the project task chain.
3. All tasks in the network will be converted to critical (optional step).

The first two steps outlined above are relatively simple. Just add a buffer at the end of the feeding chain and one at the end of the project. A

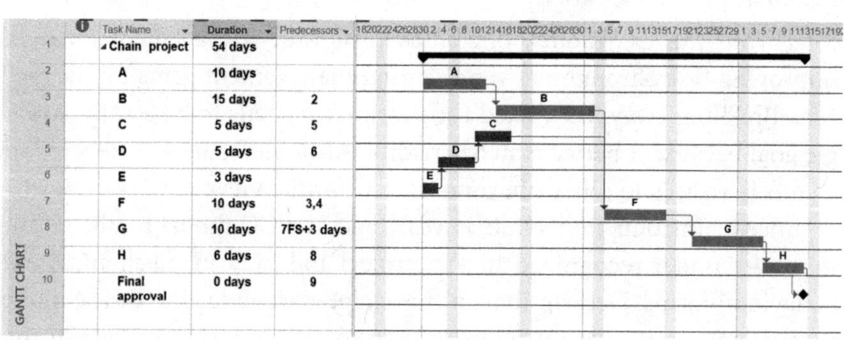

FIGURE 21.5 Two-chain project plan without buffers.

more complex design problem occurs in making all tasks critical. That can be done the right way or the easy way. The right way to do this is to use the predecessor code, SF, to force the feeder chain to start as late as possible. That same approach could be used for any task that had slack. However, personal experience suggests that this practice may not be worth the complexity it brings to the design. Some might even argue that this is overkill—leave the slack as is. Purists will argue this point, and so be it. However, the CC design rule clarifies that removing all slack is the proper setup to encourage faster execution.

Now, let us go back to how to do it. A straightforward way to accomplish moving the tasks as far to the future as possible is the following:

1. Use the Information option for each task
2. Select the Advanced tab
3. Select the Constraint type
4. Select the option "As late as possible"

Once this has been done and the buffers are attached to each chain, the result is a reasonably legitimate CC compressed schedule structure (see Figure 21.6).

Looking at this structure, the feeling of speed seems to jump out. Work rushes down each chain. One planned schedule gap for the chain starts at task G (FS+3). All tasks are meant to start as quickly as the prior task is completed, which means the associated resources must be managed

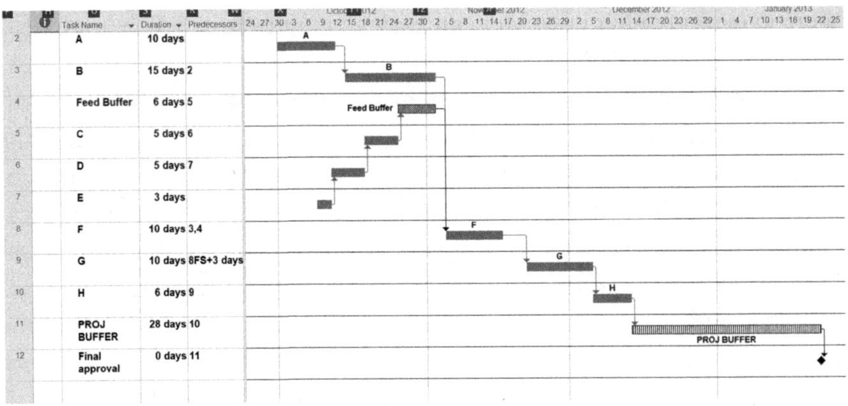

FIGURE 21.6 Two-chain model with buffers.

promptly. When a resource is scheduled for a task, no other work is assigned to these workers. No multitasking. As stated earlier, this structure has no magic, but notable operational differences exist that are both visible and subtle.

As a task overruns its scheduled duration, the amount of the overrun is extracted from the associated buffer. If all goes as planned, the buffer will be exactly consumed at the last task. The completion prediction is perfect! No, that is not the way it will work! The challenge during execution will be to monitor the remaining size of a chain buffer versus the remaining work to be executed. There is no defined schedule for these tasks. Just go down the chain as fast as possible. Some tasks may take less than the estimate, while others overrun. Every effort is made to move on to the next task as soon as one task finishes.

From a raw mechanical view, it is tough to ignore the logic of this approach, so the question that follows is, "Why has CC not become popular as a planning and control technique?" If you can answer that question, you can also answer many similar questions regarding why superior tools and techniques have not been incorporated into the standard tool and process mix.

21.6 EXECUTION PHASE

As indicated above, the status of a CC plan is heavily dependent on buffer status because there are no fixed task calendar schedules. Removing a fixed calendar date for status comparison seems to be one of the major stumbling blocks in embracing this model. The industry has a locked-in cultural bias toward "scheduled" work, meaning assigning a date to compare for control reasons. Most have never experienced a senior executive who, when looking for status, would ask for anything other than a calendar date.

This textbook section has focused heavily on recognizing task variability and better techniques to manage that reality. Traditional project plans give the image of a clear definition; however, the values shown are not accurate. Continuing to provide the wrong answer to stakeholder questions is not the best strategic approach for respectability. The simulation chapter (see Chapter 20) demonstrated the probabilistic view of the project when individual task variability is considered. This suggests that the delivery models used should also recognize this. CC mechanics do not show the same histogram view of reality, which is desirable, but it attempts to produce a more accurate completion estimate. Possibly, the combination of simulation and CC has the potential to deal with this status issue in the future.

The examples shown here have attempted to illustrate the cycle time reduction and the issues with status tracking when migrating to the CC model. Other supporting tools, such as simulation, can add value to this new view. One high-level view of CC is that it is designed to "shoot" at a fixed target at a fixed target date, but it does not have much capability to show quantitative time beyond the buffer status view.

Conversion of the traditional project view into a CC model would have significant positives over the current traditional approach. The lack of willingness to do this seems more of an unwillingness to change ineffective traditional methods and convert to the "Mysteries" of buffer management for status tracking.

21.7 CC DESIGN CHECKLIST

The following steps summarize the fundamental mechanics in stabilizing a CC model plan (Kendall et al., 2005):

1. Task estimates do not have padding. They are planned at some probabilistic completion level, such as a 50% probability that the task can be completed at the estimated time.
2. Project team members are dedicated to a single prioritized task until it is completed and challenged to execute this as quickly as possible with minimal disruptions from other activities. Periodic status reports are required to indicate the time remaining for an active task. Every effort is taken to eliminate delays and work procrastination.
3. Multitasking is eliminated by assigning workers to tasks in priority order and completing that task before moving on to a new task. Industry experience suggests that multitasking creates inefficiencies amounting to 30–40%.
4. Managing tasks by the due date is not followed. Workers and tasks are not measured based on scheduled completions. The management approach is to pass on the task to the next activity as quickly as possible. *This is a track relay race metaphor.*
5. There are no defined task completion dates. Just move along as fast as possible.

6. The longest sequence of dependent tasks can be seen more clearly by considering resource and logical dependency. This longest task chain sequence may cross logical paths in the plan network. So, the management view deals with resource status across the chain, not critical path schedule management.
7. Buffers are a vital mechanism to manage desired schedules. These will be defined to protect various aspects of the project, including the project completion date.
8. Project status will be monitored through the evaluation of buffers.
9. Task completion is viewed as passing the baton to the next task in the work chain. When a task is nearly completed, the next task's resources are queued to get ready to go immediately after the preceding task is complete. In this manner, the chain has little delay, and early completions can speed up the process. Every effort is made to create this culture in the project team.

Recognition of resource constraints requires a staggered introduction of multiple projects into the system, and resources for these efforts will be allocated on a priority basis. This method improves the completion of projects in priority order, increases the outcome predictability for each project, and increases the effectiveness of critical resources by minimizing multitasking. Shorter project cycle times improve the overall delivery of the project set.

21.8 BUFFER MANAGEMENT

Because every activity in the project is scheduled with no safety padding, some tasks have a high probability of overrunning their estimated time. Embedded protective buffers are used in various places to control the overall schedule from overrun. Placement location logic for the buffers will be discussed below, but for now, let us look at the basic mechanics of a buffer. Imagine "chains" of tasks in the plan. Each task is sized at a 50/50-time estimate (essentially a median value). All buffers aim to protect that chain from overrunning its overall schedule. Of course, the complexity trick is correctly estimating how large such a buffer should be. For this discussion, a typical number will be used, but accurate estimation can be improved through experience with specific project types. Figure 21.7 provides a

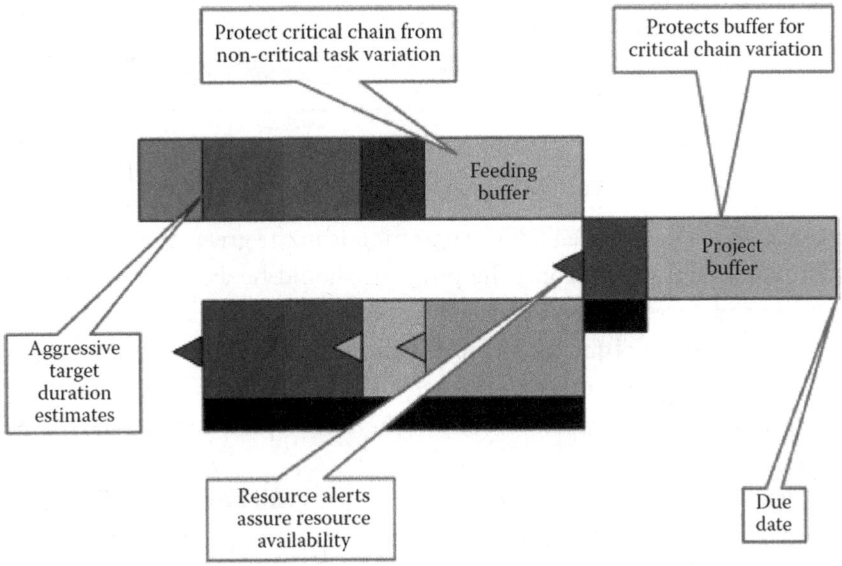

FIGURE 21.7 Buffer logic.

good overview of buffer logic where feeding and project buffers are used (Patrick, 1998, p. 6).

A feeding buffer represents a set of noncritical path activities that must be protected from impacting the critical path. A project buffer inserted at the end of the project critical path protects the project from schedule overrun. In this operational mode, the focus for management moves to buffer status because some tasks are now expected to overrun. There is no value in looking at the plan versus actual task performance, which has become one of the main cultural changes associated with this implementation. If appropriate buffers protect the various chains, the overall project will finish on or before the planned schedule. Assuming this model works as advertised, the CC model would greatly interest all organizations. Growing evidence shows it accomplishes the desired result when properly and adequately implemented.

One of the typical ways to view buffer status is to divide it into three equally sized regions (see Figure 21.8).

The first third of the buffer could be color-coded as the green zone, the second third as the yellow zone, and the final third zone as the red zone. The associated protection buffer is "eaten as the chain duration overruns." So, a ten-day overrun would extract that number of days from the buffer.

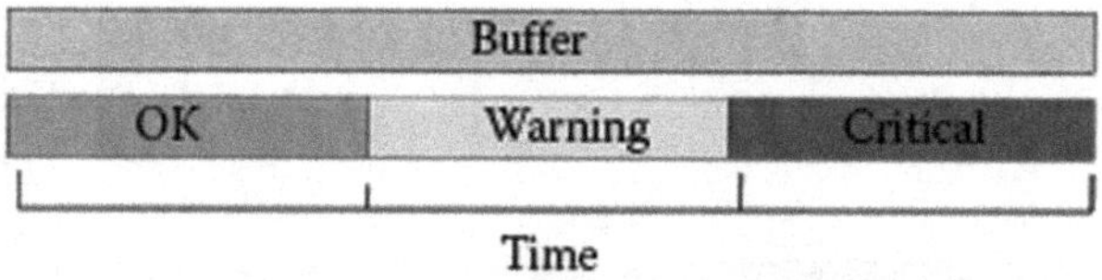

FIGURE 21.8 Buffer "Burn Rate" status scheme.

No action is taken if the buffer penetration is in the green zone. If penetration enters the yellow zone, the problem should be assessed for possible review and corrective action. Finally, if penetration enters the red zone, a careful review is required to protect the project schedule. Action plans should involve finishing uncompleted tasks in the chain earlier or accelerating the remaining work to correct the buffer penetration rate (Retief, 2002).

Defining the project and feeding buffer sizes will be a new skill requirement for the project team to master with previous comparative data (Geekie, 2008). Published research is available on this topic, but it will be necessary to test various methods before settling on a local approach. Leach (2005b) offers detailed explanations for three methods in *Critical Chain Project Management*. Reasonable estimation of buffer sizes is a critical element of successful implementation. The first time this model format with multiple buffers appears in a project plan, it will likely necessitate an explanation for various stakeholders, and it may even be necessary to prepare a high-level "shadow" plan showing only key completion dates. So long as the project team meets those key dates, the lower-level view should not be required outside the project team.

Conceptually, buffer sizing is a probability and statistics-based issue. The primary methods follow a similar logic to the one in the introductory CC plan. Simulation techniques can also provide insights into the variability of a particular chain. However, as with most items in the project management domain, the forecasting accuracy of execution data is the Achilles' heel of inaccurate quantification. So, until there is a more tested local strategy, Leach (2005a) recommends utilizing the "Half the Chain" method for introductory buffer sizing.

Many organizations will likely reject such an approach because it seems too simple, and a 50% chain buffer size appears too large; however, this has some mathematical basis resulting from how the time estimates were initially formed at the median size. Remember, the role of the buffer is to protect critical points in the chain from delay and approach their usage with that mindset.

21.8.1 Buffer Types

So far, the model discussion has focused on the project and feeding buffer roles, but other buffer types are defined in the formal CC model. Each buffer type serves a protection-type role—protecting the project from a delay. This section briefly summarizes the five buffer roles Goldratt (1997) outlined in the formal model.

Project buffers: As initially shown in Figure 21.4, this buffer type is located at the end of the critical path to protect the project completion date from the compressed estimates in the associated chain activities. This buffer protects the project because any total overrun of activities in the critical path will affect the project schedule. As a rule of thumb for an initial strategy, the project buffer should be half the size of the chain it protects. Even with this size, the project should still be 25% shorter than the traditional equivalent.

Feeding buffers: These buffers are placed wherever a noncritical task chain activity joins the primary critical chain. Their role is to protect the primary chain from interference. The sizing of the feeding buffer is typically recommended to be half the related chain duration (Goldratt, 2007).

Resource buffers: These buffers are designed to protect the project from resource shortages, so they are placed before tasks that utilize scarce resources. This is a protection for resource gaps.

Drum buffers: A drum resource is a system constraint and is viewed as a potential bottleneck to the schedule. For this reason, projects are synchronized around these resource constraints, otherwise resulting in schedule delays. This buffer is placed so that critical resources will be assured when needed. In that respect, the drum buffer is like a feeding buffer. They are placed in the project schedule immediately before the activity using the drum resource. This buffer protects the activity's start date. These resources will be allocated on a priority basis, and buffer attempts will be made to ensure that the priority project does not have to wait for that resource.

Capacity buffer: When synchronizing multiple projects around a drum resource, a capacity buffer is created between the tasks of different projects on which the drum resource is needed (Retief, 2002). Figure 21.9 shows how this buffer would be applied.

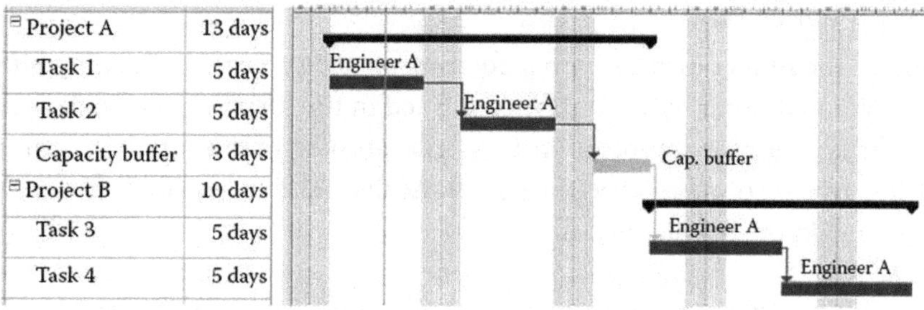

FIGURE 21.9 Drum buffer between two projects.

Reader Note: Experience indicates that implementing resource, capacity, and drum buffers often is viewed as having more complexity than benefit. For the initial implementation phase, it is recommended that the resource capacity issue be concentrated on an external disciplined process to manage these buffers outside of the plan task structure. Their logical role is valid, so this is purely an operational recommendation for initially dealing with the problem.

The core status assessment variable for each buffer type is the rate and amount each buffer consumes relative to project progress. Assuming the task variation throughout a chain is uniform, the project buffer would erode at the same rate as the completed chain. If the buffer sizing is perfect, it will be entirely consumed when the final task on the chain is completed. Monitoring each buffer's status is now up to the PM's knowledge and skill to determine the corrective actions necessary to recover if the remaining buffer size is deemed inadequate.

A second aspect of CC status tracking is to review the remaining duration for all active tasks. Resource and completion status will be monitored for all tasks in progress based on the number of days estimated to complete respective chains. If the remaining duration value on an active task remains static for a short time, a warning is triggered to review that activity. Given the model design, this logic is caused by the fact that no active task should stay static. No progress means that something has gone counter to the plan.

21.9 RESOURCE ALLOCATION

Once the CC plan structure is established, the problem becomes resource allocation and allocation conflicts. In Figure 21.10, resource "Y" was initially scheduled to perform two tasks simultaneously, so there is a conflict regarding which activity to perform.

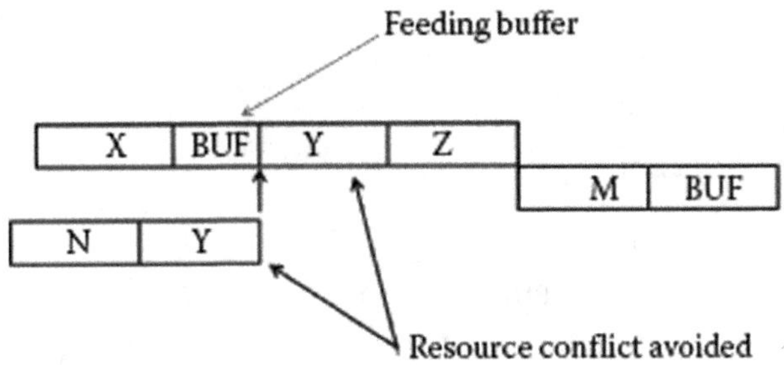

FIGURE 21.10 Resolving resource conflicts.

To resolve the conflict, the feeder branch is planned to start earlier and now becomes part of the longest chain. Also, a buffer is inserted after activity "X" to allow time for the feeder chain to complete.

The placement of resource buffers helps the PM to focus on that aspect of the project. These allocations become a communication target between the project schedule and the team. As work progresses, the resources document actual time estimates for completing the task. When a predecessor task is reported, it will be completed in five days. Hence, Goldratt (1997) recommended that the resource alert process be triggered to inform the successor task resources to prepare for their work to start at that planned time. In the track relay race metaphor, this is analogous to getting into their lane and preparing to accept the baton. This mechanism represents a dynamic countdown for the successor. If the predecessor reported two days later that there were still five days remaining, this information would be passed on to the successor. Note that this process heavily focuses on the resources needed and preparing to start the next task. The cartoon picture in Figure 21.11 illustrates the resource alerting process.

The final descriptive step of CC involves tracking and managing the schedule. The theme of this step is *buffer management.* Earlier discussion described the buffer's green, yellow, and red zones to measure status. This status evaluation is more like that of a moving train, with the goal of not being late. More quantification sophistication can be added by calculating the "*burn rate*" of the buffer so that if one is decreasing faster than task completion, it will trigger a review of that chain to see what needs to be done. The buffer management process highlights potential problems much earlier than they would ordinarily be discovered using the typical status techniques because buffer status can be reviewed anytime. Also, there is more ability to have improved insight into trends.

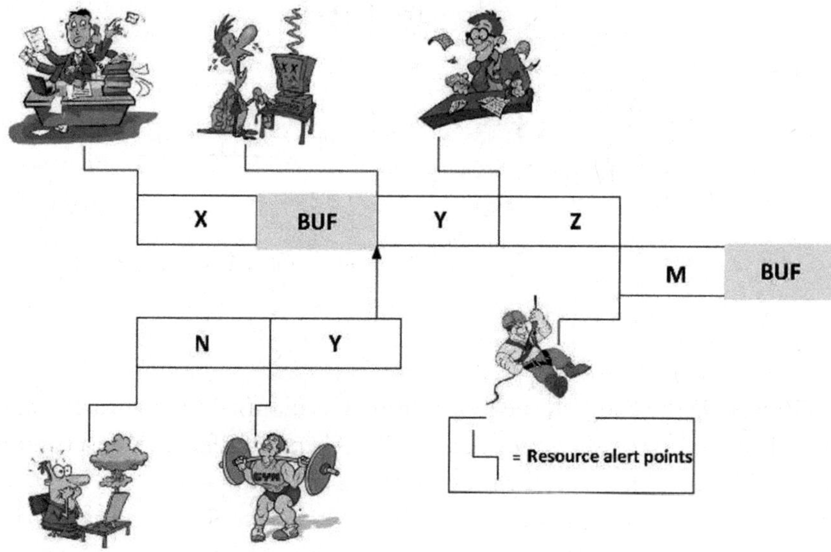

FIGURE 21.11 Resource readiness alerts.

One of the unfortunate consequences of this model is its negative impact on the formal process of using traditional Earned Value (EV) as a status metric. Chapter 22 shows a customized approach to modify the standard approach into a customized modification using variable task times.

21.10 IMPLEMENTATION CHALLENGES

There is evidence that successful implementation of the CC model process will improve project throughput, and such improvement has been observed in multiple documented cases. However, this method's operational and cultural impact on the organization is significant, and its implementation makes the conversion more complex than just changing the planning structure. There are three significant areas where complex conversions must be dealt with: organizational maturity, leadership challenges, and skill requirements. The sections below will describe some of the main issues in these areas.

Implementation of this model requires the project team and other stakeholders to be versed in the method's fundamentals, which will require a new way of looking at the project plan and subsequent status. So, for the transition to work, revisions will need to be made to the existing planning methodology, which historically has been a complex item to standardize. Once the new plan format is in place, the status-tracking process resource allocation methods will have to change. Some stakeholders would need to be enticed to try the method with a promise of improved completion times. Unfortunately, there are subtleties involved that will require more disciplined cultural change in resource management. Mature organizations with proven track records would be the easiest to convert because much of the operational discipline already exists. The less disciplined an organization is regarding overall project planning and resource allocation, the more noticeably tricky this change will be. PMs with a proven track record would likely be the best trendsetters for this change. Stakeholders and team members tend to follow creative leadership in new territory. Finally, mature supporting processes like Integrated Change Control (ICC), risk management, issue/action management, and communication processes are essential for the success of CC (Leach, 2005b).

As indicated earlier, the concept of extensive buffer use and the related sizing of these will cause critical responses from traditional managers. Historically, PMs have been discouraged from visibly adding buffers into project schedules. Hidden buffering was accomplished by padding individual tasks and hiding them from sight. Consequently, all players did not recognize buffering that existed in the traditional plan. Even a visible risk contingency fund was questioned, and we have never observed a visible scope reserve in an internally organized plan. These items are just glossed over, yet all know these factors will come into play later. The CC model demands a culture of lean tasks with visible buffers to handle overruns. Even though risk reserves and scope change are not formally recognized as buffers, they serve the same role. If something as obvious as these are not recognized in the project plan, it suggests that introducing the CC buffers would be of even less interest.

Seeing visible buffers in a project plan flies against the standard operating process in most organizations. The irony of this is somewhat confusing. Anyone around a project realizes that change requests are normal, and they add work to the base plan. If one believes that risky events happen, one should not quibble with the concept of recognizing these events. So, if the existing plan is supposed to represent what will happen, where do the required extra funds come from for these two events—task padding became the answer. The impact of these collective items is spread across the task list by adding extra time to those tasks. The project plan is to have hidden buffers on each task. Goldratt (1997) tried to sell the industry the negative static view of the traditional model, but he failed, and the original practice remained. When implementing CC, managers and stakeholders will be challenged to accept not only the existence of the chain concept and its buffers but also the acceptance of using median task estimates and visible buffers. Patrick (1998) makes the following comment regarding the nature of buffer use:

> Once developed, assessment of the full schedule, including the contribution of the buffers to project lead-time, provides a clear view into the identified potential schedule risk for the project. In non-CC environments, when a reserve is included, it is often hidden either in the management reserve or in other less visible sources. The common practice of keeping these components off the table hides their true impact and implications. The open and explicit communication of buffers allows a clear assessment of what could happen 'in the best of all possible worlds,' versus what might happen if individual concerns accumulate to affect project performance.
>
> (P. 44)

A third cultural implementation resistance issue lies in the approach to task status tracking. CC tracking moves the process from milestones or comparing activity calendar completion to measuring buffer status. Frequent task duration overruns are expected because task or activity estimates are at the 50% probability level. Because of this modified task approach, the "due date" concept loses meaning. It refers to "as quick as you can finish the task" (Patrick, 1998, p. 5). Stakeholder understanding and agreement on this new view are crucial to success. Also, securing stakeholder agreement on an acceptable project reporting methodology must be resolved during the project planning phase. Leach (2005b) and Patrick (2001) both recommend using a "Red-Yellow-Green" buffer status approach for reporting. Figure 21.12 illustrates another schematic approach for adding more granularity to the status view.

Compared to traditional project management, the philosophical status-tracking changes outlined here shift away from "what was previously done" and move the focus to a metric representing a more forward status look through the chain. Also, the various buffers offer insight into potential resource roadblocks. In other words, CC has more of a future look than a historical one.

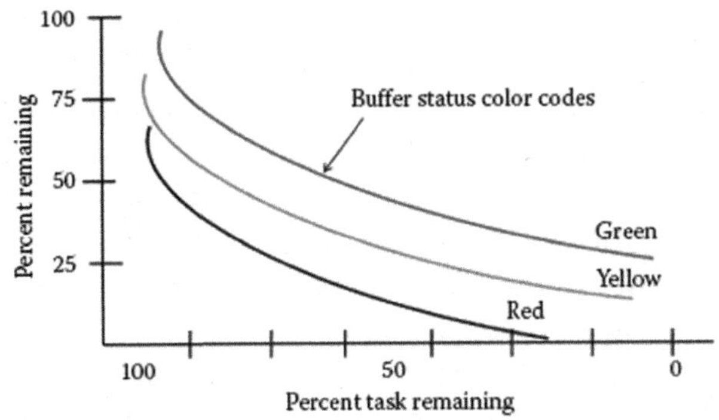

FIGURE 21.12 Buffer consumption reporting.

21.11 CHANGING THE CURRENT ORGANIZATION

Organizations with established Project Management Offices (PMOs) or standardized project management policies and procedures may not have the flexibility to implement CC. For example, using a 50/50 estimating model may be a losing proposition to some organizations. Why plan for tasks to overrun? It looks like you do not know what you are doing. Once you see the logic of the CC model, this attitude will be hard to understand, but it will exist. Leach (2005b) asserts that formal training must be a part of the implementation process, and below is his statement on that aspect:

> assuring that the CCPM approach permeates all policies, procedures, and measures of the organization and is formalized into training programs to ensure that new people are properly indoctrinated into the organizational process. In the end, CCPM should not be an additional thing. It should just be how we do business around here.
>
> (P. 193)

There is an adage that says, "If it ain't broke, don't fix it," and this attitude often results in resistance to change. Some project practitioners may argue that CC is just "rearranging stuff." To counter that statement, try to influence the notion that the starting place for CC is a good critical path network that has been effectively resource-leveled. Using that as a base, the CC approach enhances the ability to optimize the schedule. It does this through an improved method of handling task variability and a creative approach to project monitoring and control (Kendall et al., 2005).

Organizations suddenly faced with the challenge of estimating activities in the 50% probability range may counter with, "We already use overly aggressive estimates and always work to meet or beat them" (Leach, 2005b, p. 207). That response is not a reality, given task variances. A reserve in CC is a padded task in the traditional model. If an organization was factually doing such restricted estimates, the overrun rate would be evident, and there would be no buffers in place to protect the final answer. Therefore, they are arguing against themselves.

Mature organizations often develop existing processes with project templates to aid in plan development. One such example is an estimating technique or model that is "accurate." This might mean that it is seldom overrun and has significant internal buffering built in. A standard project life cycle template based on a traditional network structure might hold the same. In such situations, there is likely to be resistance to change approaches for an environment that is working adequately. An organization in trouble might be much more receptive to new ideas.

The issue of status-tracking changes can have both political and cultural resistance points. The most likely scenario is the existence of a standard tracking and reporting system that everyone understands. Its format is embedded in the corporate culture. Second, there could be a situation where the customer dictates a contractually obligated traditional approach to use a traditional approach for an external customer. These would be very tough areas in which to start CC. A third minority potential resistance would come from the scenario where traditional EV reporting was required. CC has destroyed the fixed task view that traditional EV requires. Chapter 22 will describe a modified variable task view EV approach, which might also be modifiable to fit the CC process. A fourth resistance scenario involves customer-required milestone reporting, which is not the usual way CC models look, but this can be accommodated by inserting the milestone at the end of the buffer. This might be an advantage to CC because the reporting requirements appear high-level, with interest only in the milestones rather than a more detailed task view of the project. Christ (2001) relates the following from his own experience, but more can be found in his article:

> There was an assumption in place that since the customer required Earned Value Management (EVM), no other method of management [approach] was authorized or acceptable. When queried, the customer indicated that factory management methods were

> company business, but the customer wanted data in EVM report format. This response opened the door for CCPM schedules in conjunction with EVM reporting.
>
> (P. 2)

All the CC experts believe that changing an organization is complex, and CC-created changes will be equally challenging. Classical human nature regarding change in their work process will be a complicating factor. "Low tolerance for change is often rooted in a fear that one will not be able to develop new skills and behaviors that are required". Leach (2005b) states that resistance to change is a characteristic of any stable system. One might argue that CC could be more challenging to sell in stable, mature organizations that were doing well. Therefore, they are not motivated to make what might be viewed as a radical change. "Regardless of the organization, there will likely be negative reactions from the various entities with these new approaches to planning and control" (Leach, 2005b, p. 203).

On the counter side of this argument, Hyang R (2025) comments "that in certain instances, employee resistance may play a positive and useful role in supporting organizational change" (p. 26). Insightful and well-intended debate, criticism, or disagreement do not necessarily equate to negative resistance but may be intended to produce a better understanding and additional options and solutions. With proper selling, the CC theme will resonate with many organizational groups seeking faster project completion.

From a leadership perspective, PMs should respond to critics in a positive, constructive fashion. It is important to communicate how the local project links with organizational success (Boyles, 2021). The case for CC must also be presented couched the right way. *It is NOT a technology initiative. Instead, it is a management technique designed to deliver defined value quicker than would be done otherwise.*

CC attempts to speed the delivery of the project without sacrificing requirements, increasing cost, overstressing resources, or changing any of the underlying base variables. It focuses on finishing the project faster and delivering its value quicker. In addition, there are other less visible benefits from this goal. A shorter cycle time can positively impact the rate of scope change observed. Leach (2005b) states that PMs complaining about scope creep admit to an ineffective project change control process. Although seemingly just a process issue, project scope control is recognized as one of the most important factors for project success. By completing the project sooner, the level of scope change should be less.

21.12 CC PROCESS IMPACT

The previous discussion has outlined some more significant process changes related to implementing CC in an organization. It would be a disservice to suggest that all players will understand these and accept them willingly. It will require considerable work to translate these processes into acceptable ideas for implementation. This section will offer brief comments on the most obvious of these.

Creating an effective traditional project plan requires sufficient time, and adding the CC model will not shorten that time. Adding a CC structure requires additional discipline in evaluating resource capacity and buffer requirements, as well as looking at multiple project resource impacts. CC requires more discipline in these and other areas of the life cycle (Kendall et al., 2005).

A well-thought-out Work Breakdown Structure (WBS) is important for various reasons, as outlined in Chapter 6. This artifact contains important information to help organize, integrate, assign responsibility, measure, and control projects (Leach, 2005b). A WBS also serves as a planning artifact that helps identify work units performed in parallel, contracted, and allocated to lower-lev.el skills and a host of other decisions to reduce overall project time (Kendall et al., 2005).

During schedule development, the project team must be permitted to manipulate the structured work unit plan and establish the necessary feeding and project buffers. This includes a disciplined process of assigning resources to all tasks and having tight control over those resources to work on one project task at a time. Management must support this overall process discipline, including resource allocation, predecessor relationships, buffer allocation, and the CC model, itself. Patrick (1998) reminds us that the scheduling mechanisms provided by CC require the elimination of task due dates from project plans.

As indicated earlier, the calculation of buffer sizes will be spotty in the early stages of implementation. Regardless of the methodology used, the buffers should initially be calculated based on rough rules of thumb.

Setting an arbitrary delivery date without validating the time and resource implications invalidates any plan. An exception might be necessary in some cases as some external driver, such as a regulatory requirement or market date, may dictate a predefined date. When a fixed delivery date is used, it must be prioritized against the existing workload to test for feasibility.

Using a strategic project portfolio view will help decrease the severity of resource contention. Projects must be prioritized in terms of multiple factors beyond just benefits to minimize this issue. Leach's (2005b) view is more direct:

> If management does not adhere to a priority list, the multi-project system will not work. It is a simple choice. Behave to double throughput, or do not. Once they see the comparable result and project teams doing much better at this than many thought. After all, when the system makes more money, people's jobs are protected, and often they make more money too.
>
> (P. 162)

The organizational methods used to select and initiate new projects can be a source of interruption or resource contention for ongoing projects. If the new project is placed as a higher priority than some of the ongoing projects, the schedule of the ongoing projects will need to change quickly because of the decreased CC resource allocation scheme. Leach (2005b):

> It is necessary to keep in mind that the worst possible priority decision is not to make a priority decision but to encourage everyone to do his or her best and hope the problem goes away. This inevitably causes bad multitasking and poor performance on all the projects.
>
> (P. 161)

The practice of "pulling" an already committed resource from an ongoing project is a significant scheduling challenge for any project. However, it can be a disaster for a CC-driven model effort. Upon receiving a notice that a task will be finished soon, resources for the successor task must be prepared to finish up interruptible (noncritical) work and get ready to start work on the assigned active critical project task. The new view must change from "When is my task due?" to "When will my next prioritized task start?" Note how this represents a change from working to a flawed fixed task calendar to focusing more on accelerating the completion.

The change management process is responsible for protecting project resources from unnecessary or significant changes in scope (Patrick, 1998). No project can be successful when new, approved work is being added to the requirements without considering the impact that this has on existing work. It is common for organizations to "pull" project team members away from project work to address some important operational issues, and sometimes this is unavoidable. However, once again, failure to have the required resources at the right time and in the right quantity will adversely affect the schedule and defeat the model delivery goal. Management has the responsibility to make a priority call on resources, but the PM should attempt to quantify the impact of an unplanned resource reallocation and not just let these decisions go quietly.

The belief that one can handle multiple tasks concurrently (i.e., multitasking) represents one of the "merit badges" of today's work environment. This is often cited in job descriptions and is valid for project teams. This was once considered a productivity multiplier, but management and workers are beginning to recognize the overall negative limitations of multitasking (Anderson, 2001). Research has confirmed this negative impact. There is increased job complexity and time lost when workers switch between complicated tasks. Leach (2005b) states a similar conclusion:

> Practical applications of CC have demonstrated the most significant gains in multi-project enterprises. This is because those environments usually require everyone to multitask much of the time. Eliminating much of the bad multitasking has the greatest (positive) impact on overall enterprise project throughput.
>
> (P. 163)

When one considers the productivity degradation effect of multitasking, it should be clear that assigning resources to jump back and forth across task or project boundaries is not an efficient mechanism. Replacement of a more systemic process to allocate resources with synchronization, combined with their management in a "relay race" behavior, will go a long way to reduce this issue and help accelerate project completion (Patrick, 2001, p. 8). Leach (2005b) recommends that one of his success factors is to limit multitasking. Similarly, a more orderly task allocation process is a stress reduction technique for team members.

In the CC mode, team members must have clear task priorities communicated. There are two views of project resources: those performing active tasks and those performing noncritical tasks. The ones that must be most carefully controlled are the active task critical group because this group most directly determines how long the project will take. The operational process must ensure that critical task-assigned resources are available when the preceding task is done without relying on a predetermined predecessor completion due date. The underlying assumption is that the work required for a task will take as long as it takes, no matter the schedule estimate. Once allocated to a priority task, the resource will work on it until there is nothing more they can do. There should be minimal distractions until they are complete (Patrick, 2001). In an ideal case, this means meetings, email, or anything else that takes away focus on the assigned task. The resource view outlined at this point deals with the required management process rather than the technical skill needed to produce the product.

The concept of variable time scheduling means it is impossible to pick the exact execution time for a task. Too many variables are involved to make this a deterministic value, but the 50/50 estimating approach will be a significant philosophical shift for many. Task estimators must be coached and encouraged not to develop task estimates with built-in safety. Completion targets must be achieved through the buffer rather than padding the task estimate. A 60% probability time represents a slight compromise that might be used as a beginning strategy because the 50% point represents a significant organizational cultural change. Leach (2005b) describes the tendency of organizations to believe that all 50/50 tasks are underestimated because if this is not the case, then they are concerned that something could go wrong to prolong the task. This is often the case in an organization with extensive multitasking and interruptions. Regardless of the rationale, there is a strong tendency to overprotect an estimate by adding significant padding to its expected duration. Team members and work unit owners will have similar biases left over from traditional management practices. This cultural item will be a leadership challenge, as it shows everyone that meeting a particular padded activity time estimate is no longer the criterion for success.

Once the decision is made to move forward with CC principles, it should be tested on a relatively small project with an experienced PM who believes in the new method. Every reasonable effort should be made to gain commitment from the team, and careful metrics should be kept when evaluating the results. The famous 1930s Hawthorne lighting experiment illustrated the value noted when individuals felt special because they were involved in something new (Perera, 2024). That same concept can be a motivator for CC introduction. There are some similarities between this implementation process and some of the reactions of the Hawthorne study.

Task estimating will likely be the most challenging activity in the implementation and the first opportunity to either commit or resist. Newbold (1998) recommends running the team with a "Pit Crew Mentality" (p. 195). Everyone must understand how their work contributes to the track race and attempt to translate this into the new process.

CC status reporting is based on team members accurately reporting task completion status. This is done not as the traditional percentage complete but focuses on the estimated time required to finish the task. Actual completion of tasks, as well as a forecast for completion, must be communicated on time. The traditional once-a-week status reporting will not work well for this requirement. "Activity reporting and resource management need to be a real-time mechanism" (Leach, 2005b, p. 201). Status reporting and resource management become tightly linked in the CC operational environment because status information controls the next step in the various project chains. Patrick (1998) details two simple steps required to accomplish this:

Step one: Ask the responsible work unit owner how much of a warning they need to finish up their other work (noncritical if all is going well) and shift to uninterrupted work so they can drop what they are currently doing and start work on their critical task.

Step two: There must be discipline in providing appropriate status updates for each active task. When the estimated time to complete task A matches the warning needed by the task B resources, the resource management alert system will trigger and communicate this to that resource. They then need to plan on shutting down what they are doing and prepare for soon-to-be-completed predecessor tasks.

This advanced warning process is called a "resource alert" in the CC vernacular. Team members and their managers may resist the required increased frequency of reporting compared to traditional status methods;

however, the importance of this must be stressed. Remember, the new culture must resemble a track relay and a pit crew team mentality. The traditional model has been characterized as waiting at the bus stop for a loosely scheduled bus to arrive. We are now trying to speed up the bus schedule. One should be able to see from this dialogue that the resource alert process is used in CC to produce a smooth workflow. It and the frequent status reporting process are key to achieving CC success. The only dates in a CC schedule are launch dates for chains of tasks and final due dates associated with related deliverables (Patrick, 2001).

21.13 ORGANIZATIONAL CHALLENGES

This chapter has presented a project model with great potential, although culturally challenging to implement. CC brings with it new processes and related concepts. Before leaving this subject, it will be good to summarize eight of the most significant challenges that organizations will find. The review summary items are the following:

1. *High-level management support*: As with all organizational changes, the support of senior management is paramount.
2. *Cultural change in managing teams and projects*: CC changes how project activity is pursued. The entire organization must understand and work with the new paradigm.
3. *Status reporting methods*: Traditional status reports must be replaced, and all stakeholders must be educated about the new approaches. There will likely be a need to compromise regarding subproject completion reporting. There are many related status-oriented changes embedded in the CC. It is important to remember that effective communication with the stakeholder community is a prime goal for the PM. CC will not be a transparent change in this regard.
4. *Translate estimating techniques to 50% probability*: Taking away time padding will be a significant cultural problem because of the stigma of time overrun in the traditional view.
5. *Task overruns are now the norm*: Traditional status reporting looked unfavorably at time overruns. In the CC model, they are expected. Management and other stakeholders must understand this new phenomenon.

6. *Team evaluation*: In a relay race, the team wins, which is how CC must work.
7. *Resource allocation and project priorities*: Resource alert and formal project prioritization are required to manage the workflow process. Both issues require more discipline than exists in the typical organization.
8. *Multitasking avoided*: This implies that once a resource is moved to a task, it will work on it until it is completed—no jumping around to other tasks.

21.14 CC IMPLEMENTATION STRATEGIES

Two seemingly viable scenarios exist to introduce the CC model into an organization. The first scenario would be to present the CC model as a challenge to an existing high-performance, recognized, and established team routinely assigned challenging or high-priority projects. In this case, the team would not likely be intimidated by the change in control approach, and their ego could be a stimulus to move higher in respect. This would be similar to the behavioral concept of Hawthorne's experiment but in modern times. Offering this group the opportunity to utilize the new method to reduce their delivery time and increase their effectiveness and value to the company would provide the level of challenge that such teams often thrive on. This strategy is like introducing a new tool or technology. Successful project delivery by this pilot team would serve to demonstrate improved capability, thus paving the way for broader acceptance.

A second implementation scenario is introducing the concept in a project with an experienced and respected PM and a technically proficient team, but not one steeped in traditional methodology. Many of the behaviors described in the CC model would possibly appeal to such a group because they would recognize the logic behind the model. This second type of environment would not be so inclined to protect an existing approach, and this would avoid a defensive reaction often found in a traditional team.

21.15 CONCLUSION

The CC model represents an exciting new option that allows organizations to increase the number of projects that can be done with the same resources and reduce the average duration of projects. This approach offers interesting insights into the existing slow project completion problems.

Even though the processes embedded in the CC model may seem unusual, they are just a logical extension of current project management practices. Focusing on these differences may be academically interesting, but one should look at the concepts and see how they might be implemented before taking on the formal model.

This overview of the CC model concepts examines the basic mechanics and some operational complexities associated with the model design. [21C3] As stated, this is not a technical change but a behavioral one, which is the organizational area to focus on. It should be apparent from the description that potentially excellent project management ideas concerning methods for improving project throughput are embedded in this model. Indeed, buffering and restricting time estimates could be implemented in some fashion, even in the traditional view of projects.

Many organizations today are searching for better ways to achieve breakthroughs in project development cycle times to stay competitive. There is a need to complete more projects through their organization per resource allocation unit. This goal must also often be achieved without increasing the number of people allocated to projects. The availability of skilled resources will always be a project constraint in good and poor economic times. In healthy periods, the aggressive requirement outstrips demand, and in tough economic times, executives are reluctant to hire even though the demand for new projects remains.

Out of all the project management schemes proposed today, the CC model logic significantly improves project delivery strategy. It is based on an improved reality view of the process. The uniqueness of the CC concept is that it hits at the heart of why projects take too long to execute. The best traditional project management process can be implemented, but as long as padded estimates and multitasking remain the norm, the results of excessive time will not change significantly. CC projects in mature support organizations have verified the concepts outlined here, and throughput improvement in the 25–30% range has been demonstrated. This is a problematic statement for management to ignore.

There are more process ideas to deal with as we seek techniques to improve the modeling of variability and reality in the management process. The next chapter will use this project model view to introduce a modified look at status tracking using EV techniques.

> ***Author note*: A representative CC schedule is available from Patrick (1998) (see https://www.academia.edu/17351333).

CASE REVIEW PROBLEM

This sample project has a somewhat abstract traditional project structure with multiple feeding chains. Assume the task estimates are padded. Convert these to 50% of the values shown, then restructure the plan to a CC structure using rules from the chapter. Utilize primary chain and feeding chain buffers. The project WBS is shown in Table 21.1.

The associated plan task data for a traditional plan is shown in Table 21.2.

Compare the CC plan with the initial traditional plan. If a study group is available, use this plan to discuss the CC model further.

TABLE 21.1 CC Problem WBS

WBS	Task Name	Duration	Predecessors
1	CC Start Example	113 days	
1.1	Summary 1	41 days	
1.1.1	a	12 days	
1.1.2	b	18 days	
1.1.3	c	11 days	3,4
1.1.4	d	12 days	5
1.1.5	Review 1	0 days	6
1.2	Summary 2	23 days	
1.2.1	e	11 days	
1.2.2	f	12 days	9
1.2.3	g	12 days	9
1.2.4	Review 2	0 days	11
1.3	Summary End	113 days	
1.3.1	h	20 days	
1.3.2	l	13 days	14
1.3.3	j	14 days	15
1.3.4	k	14 days	16
1.3.5	l	25 days	17
1.3.6	m	12 days	18
1.3.7	n	15 days	19
1.3.8	Project close	0 days	20
1.4	Summary 3	49 days	
1.4.1	o	18 days	
1.4.2	p	11 days	23
1.4.3	q	12 days	24
1.4.4	r	11 days	23
1.4.5	s	12 days	26
1.4.6	t	8 days	25,26
1.4.7	Review 3	0 days	28

TABLE 21.2 Startup Problem Table Data

ID	WBS	Task Name	Resource Names	Duration	Total Slack	Pred
1	1	Car Problem First Cut		315 days	0 days	
2	1.1	Planning & Project Initiation		70 days	0 days	
3	1.1.1	Charter	Business Analyst, Designer	10 days	0 days	
4	1.1.2	Stakeholders Identification	Business Analyst	5 days	300 days	3
5	1.1.3	Scope Definition	Business Analyst, Designer, Electrical Engineer, Mechanical Engineer	15 days	0, days	3
6	1.1.4	Schedule Development	Designer, Electrical Engineer, Mechanical Engineer	15 days	0 days	5
7	1.1.5	Risk Assessment	Business Analyst, Designer	5 days	0 days	6
8	1.1.6	Budget Definition	Business Analyst	15 days	0 days	7
9	1.1.7	Management Charter Approval	Business Analyst	5 days	0 days	8
10	1.1.8	Set baseline	Business Analyst	5 days	138 days	9
11	1.2	Engineering		112 days	0 days	
12	1.2.1	Body/Engine Draft Design		45 days	0 days	
13	1.2.1.1	Initial Draft	Designer [200%], Electrical Engineer [200%] Mechanical Engineer [200%]	35 days	0 days	9
14	1.2.1.2	Integration analysis	Electrical Engineer [200%], Mechanic [200%], Mechanical Engineer [200%]	10 days	0 days	13
15	1.2 2	Mechanical Engineering		80 days	148 days	
16	1.27.1	Engine Design	Designer [200%], Mechanical Engineer	80 days	165 days	10
17	1.22.2	Mechanics Design	Designer, Mechanical Engineer [300%], Mechanic.	30 days	138 days	DFS-10 day
18	12.3	Electrical Engineering		57 days	138 days	

(*Continued*)

TABLE 21.2 (Contined) Startup Problem Table Data

ID	WBS	Task Name	Resource Names	Duration	Total Slack	Pred
19	1.23.1	Electrical design	Designer [200%], Electrical Engineer [300%]	30 days	138 days	17
20	1.23.2	Wiring	Electrical Engineer [200%], Designer	15 days	160 days	19
21	1.23.3	Others Elect	Electrical Engineer, Mechanic [200%]	27 days	138 days	19
22	1.2.4	Structural Engineering		67 days	0 days	
23	1.2.4.1	Body	Designer, Mechanic, Mechanical Engineer	30 days	0 days	12
24	1.2.4.2	Other Body Parts	Designer, Procurement	30 days	165 days	12
25	1.2.4.3	Design integration review	Procurement	10 days	138 days	4,17,21,24
26	12.5	Body/Engine Structure		30 days	0 days	
27	1.25.1	Chassis	Metal [3.51], Mechanic [300%]	20 days	155 days	23
28	1.25.2	Other Structures	Fiberglass [2] Mechanic [200%] Metal [3]	10 days	Odays	23
29	1.25.3	Mechanics Parts	Fiberglass [10], Metal [4], Procurement, Tires [8]	20 days	145 days	28
30	1.25.4	Electric - Electronic Parts	Mechanic [300%] Procurement, Wires [7]	10 days	Ddays	28
31	12.6	Engineering Final review	Procurement	10 days	0 days	30
32	13	Logistic & Procurement		45 days	0 days	
33	13.1	Part Ordered	Business Analyst, Designer Procurement	15 days	0 days	31
34	13.2	Part Delivered	Procurement [200%]	10 days	0 days	BFS+20 days
35	1.4	Manufacturing Assembly.		18 days	0 days	
36	14.1	Station Mechanic	Mechanic [200%] Mechanical Engineer	6 days	0 days	34
37	1.4.2	Station Electric	Electrical Engineer [200%], Mechanic [300%]	5 days	0 days	36

38	1.4.3	Finishing	Electrical Engineer [200%], Mechanic [300%], Mechanical Engineer [200%]	7 days	0 days	37
39	1.5	Testing		26 days	0 days	
40	1.5.1	Components Testing		10 days	0 day	
41	1.5.1.1	Electronic Components Testing	Business Analyst, Tester [300%]	10 days	72 days	38
42	1.5.1.2	Mechanic components Testing	Mechanical Engineer, Tester [200%]	10 days	0 days	38
43	15.2	Test drive	Driver, Tester [200%]	8 days	0 days	42
44	1.5.3	Safely test	Deriver Mechanic [200%], Tester [200%]	8 days	0 days	43
45	1.6	Formal Closing		21 days	0 days	
46	1.6.1	Documentation	Business Analyst, Designer	7 days	0 days	44
47	1.6.2	Close Records	Business Analyst [200%]	7 days	0 days	46
48	1.6.3	Formal Shutdown	Business Analyst [200%]	7 days	0 days	47
49	1.7	Project Buffer		35 days	0 days	48
50	1.8	Close Project		0 days	0 days	49

REFERENCES

Anderson, P. (2001). Study: Multitasking is Counterproductive. https://www.cnn.com/2001/CAREER/trends/08/05/multitasking.study/

Boyles, M. (2021). How to Implement Organizational Change Successfully, Harvard Business School, https://online.hbs.edu/blog/post/implement-organizational-change

Cherry, K. (2021). What is Parkingson's Law, https://www.verywellmind.com/what-is-parkinsons-law-6674423#:~:text=The%20main%20premise%20of%20Parkinson's,is%20allotted%20for%20its%20completion

Christ, D.K. (2001). Theory of Constraints Project Management in Aircraft Assembly. https://files.znu.edu.ua/files/comp.books/books/goldratt_boeing_toc_pm.pdf

Geekie, A. (2008). Buffer sizing for the critical chain project management method. *South African Journal of Industrial Engineering*. https://scielo.org.za/scielo.php?script=sci_arttext&pid=S2224-78902008000100006 (Accessed September 2, 2025).

Goldratt, E. (2007). Critical Chain. http://www.goldratt.co.uk/resources/critical_chain/index.html (Accessed September 3, 2025).

Goldratt, E.M. (1997). *Critical Chain*. Great Barrington: The North River Press.

Hyang, R. Mqy (2025). Organizational Change Mangement, *Journal of Organizational Change Management* (2015) 28 (3): 356–368. https://doi.org/10.1108/JOCM-07-2014-0130, (Accessed September 2, 2025).

Kendall, I., Pitagorsky, G., & Hulett, D. (2005). Integrating Critical Chain and the PMBOK® Guide. http://logmgt.nkmu.edu.tw/news/articles/criticalChain-PMBOK.pdf

Leach, L. (2005a). EMV and Critical Chain Project Management. http://pmchallenge.gsfc.nasa.gov/Docs/Presentations2005speakers/Day%202/newideas/Larr

Leach, L. (2005b). *Critical Chain Project Management*, 2nd edition. Norwood, MA: Artech House. p. 2353.

Moralis, S & Dinan, S. (2022). The Myth of Multitasking, Psychology Today, https://www.psychologytoday.com/us/blog/the-therapeutic-perspective/202202/the-myth-multitasking

Newbold, R.C. (1998). *Project Management in the Fast Lane: Applying the Theory of Constraints*. Boca Raton: St. Lucie Press.

Patrick, F.S. (1998). Critical Chain Scheduling and Buffer Management: Getting Out from between Parkinson's Rock and Murphy's Hard Place http://www.focusedperformance.com/articles/ccpm.html

Patrick, F.S. (2001). Critical Chain and Risk Management: Protecting Project Value from Uncertainty. http://www.focusedperformance.com/articles/ccrisk.html

Perera. A. (2024). *Hawthorne Effect: Definition, How It Works, and How to Avoid It*. Simply Psychology, https://www.simplypsychology.org/hawthorne-effect.html

Retief, F. (2002). Overview of Critical Chain Project Management, Critical Chain Synposium. https://auth.scribd.com (Accessed September 2, 2025).

CHAPTER 22

Variable Time Status Model

Project N Motion Videos for This Chapter

22VAR	Project Variability	Variability Overview

The goal of this chapter is to produce a modified look at project status using variable task times in the project plan. Essential components of the Critical Chain (CC) and EV are used in this plan structure. Overall, project performance is compiled from individual task measures using a customized approach. In this view, tasks are assumed to be variable, and CC-like buffers are used to protect against overruns. A modified method for calculating EV parameters will measure task performance. This plan structure unveils one of the shortcomings of the standard EV model but also shows that the underlying measurement theory still has value if one understands the core approach to measurement. Said another way, a valid performance interpretation can still be made from the same calculations if you know what you are measuring.

We are warning readers about this chapter's content and approach. Up to this point in the text, the material presented has been focused on delivering a standard industry view of the chapter topic. This customized model description will use that core background, but this standard

DOI: 10.1201/9781003532989-27

process is used when creating a modified view of project performance. [22VAR]

22.1 INTRODUCTION

Utilization of variable time task estimates invalidates the traditional project control model linked to fixed task schedules. The modified custom model view described here utilizes portions of standard EV theory and automated support to produce a new method for evaluating project status. This text references the invalid model theory of fixed task times, yet industry practice ignores that view, creating measures that have no actual meaning. This customized model will show a technique to recognize task variance and still produce valid performance measures.

The material presented here is considered advanced in several dimensions. First, the logic of changing the estimating practice to recognize variable task time is similar to the CC theory. In addition, recognizing variable task work chains changes the ability to calculate standard plans versus actual status measures.

Changing the method of examining project performance in this manner will be an excellent test to see how well you have absorbed the previous standard CC logic described in Chapters 14 and 21, demonstrating the type of redesign logic needed to improve the traditional model. A comparison of this method with the standard MSP-generated status measures would demonstrate why the current automated values have little chance of being correct.

Current project planning and control practices have conceptual flaws in their underlying logic. These flaws are subtle in origin but permeate the life cycle in their impact. This section aims to highlight the root cause of the standard measurement problem and a method to improve the process from this design. The process described does not resemble any known industry method and is offered here as a thought process that contains more valid perspectives regarding project status measurement. Industry surveys point to hundreds of reasons for poor project outcomes. No single change in the standard process will fix all the items mentioned, but an improved view of project status offers the potential for improved identification of performance areas.

Previous sections of the text have been critical of the traditional model view that shows project plans with static deterministic durations when there is every reason to expect variability. This assumption negatively affects the validity of the project completion estimate and, more subtly, the accuracy of project status measurement. Unfortunately, this model design flaw inhibits producing accurate status or completion forecasts. It is essential to recognize that the modified view described here is counter to prevailing usage, so it is necessary to define how this process is compiled.

22.1.1 Design Assumptions

Before dealing with the measurement process, a comment on three core design assumptions is necessary. First, it is assumed that the goal of a project is to complete the target as expeditiously as feasible, and the associated project plan should reflect this goal. CC theory clearly describes various mechanics necessary to support this theme. Second, using an initial project baseline for performance measurement has profound performance measurement logic and flaws, given significant scope change. Third, recognition that tasks are variable eliminates the ability to define start and finish times, which wipes out the validity of comparing because the tasks are now essentially viewed as floating through time. Changing the standard model this way also requires changing the performance measurement process. This modified view represents the starting place for the material presented in this chapter. However, rejecting segments of the prevailing theory does not mean that all of it is wrong; it is just how it is packaged. The classic roles of a Work Breakdown Structure (WBS) and the task network logic are still fundamental to the planning process, and every effort will be made to salvage existing automation support as a computation aid.

22.2 TIME ESTIMATES

The underlying view of variable task time logic described in the CC model is compelling. From this, it is hard to ignore its description of why task padding does not properly represent the project variation. The custom modification shown here will start with accepting the base view outlined by the CC model. The variable task time will be defined as a median value of some underlying probability value. Normal (Gaussian) or beta distributions will be most relevant to the project situation, but this is equivalent to the CC 50/50 view. The key here is that the task is variable and will float with no defined start or finish time. CC-type buffers will protect completion times.

One significant status measurement issue with this plan design is the loss of defined calendar data for the start and finish of a task. In the modified model, tasks are "floating" through time, so their status must be measured differently. On the positive side, more needed reality is added to the plan view.

Another reality factor that is not well handled by the traditional modeling approach is the addition of new tasks based on approved scope changes. The new tasks are not included when the status comparison process measures a fixed-time task against the original approved baseline. This is another reason why the traditional concept needs to represent status properly. To deal with performance properly, the performance measurement method must recognize both the task variance issue and the newly approved scope change tasks. In other words, the project plan must represent the current delivery requirement and contain a method to measure the total work performed to the status point. Also, a new model needs an improved method for forecasting completion dates and costs. This new view of the project may be complex to envision initially, and it may be helpful to use a set of moving Gantt bars to simulate this task variation. If one tries to make a project plan more realistic, some recognition of these two reality areas is needed. The CC model attempted to salvage the general waterfall structure view and use buffers to correct for the task view, but it did little to quantify task performance or completion forecasts.

22.3 THE MODEL

The first step is retooling the planning model to recognize that task estimates must be viewed as variables, and padding is the wrong patch for this. Goldratt's (1997) CC work psychology logic will be the basis for the new structure. Buffers will be simplified as much as possible, primarily focusing on a project buffer. Other buffer types can be added as the concept is accepted, but the assumption is that excessive use of buffer types might limit acceptance. A separate, improved, robust resource management approach will be used to lessen the need for additional buffers.

Goldratt describes the expedited delivery process as removing excess task time by challenging the team with estimates based on a 50/50 median estimate basis. This concept goes against the current project culture, where task overage is viewed as a negative performance. In the variable time view, task overruns are to be expected, so the cultural overrun view will have to be changed. As the previous CC model description has illustrated, there are compelling reasons to consider this planning approach based on execution efficiency.

As a starting point to show the underlying logic details, the same three 50/50 sequential tasks used in the CC model chapter will be used as a

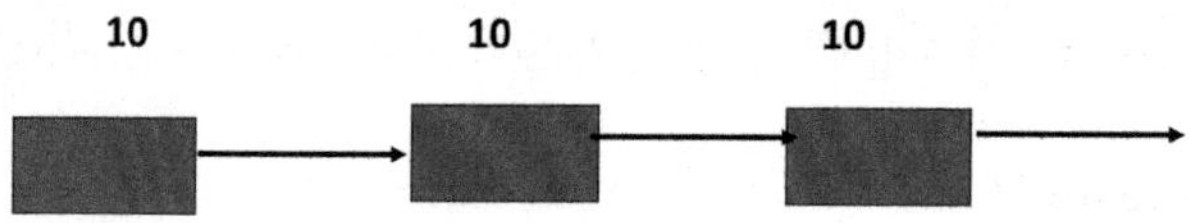

FIGURE 22.1 Sequential compressed chain tasks.

starting point. Three traditional sequential tasks are padded with estimates of ten days each. The median schedule shows the same buffer protection logic. Figure 22.1 shows the schematic task view.

The CC model specification states that all tasks in the plan need to be converted to zero slack, which is sound but complicates the plan design. The demonstration example will obey this model requirement but might also be relaxed for initial testing. From a work psychology viewpoint, this design view introduces the "bus schedule" mentality, which is undesirable.

At this point, we need to take a step back and look at this modified plan view. The tasks now have the perspective of a track race, and it is appropriate to seek out work chains that can be handled via the CC buffer view. We expect the median-sized tasks to overrun but will keep the anticipated start dates in a plan view even though this is not a technically accurate date. One of the main visible performance views is the status of buffers. EV logic will evaluate individual task performance compared to its median value.

22.4 MODIFIED PROJECT CULTURE VIEW

The description above is sufficiently scary and needs some discussion on operational views. This model modification reflects certain reality factors but also changes the simplified traditional culture of plan versus actual status measurement. This significant gap will be hard to fill because there is now a simple answer to the situation.

At this point, it is also essential to introduce a somewhat different view of status. The idea that all project tasks are critical is a misnomer. Many items in the overall work package fall into the category of "time available." These might be documentation, internal training, reading emails, etc. These are legitimate work requirements that should be reflected in the project plan but should not be considered critical. These are the tasks to be worked on while waiting for the variable task to be finished. This changes how work is assigned compared to the traditional predecessor method. Many of these could be categorized as "time available" with less of a hard lock to a specific predecessor. In addition, the no multitasking rule stays in effect for active tasks. This modified model approach will keep the CC view of the "runner with the baton" metaphor.

Each of these process items will be used, and active training in the method for the project team is an integral part of the effort. This must be sold as a "work smarter, not harder" logic. In addition, the modified management practice must support this view with a compatible resource management process to cover the omission of complex CC buffers in this implementation. Cultural changes in the areas of multitasking and resource management are two essential success factors.

Now, for the million-dollar question, "Is life in this model so hectic that it is unlivable for the project team?" The answer is no, but there does need to be a stronger focus and discipline on active work targets and completion. A project's benefit is not realized until it is finished. That goal needs to be part of the local culture. The role of project management should not be to make the worker's life difficult, even though adding more discipline to the process may invoke that feeling. In this case, the improvement comes from focusing on the right targets with the right execution strategy. Worker motivation needs to come from successful completion. Performance bonus programs would not be a bad strategy to enforce the importance of time.

This relay track race view brings the need for a change in how a project is viewed from a schedule communication point of view. Rather than judging performance by looking at a scheduled task finish time, the process must change. Simulation modeling has shown one method of demonstrating how project completion might be modeled. Tenants of that approach are certainly candidates to improve that aspect of completion forecast communication and could be added to the methods outlined here.

The new model's primary operational status-tracking logic now converts to the status of the various buffers, as the CC model has described. So long as buffer usage is within bounds, the theory suggests that the execution process is in control. Classic statistical control techniques from the 1980s looked at manufacturing processes this way and devised control bounds. The buffers serve the same purpose. This new status measurement role means that buffers become the most suspect element in the model. The operative question is, "What do they represent, and why are they this big?" Significant literature is available to describe how to size buffers and how to use their current size to measure project status. However, this topic will require some time to develop historical data to help justify any sizing strategy. Do recognize that the buffer's accuracy directly affects the plan's forecast accuracy and its work chain's associated status. There is a subtle value in separating task estimates from their buffer, even if the buffer value is wrong. The separation allows a more direct view of the execution performance for the task. If the task is viewed in padded form, one could only conclude that the task execution performance was not as bad as the bloated estimate said it might require. That is not a valid performance measure and must be eliminated from the culture.

Multiple internal process issues must be revised once one commits to the variable task chain theory approach. These involve changes to the associated execution mechanics and their management approach. Overruns are now learning events to guide improvements in the internal estimating for the next project cycle. There is a personal belief that the resistance to looking at the planning process this way is often due to a lack of internal management maturity and a lesser understanding of the importance of resource management. The goal here is to emphasize how schedules are compromised and what it takes to close those performance gaps. Raytheon and other organizations have committed to this approach and advertised its value, but the process described needs to be observed in most projects. Also, the use of EV is shown here as a thought-generation idea for improved status measurement.

22.5 SAMPLE PLAN

Figure 22.2 shows an introductory example of this planning approach. Each task represents a median task estimate with a 30% chain buffer value.

This set of sequential tasks is estimated using the median estimating logic, and a project buffer is attached at the end of this chain to protect the predicted 4/1 completion. In this example, the buffer size is set to 15 days based on previous project experience.

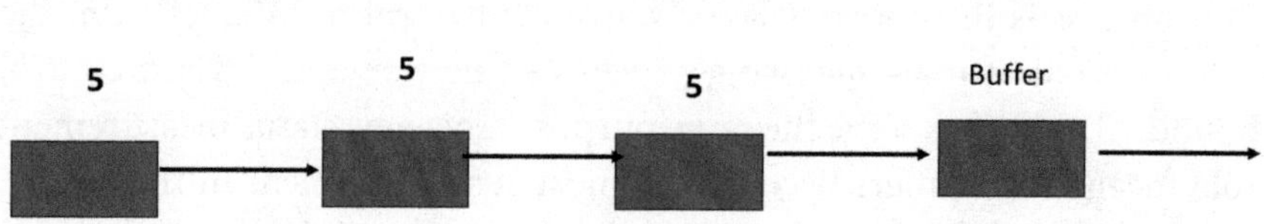

FIGURE 22.2 Median task estimate project schedule.

22.5.1 Dual Chain Model Plan

The WBS in Figure 22.3 shows a two-chain structure model. A summary task group collection is represented by units A through E. Then, a feeding summary sub-chain is defined by work units H, I, and G. Two protection buffers will be attached to these two chains. The buffers are optionally shown in separate areas to better segregate the work protons.

In this example, a feeder chain buffer will be needed to protect that portion of the plan, and it will be the core status-tracking element for that segment. Figure 22.4 shows the associated plan constructed according to the tenets of the modified model.

This plan has been designed to provide some worthwhile design examples that mechanically illustrate the role of protective buffers and how a project status can be derived from that view. It is not necessarily designed to show optimum work package sizes as it violates that sizing rule. The following non-standard layout modifications in the sample plan need to be explained:

1. WBS numbering is simplified to make references to tasks easier to follow.
2. The task sizes shown are a little too large according to work package theory but help to shrink the plan's physical size.
3. Task values have been selected to help highlight the EV process.
4. All task estimates are assumed to be median based on a preferability distribution.
5. The project completion variables are protected by a project buffer of 60 days and $15,000.
6. A "feeding" chain buffer at WBS 7.1 of 12 days and $9,000 is inserted before task D to protect the WBS 6 summary task group from impacting the main project chain, which would be the critical path

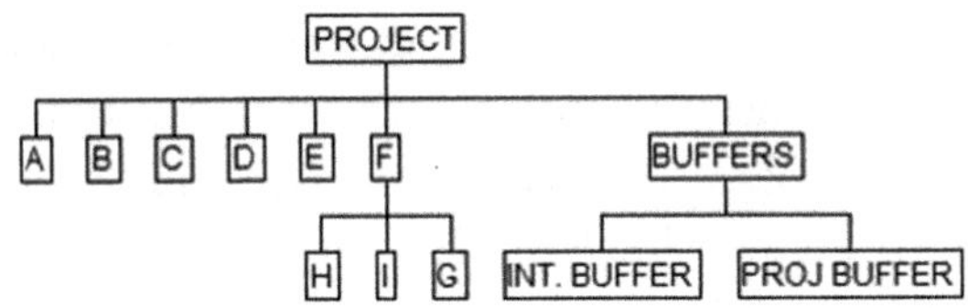

FIGURE 22.3 Dual chain structure.

ID		WBS	Task Name	% Complete	Duration	Cost	Start
1		0	PROJE	0%	373 days	$330,000	1/1
2		1	A	0%	110 days	$200,000	1/1
3		2	B	0%	66 days	$35,000	6/3
4		3	C	0%	48 days	$10,000	9/3
5		4	D	0%	40 days	$10,000	11/8
6		5	E	0%	49 days	$10,000	1/3
7		6	F	0%	70 days	$41,000	7/17
8		6.1	H	0%	30 days	$14,000	7/17
9		6.2	I	0%	20 days	$15,000	8/28
10		6.3	G	0%	20 days	$12,000	9/25
11		7	BUI	0%	161 days	$24,000	10/23
12		7.1	IB	0%	12 days	$9,000	10/23
13		7.2	PB	0%	60 days	$15,000	3/13

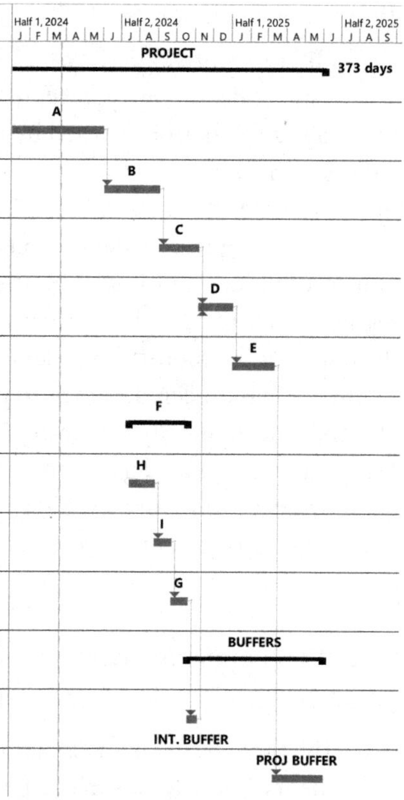

FIGURE 22.4 MSP plan (with protection buffers).

in a traditional project. This is a crucial mechanic to understand in the plan design structure.

7. All buffers are packaged into the WBS label 7 summary group. This is an optional layout to isolate and focus on buffer status later. Typically, buffers would be physically inserted at the end of their respective chains. This is logically done here but physically separated for focus (the predecessor list is not shown here).

8. Summary WBS 6 illustrates a significant task structural planning change that follows the core theory of a CC. Typically, a plan is structured to show a critical path and then show non-critical feeder segments linked to it. The *formal CC theory specifies that all tasks are to be started as late as possible, making the entire task network critical.*

The project start date is set for January 1, and the status review date is March 14. Note the vertical line running through the Gantt bars at the edge of task D to help visually focus on the status. This helps to see the tasks that compose the status population. The buffer sizing parameters are as follows:

A primary buffer of 60 days and $15,000

Feeder buffer of 12 days and $9,000

It would be instructive to review Figure 22.4 Gantt bar view of this plan, to see how the buffers are linked to their respective chains.

Following CC theory, the WBS 6 summary task group has been forced to start as late as possible and protected by buffer 7.1. This design approach is non-traditional and is worthy of some explanatory comments. Note that the tasks in the WBS 6 group would usually be shown with no predecessor for work unit "H." This would cause that entire chain to fall. However, forcing the entire chain downstream to mesh with the primary chain removes all the feeder chain slack. This significant cultural work design philosophy change was explained earlier as part of the relay race metaphor. Because all tasks are now considered critical, it is easy to see how this changes the work dynamic from the traditional mode of defined slack time and padded tasks. This is the most visual mechanical example of chain culture shown thus far. If the execution goal of a project is to minimize wasted time, this view reflects that perspective. As we progress through this example, following this design logic and seeing why rejecting the traditional structure has merit is essential.

22.5.2 Status Calculation Process

Using software to evaluate the task performance has computational advantages for producing two core standard EV parameters, Cost Performance Index (CPI) and Schedule Performance Index (SPI). Table 22.1 reviews the dual notation previously described in the EV chapter.

TABLE 22.1 EV Dural Notation Parameters

PV	BCWS	Current baseline cost value to status point
EV	BCWP	EV calculation up to a status point (work performed)
AC	ACWP	Actual cost of planned work performed up to status point

CPI and SPI parameters will be used here to evaluate the performance of a median-sized work unit; however, other EV parameters are available to translate and show different forms of status measures.

The sections below will move through the work units to generate performance measures for each. This segment of the status measurement process identifies performance anomaly areas to investigate, but the primary role of these parameters is to improve the estimating process for future projects. Significant deviations can also signal that something is going wrong with a task. For example, if a work unit's performance exceeds some critical level, such as a 3.0 value, it would invite a more detailed review of the cause. A defined threshold review value becomes the key to deciding when to look at the deviation as more than an estimated variation.

The new model quantification process utilizes as much traditional calculation software as possible. Existing software can produce EV parameters if the current plan is treated as a macro goal for task timing purposes. A cleaner status measurement approach would be to have software custom-designed for this view. That would allow a cleaner approach to task selection at the status point. They are evaluating all work units that were in action before the status point. We will take the calculated results for the demo model view using the task view later shown in Figure 22.6. Given the floating task view, we can't guarantee which task will be at the status point, but data from actual work performed will be used. This task schedule assumption needs to be evaluated based on experience with the model usage. Standard EV parameters and their actual data will be produced from these target tasks.

The first calculation gap to recognize is that the existing project software will not produce values that fit standard interpretation based on the variable task time. The traditional software assumes the tasks are fixed for the calculation process. Also, buffers are not appropriately recognized in the existing software, so the existence of a buffer in the demo plan would be treated as just another task. Tasks consume time and cost, while the role of a buffer is to hold their value and use portions to cover chain task overruns. As initially shown, the buffers add schedule time to the plan, but once the project starts and tasks are overrun, the buffer is being "eaten" to preserve the planned schedule. This buffer consumption process allows the project task dynamics to occur while keeping the chain schedule intact. Monitoring the buffer "burn rate" provides insight into its adequacy to protect the schedule. Consider this process carefully, as it is fundamental to the model concept. Also, note that this mechanic does not forecast completion; it only serves to protect a defined completion. If someone asks when the project or chain will be completed, the technical answer is that an estimated buffer size protects the date shown. But that does not predict the actual completion. This interpretation requires further discussion as we examine the sample results. A desired additional status view would provide more quantitative insights into the actual chain performance. At this point, it is time to outline how this model approach changes the traditional status calculation process.

22.5.3 Work Unit Performance Measurement

The following steps are needed to compute EV parameters for the defined task:

1. All tasks that are active before the status point are evaluated. For the sample plan, only a portion of Task D and all of Task E are outside the status domain (see Figure 22.4).
2. Actual status data parameters for Actual Cost (AC), Earned Value (EV), Planned Value (PV), and %Complete are recorded for each active work unit. This populates the EV formulas for CPI and SPI.
3. Process the task EV parameters using the standard MSP process described in Chapter 14.

The project plan shown in Figure 22.4 is used to illustrate the EV quantification process. A defined status point is used to define assumed active tasks. CPI and SPI will evaluate those tasks' performance based on their median estimate. These steps represent a significant divergence from the traditional use of EV. They will require an alternative interpretation of the parameters because there is no fixed task duration or calendar date.

22.5.4 Model Review

Within this scheduling envelope, envision a chain of tasks with an associated buffer. Tasks are floating along their defined chain and naturally tend to overrun because they are estimated with median values. The plan design goal is to size buffers to protect their defined chain schedule. The goal for feeding buffers is to keep that chain from invading the primary chain schedule, which in most cases represents the traditional critical path. Likewise, the primary project chain buffer protects the total project completion date. This buffer will have a target date, which is a target value not to exceed. Chapter 20 described mechanics to simulate project duration and produce a quantitative probability view of completion. Given the distrust of buffers in project plans, this view is more appropriate for external discussion of completion. If the view of schedule completion still needs to be clarified, the upcoming numerical example mechanics should help with the demo specifications.

One of the significant logic gaps this task view creates is rationalizing buffers that look like task padding when the same idea has been avoided in traditional plans. This form of overrun protection is not generally approved but logically justified based on the logic of compressed task estimates. There is also the problem of buffers in the automated software, as they are also not recognized as having different roles in the plan. These issues require that manual evaluation be used to evaluate the project status.

Status measurement is assumed to involve the currently active task work plan, not the initial baseline plan. The logic for this view was outlined earlier but needs to be mentioned again. Measuring the status against the original approved plan is comparing apples to oranges. Assuming scope change because the original plan was approved means the current plan contains new tasks that must be evaluated. Any EV comparison from the original baseline plan would have no logical assessment. This view contradicts a standard methodology that describes status tracking against the approved baseline. If we were to redefine the term to include the currently approved work plan as the baseline, then we have an agreement.

The following three actual status data items are required to drive the status-tracking process:

- %Complete
- Actual cost
- Actual duration

The %Complete data to the task status adds the EV perspective to the evaluation. All active tasks producing work before the status point will be used in the review. These can be made via the MSP calculation algorithm

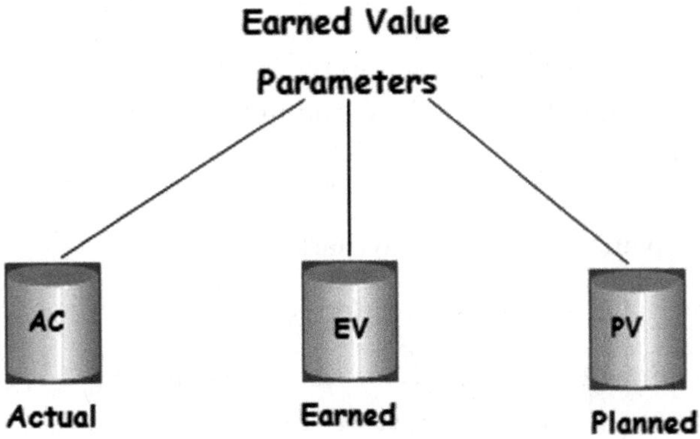

FIGURE 22.5 EV status data.

using these data values to produce EV parameters. The software produces the BCWS/PV value for finished and partially completed tasks. The Gantt bars and associated task data can visually track the resulting process, but the software simplifies the calculation process.

Recall from Chapter 14 that the EV parameter formulas are driven by the variable "AEP." Figure 22.5 shows the three data elements required to produce the EV parameters.

The EV parameters are generated directly from these three data items. Translation of these formula terms into the software model is summarized as follows:

- Actual Cost (AC)—The actual cost incurred for the planned work packages (tasks) up to the status point.
- EV—This metric is labeled as BCWP in the classic notation, and it is a measure of completed work units and partially completed portions of work units up to the status point. For example, a 50% complete work package would receive 50% of the planned value as earned.
- Planned Value (PV)—This metric is labeled as BCWs in the classic notation, and it is the sum of planned baseline cost for all or partial work units scheduled to be accomplished up to the status point.

With the addition of a third evaluation variable, it is now possible to produce quantitative BCWS/PV and BCWA/EV measures that drive the EV formulas.

22.5.5 Software EV Generation Process

Multiple steps are required to set up the software to generate EV values. A full description of these can be found in Chapter 14, but the core MSP essentials are as follows:

- Collect plan data for the active tasks
- Set the status date—12/14 is selected here
- Open output table columns for %complete, AC, EV, PV, CPI, and SPI (see output of these in the results Table 22.2)
- Enter actual values for all active tasks defined before the status point

MSP will then compute the related task EV parameters based on the actual status data. Status interpretation must be manually performed because non-standard issues are recognized when evaluating floating duration task status. Also, the MSP raw data will need to be extracted into another data media, where it can be more freely manipulated. The major logic gap between the coded software and the chain view lies in the role

TABLE 22.2 Raw EV Status Values

WBS	Task Name	DYR,	% Comp	Baseline Cost ($)	Actual Dur.	Actual Cost($)	BCWP/ EV ($)	BCWS/ PV ($)	CPI	SPI
0	PROJECT	349	67	294,000	258.28	296,000	251,000	271,041	0.85	0.93
PROJ	SUMMARY									
1	A	100	100	180,000	110	200,000	180,000	180,000	0.9	1
2	B	60	100	30,000	70	35,000	30,000	30,000	0.86	1
3	C	40	100	10,000	43	10,000	10,000	10,000	1	1
4	D	40	10	10,000	4	10,000	1,000	10,000	0.1	0.1
5	E	49	0	10,000	0	0	0	2,041	0	0
prim.	CHAIN	289		230,000	227	255,000	221,000	232,041	0.87	0.95
6	F	77	100	can	93	41,000	30,000	30,000	0.73	1
6.1	H	30	100	10,000	30	15,000	10,000	10,000	0.67	1
6.2	I	22	100	10,000	38	14,000	10,000	10,000	0.71	1
6.3	G	25	100	10,000	25	12,000	10,000	10,000	0.83	1
FEED.	CHAIN	77		22,000	93	82,000	60,000	60,000	0.73	1
7.1	FFEEDER BUFFER	12		24,000	0	0	0	9,000	0	0
7.2	PRIMARY BUFFER	60		9,000	0	0	0	15,000	0	0

and related interpretation of buffers. The individual task measures can be interpreted similarly but with a median duration base. In other words, a CPI of 2.0 still means the task overran by the plan value at 200%, but the plan value is now the median rather than a fixed view. This then requires an interpretation of how reasonable that overrun is. Also, the interpretation for "reasonable" overruns is more of an estimating review than a traditional performance review. We now uncover another reality-based change. Status measurement is more of a signal than a pure quantitative view. We admit that tasks vary and do not measure performance on normal variance, but we use the excessive value as the management triggers. This is another cultural change to deal with. The project team will use the internal measures as technical task measures, leaving the broader status measures of chains and simulation models to the external audiences. This status view must be compiled in segments as it is not produced automatically in the current software. Review the original plan to see the relationship between primary and feeding chain task names. These new views will be described further in the sections below. Table 22.2 was created with MSP and directly downloaded to Excel for manipulation and data testing. Both classic and modern EV notations are attached to data table headers.

As seen from this raw data list, there is no simple answer to this project's status. Data contained here has characteristics of present, future, and past perspectives of the activity, which is helpful for different audiences and perspectives.

22.6 EVALUATING THE PROJECT STATUS

A modified status will be produced using the calculated data. The following status measurement can be made by the following:

1. The current buffer chain status regarding overruns of cost and schedule can be computed by evaluating actual chain value overruns compared to the median plan values. Task overrun values will need to be deducted from their associated buffer. Then, an assessment will be made to evaluate that buffer's adequacy and protect that chain's plan value.

2. Producing a summary status for the overall project involves an assessment of the primary chain buffer. Simulation modeling of this is deferred to the next chapter for details. The key project-level

questions to resolve are: "Will the project finish on plan schedule and provide a more quantitative view of that status?"

3. Task performance examination involves adequacy of the variance information compared with the original plan that generated the median values. Excessive EV parameter values also trigger a detailed management review of that task.

The first quantitative view of status revolves around the computed values of CPI and SPI. These values will indicate the level of individual execution performance for each task. Buffer status represents a companion view of schedule performance.

Note that all plan tasks in the WBS 6 summary chain are completed at status time, so this grouping makes a good starting place to examine status. Manual summary data for this group is shown in Table 22.2, labeled "FEED." Note that these values are the same as the software generated because all these tasks were 100% completed before the status date (see WBS 6). So long as buffer data does not get comingled with the task chain data, the computer-computed values will be accurate. Still, the calculation rule is to manually separate the two types and review the calculations for any tasks that were only partially completed. The feeding chain has an SPI value of 1.00 because all tasks in the chain were completed before the status date (that does not mean that they were completed on schedule). Individual task EV calculations in this group can be verified by manual calculations of the EV columns arranged in the order described in the EV chapter (i.e., AEP). The standard EV "AEP" formulas can be manually used to check the calculation values for CPI and SPI. The feeder chain, WBS 6, CPI value of 073 indicates a significant cost overrun for that group (e.g., 27% of the median estimate). This is considered a reasonable variation from the median value, so no visible action is required. The EV parameters indicate performance issues for the primary chain summary data, but we now move down a layer to look at performance trends in the chain, which is not yet fully completed. Note that Task D's performance does not fit the trends shown for the previous tasks (i.e., high performance of the others versus only 0.1 for D). This is our first reminder that just looking at a chain task buffer status may not be sufficient for evaluating the future. For now, we stay at the higher level.

WBS 6 feeder chain offers the best calculation example because all the tasks are finished, and the EV parameters can be used to show how they might help evaluate performance, even though current data can never completely forecast the future. In this case of serial task performance, there is consistency in the CPI values, with an average overrun of 27%. Even though this might be considered a significant deviation from the traditional view, it does not mean the same for a median variance based on a probabilistic estimate. Here, we have both consistency and reasonable variation performance. This would be considered a stable chain. One may still quibble that no work is performed at less than the negative median value. Technically, we should see values on both sides of the median. If this becomes a trend, one can assume that the estimating process is perfect and procrastination delay issues are minimal.

The status review generation mechanics will evolve downward from the buffer level status to increasingly lower-level task views. Let's now dive deeper into the data to see what these results indicate.

22.6.1 Chain Evaluation Step

The project's primary chain buffer is labeled WBS 7.2. Tasks A, B, and C are 100% completed. Task D is only 10% complete, Task E is not scheduled, and work has yet to be performed at this status point.

Values for the feeding chain, summary group F, are shown in Table 22.2. Note that this chain and all its subtasks are completed by the status point, with CPI and SPI values of 0.73 and 1.00, respectively. This translates to a cost overrun of 27% for the chain based on median estimates; however, this is still not viewed as a negative performance, given the expected range of a probabilistic task. Also, the consistency of the CPI values in this chain indicates that the process is proceeding in a stable form other than the constant overrun values, which are worthy of review. This level of data granularity offers an opportunity to understand better what is going on with the process. Are the estimates inconsistent, procrastination lingers, or some other factor? In general, this process is not out of control from a task performance point of view, but there are still issues to uncover here.

Because the tasks in the feeder chain are completed by the status date, the buffer size can be audited. The baseline plan duration for this chain was 77 days, while the actual duration was 93 days, which means the chain overran by 15 days. Because there were only 12 days allocated for the feeding buffer, we have a chain overrun into the primary chain, which is not a desirable result. Also, note that the EV SPI parameters failed to show this problem because of how the status was calculated (i.e., the planned value equaled the actual value). This highlights the need for a frequent review of buffers during execution. Part of the monitoring process should be anticipating such events and taking corrective action; the text material pointed out this type of review. As seen from this example, buffer sizes are not guarantees of schedule or cost protection but more targets to monitor.

As outlined in the previous discussions of summary group 6, all the software-computed CPI and SPI Manual calculations for the chain are calculated for the primary task chain. The computed summary group values for CPI and SPI are 0.87 and 0.95, respectively. These values indicate an approximate overrun of 15% for cost and less for schedule. These values are considered normal measures for the anticipated median variation, although all overruns are still in the positive direction (i.e., no EV values greater than 1.00). There are slight differences between the manually computed values and the software summarized, likely caused by MSP internal calculation mechanics. One troubling status metric is shown in the parameter values for Task D, which is only 10% complete. These parameter values are not within any explainable bounds, and because the task is at the end of the chain, there was no warning for this. Because of this location, the overall chain performance at the status point is not significantly impacted. Murphy's Law says that things can go wrong at the worst possible time, and this is it. This project may have just cratered. Let's see if any of the status parameters could have helped now or forecast the future.

22.6.2 Cost Buffer Status

The CPI value for Task D has triggered a performance problem for the rest of that chain. The planned duration for the primary chain is 289 days, while only 227 were consumed up to the status point. The SPI values computed for the primary chain suffer from the same 1.00 syndrome found in the feeding chain. If we look at the actual versus plan data, each one had moderate overrun values. However, the potential negative performance of Task D is still unclear regarding the future outcome.

One option to estimate the remaining duration for this chain is if task E will require its planned 49 days and task D still has 90% of its planned 40-day duration (36 days). Given the history profile of the chain, moderate overruns of these two tasks would fit the previous profile. Values for these three components are needed to estimate the total work chain. The initial plan for the primary chain was 349 days (289 + 60 buffer). The estimated remaining segment of the primary chain requires 213 days versus the 200 days planned. Cost variance showed a similar profile. Based on historical

EV performance, this project looks under control with adequate buffers. However, there is an EV task value indicator that something has changed from past values, and this could be a clue that the performance of the rest of this chain will need to be reviewed further.

22.6.3 Cost Buffer Evaluation

A companion review to the schedule analysis involves the same type of evaluation of the cost buffers. There are multiple methods to evaluate buffer adequacy. A rough measure for this can be estimated through the chain's CPI value. In the case of the primary chain, the current CPI is 0.87. This indicates that project cost is overrunning the median value by 13%. So, if the median chain estimated amount was approximately $232K, a 13% EV calculated overrun would forecast that chain to cost roughly $262K, or a $30K overrun. The related cost buffer was $60K, so the initial assessment is that it looks adequate. The troublesome status of Task D still needs more evaluation.

Because the feeder chain is finished, that assessment can be handled using actual values with the EV parameters. The cost overrun for this chain was $22K (82 – 60). Between all tasks in this chain being completed by the status point, we can compare actual values with forecast values. Note that the CPI value for these chains is 0.73, which indicates a 27% median value overrun. That prediction is $16K versus the actual $22K, and this forecast difference highlights the reality involved in project forecasting and its associated modeling tools. The difference could represent variation that did not obey the probability format. Forecasting future values in a project needs to be looked at as range values and not pure arithmetic values. This project has something unusual being shown by the Task D value. Until the implications of that effort are better understood, more caution is needed in forecasting.

Traditional EV forecasting with fixed tasks assumes that the project trends are constant from the current point, and that is not necessarily true; SPI values easily produce values that signal the wrong forecast, as shown in the dome program. When using median duration values, we must recognize the added variability a forecast might have from that model view. As with CPI and SPI parameters, this is another example where experience is needed to guide which method will provide the best answer. Here is a little memory philosophy for this point. *It is better to have an approximate correct answer than an arithmetic precise answer that is wrong.*

22.6.4 Anomaly Triggers

The anomaly triggered here involves Task D with very low EV parameters. When project chain performance rolls along with similar metrics, one can often conclude that the chain is in a steady state and that the standard forecasting metrics work reasonably well. However, if there is an anomaly in the metrics, it needs to be inspected closely and avoided using standard forecasting techniques. That is the case with the sample plan. Task D has a CPI and SPI, both measured at 0.1. If further examination proves this to be only a single task issue or not caused by insufficient data, these values could signal that the project has encountered a serious problem that could continue. Assume the data for this task is correct. Remember, the chain status parameters have not generated a major warning that something was going wrong up to this point.

Task D has only finished 10% of the defined work but has spent ten times the planned cost and time of doing that portion. Using the EV forecasting value for CPI indicates that the task could require 360 days to complete for $100,000, compared to the plan values of 40 days and $10,000. This variance would almost double the project duration and add 25% to the cost. Task anomalies are not often this stark, but they must be identified immediately for corrective action. It is essential to recognize that the standard EV forecasting logic must handle this class of problem.

22.6.5 Project-Level Analysis

The macro-level project status can be measured through the primary chain buffer and EV parameters; however, other status factors are related to the family of chains in the overall project. A manually calculated project-level status summary row is shown in Table 22.2. These values are calculated from the total list of active tasks, and the resulting parameter values are 0.85 and 0.93, respectively. These values indicate that work unit performance at the total project level is within reasonable variance ranges for schedule and cost. As a warning, this example shows the standard method through which the CPI parameters may forecast the future cost value. The traditional interpretation says that a value of 0.85 indicates an overrun of 15%. The planned project median value is $271K, so a 15% calculated overrun would project the completion cost to be approximately $311K without buffer protection. If the buffers are appropriately sized, the project total would be estimated at $360K (271 + 90). The EV forecasting model will produce a better estimate if the chain tasks variance is near the median value. However, if the task variation changes, as shown in the demo plan, one must review other assumptions. These are untested processes that reflect the immaturity of this model. Chapter 20 on simulation offered another approach to providing probabilistic results for the total project schedule and cost.

22.6.6 Status Process Overview

The model outlined here breaks current cultural processes in multiple ways. The padded task culture has been broken in a less-than-subtle manner with median task values. Second, EV parameters are now introduced in a modified form with a general quantitative interpretation. The mechanics outlined require increased maturity in plan design and the associated interpretation process. Improvement in this process comes from better automation tools and experience with forecasting logic methods.

Significant progress in correcting traditional model flaws is possible as organizations see improved value in modifications such as this example. To enhance automation support, the project plan buffer structure would need some form of standardized structure to indicate which items were tasks and which were buffers. Having chains embedded in the traditional project plan structure complicates the analysis process. Machine learning capability would also be an asset in examining the raw data and uncovering parameter anomalies.

Chain buffer sizing is an integral part of the plan functionality. The buffer overflow found in the example feeder chain illustrates the result of poor buffer sizing. Goldratt (1997) initially stated that a buffer should be 50% of its chain, but that assumption is suspect as a standard rule. High-technology projects need more robust plans. The best advice is to use an initial value in the 30% range and adjust with actual results. As with all new processes, accuracy of use improves with experience. There is research literature on buffer sizing, but the local experience answer is still the best. The example plan shows that the feeder buffer was approximately 30%. This buffer was made to fail as a lesson to remember. Also, the point about periodic monitoring of buffer status is essential. One of the advantages of the chain concept is the ability to dynamically monitor the chain's progress and not rely on a lengthy review cycle. Status-tracking techniques are described in Chapter 21 on CC, which supports this. For example, an alert could be issued if 50% of a buffer has been consumed, but only 30% of the chain has been completed. These are called "burn rate" models and would be essential to operationalizing this technique.

22.7 ROLE OF STANDARD EV PARAMETERS

Beyond the status interpretation described above, other EV parameters are described in Chapter 14 on EV. These are summarized below as a reminder that they can be produced using the defined status data. These additional status terms are as follows:

EAC—Estimate at Completion

BAC—Budget at Completion

Earned Schedule (ES)—quantitative assessment of current schedule

TCPI—A measure to evaluate the potential for budget recovery

SV—Schedule Variance

CV—Cost Variance

The example variable time plan matches customized EV parameters to a more realistic view of the project plan. The computed values are not interpreted using traditional standards, as overruns or underruns do not have the same meaning. Task parameters only evaluate estimating accuracy based on the median value. There are other ways in which the project plan can be constructed, such as with less rigorous task estimates and planned delays between tasks, but the internal task and chain performance can still be measured as shown. In this case, an overall prediction of completion dates should be based more on buffer status than EV parameters, such as BAC or EAC. For example, suppose there are frequent delays expected in resource availability. In that case, the plan should try to reflect that event, but that should not prohibit an evaluation of the intra-task idle time. For example, idle time waiting for a resource is only a schedule delay, not a work performance measure.

22.7.1 Resource Analysis

The previous section focused on project status through the lens of execution speed and work unit productivity compared to the approved median duration plan. Both views are valuable performance metrics but do little to evaluate potential causal factors behind some variations. Resource analysis is a third area that focuses on resource-related topics such as allocation discipline and efficiency. In many ways, this is the most valuable status measurement process. However, as with many good management ideas, it comes with a cautionary warning. Measuring any project activity requires raw data, which requires the most granular operational details. If the goal of resource analysis is to examine allocation discipline and efficiency data to support this need, data to support this need is required, which must be included in the current culture.

22.7.2 Resource Gaps

A project plan shows work progressing from one task to the next. Models do not show gaps between those tasks, which is not reality. Anyone who has ever been involved in a project or just observed one notices "dead" times in which resources are not visible. A straightforward conclusion is that the gap between tasks directly affects the project schedule. Whether this is created by scheduling problems, lack of available resources, the priority of resources, or other issues, this resource gap affects the project. If project cycle time is in the goal set, this critical problem must be handled. Yet, this does not seem to be the case in many projects. The CC model has sensitized us to the need for this role.

Highway and building construction projects offer the best evidence of this management gap, which is equally valid in all projects. All the project delivery models imply no gap between tasks, yet reality says this frequently exists. As obvious as this statement may seem, it represents one of the management disciplines often neglected in practice. The CC model emphasized this point for good reason, and it applies to all ventures if the speed of completion is a goal. Figure 22.6 simulates a four-task work unit task chain with measured resource gaps of 5 days.

This chain shows four tasks (A, B, C, and D), with resource delay gaps between each. Assume the project is completed after task D. A model schedule calculation for this string would be 40 days, but because of poor resource allocation discipline, it is now 15 days longer. This represents a 35% schedule increase caused solely by the misallocation of resources. In

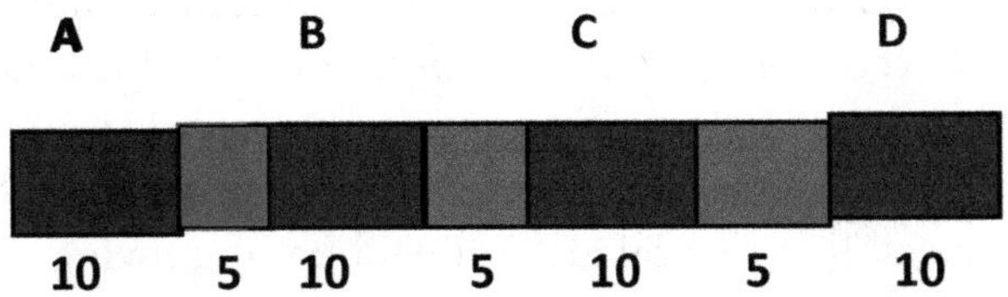

FIGURE 22.6 Task chain with delay gaps.

many cases, the status-tracking logic looks inside the task for schedule slippage when much of it is caused by resource gap issues. This variance would be measured as a chain overrun, but the low granularity of detailed task start and finish data might hide the root cause. Returning to the sample plan for summary group E (WBS 6), note that the tasks in the chain were estimated to have a total duration of 77 days, but the actual duration was 93 days. Thus, the chain overran by 19 days. That variance represents the internal overrun of the work units, but there is no easy reference to resource gap information. There is no easy reference to the information on resource gaps between the tasks. One way to retrieve this data would be to capture each task's start and finish date. Assume that actual start and finish data is captured for each work task during execution. The chain value for WBS 6 is computed to be 164 days in duration. This compares to an actual task duration for the chain of 93 days (see Table 22.2). If all had gone according to the model theory, 71 days of additional time would not be shown in the individual task data. This overrun is the resource gap for that chain! The many scheduled days were "wasted" because of resource allocation. The theme of Section V of this textbook is to highlight variability and reality in the project delivery management process. This general discussion area does not appear in any current project theory literature. There are two reactions to this. Yes, these gaps in current models are essential, and detailed data is needed to quantify the operational impact of such gaps. However, the remaining question is what level of gap is acceptable. Zero is desired, likely, but not achievable. In this demo example, we have introduced two reasons for schedule overruns: task performance and time gaps between tasks. Data is needed to assess overall performance. The critical point is that a resource gap metric represents a schedule overrun based on resource availability outside of the work. The project team often gets blamed for schedule overruns. One wonders how much of that could be attributed to resource management, which is usually external to the team. Projects that do not keep legitimate task information (non-padded) and actual task performance data cannot accurately assess the root cause of poor schedule performance. This is just a sample of one such hidden area. Having appropriate status data is needed to support the management process. Looking backward at resource gap levels and root causes is a critical determinant in improving the process. Resource gap analysis is an underutilized tool in the project status menu, and mismanagement in this area is not well recognized.

22.7.3 Use Rates

Yet, another status view involves the actual versus planned internal resource usage rates. During the planning stage, assumptions about the hours and skills required to perform a work unit are made. The skill level is generally correlated with the labor rate of the individual. If that task subsequently overruns the planned duration or cost, the implication is that some efficiency variation is indicated. That may not be the case. Is the source of the deviation from actual hours involved in the task or the associated labor rate of the worker involved? An answer to this status question is of value to the management process. Even though there is no perfect correlation between worker skill and productivity, this review is often performed as part of the project assessment. One underlying question is whether assigning a highly skilled and expensive person or a lesser skilled and inexpensive resource is better. The best way to answer such questions is to have data to support such a review. As with the resource gaps area, this question also requires data granularity beyond what many projects capture or analyze.

There are two quantitative views to this analysis:

Use variance = (Earned hours minus actual hours) X the planned rate

Rate variance = (planned rate minus actual rate) X actual hours

Assuming both parameters are negative, the cost variance (CV) would be:

CV = Use Variance + Rate Variance

The reasons for the CV component overrun are the actual time duration variation and the amount related to the worker labor rate. This calculation might show that the bulk of the overrun was related to the labor rate rather than a simple duration overrun. Also, evaluating an overrun situation might determine that the entire cost overrun could be traced to a higher labor rate used for the work rather than a duration overrun. The project team's implication regarding this fact differs from simply saying that a task overran, and the team has some deficiency. Much of this variance comes from external factors beyond the team's control.

For this type of calculation to be performed, it will be necessary to have labor-planned rate details associated with tasks in the plan. This analysis might be performed at either the task level or for the entire project. From an aggregate viewpoint, it would be easier to use an average labor rate and use that to evaluate the overall project, even though that does not support the more appropriate lower-level analysis by specific skill groups. A lower-level generic rate review might be of value as a macro measure.

22.7.4 Status Model Summary

Planning and forecasting a project's life cycle is complex, even with a well-defined task list. Having actual performance data to match against plans is key to improving future performance. Data on work estimates and resource gaps are valuable tools in the process.

The demo model improves project status evaluation based on the following enhancements:

- Tasks are planned as variable.
- EV techniques are used to evaluate results.
- Chain logic assesses performance rather than plan versus actual measures.
- A modified EV model improves performance assessment.
- Detailed task-level actual results are used to evaluate the overall resource gap picture.
- Project status is viewed as a multiple-segment grouping of factors (i.e., work chain buffers, EV parameters, and resources).
- Project status is not reported as a simple metric but a composite view based on both standard quantitative and experience concepts.

A mantra in the world of control says, "You cannot control something you do not measure." If one decomposes the items described, there is a focus on the work itself, the speed of completion, and the attributes of the workers allocated. Isn't this what status tracking should involve, rather than simply comparing the task plan versus actual data as seen in the traditional view? The conventional plan versus actual comparison has no operational status validity. This harsh statement is based on a solid view of proper management goals.

It is wrong to suggest that every project has the same level of planning and status measurement detail. The simplest of project plans could be Gantt bars showing "desired" periods to execute tasks, while a more mature view would show planned work units as described. As one looks across this spectrum of management maturity, ask what is known about status in the Gantt-level view versus the advanced type of view outlined. Sizing the management process to the project type is legitimate but should not be done by default. The modern collection of support tools has increased the ability of even small projects to exhibit more of the flavor of larger ones in terms of planning and control.

22.8 CONCLUSION

The model described in this chapter is certainly not considered the standard approach to managing a project, but it does fit a general process that the authors have utilized in the past that better represents the actual behavior of the tasks.

The underutilization of the CC model concepts is a historical mistake by the project industry. It has been rejected because of the underlying process discipline and the required significant changes in the status-tracking culture. Goldratt's (1997) view of work psychology is a compelling view of current practice, and the standard process for covering up performance variation through task padding does not represent the proper view of this activity. Project management must recognize that this is a complex activity, and a simplistic view of the delivery model does not support improving the process. The CC approach to dealing with variable task times is creative yet rejected by the industry for some reason. A culture that sees a variable problem as a fixed event is one example of this status task culture that needs to change.

This modified view of the CC model and status tracking is a significant improvement over the traditional approach. Other behavioral aspects of the CC model are also worthy of being part of the core culture.

Furthermore, it is time for the industry to understand the merits of the EV model and begin to properly use that view in defining work performance. EV parameters can be produced and used with status measurement values differently. This is more complex than the standard formula view implies and fits more along the lines described. The CC and EV models are underutilized and have value in improving the reality view of project planning and control. The project success factors outlined in Chapter 3 describe the repetition of similar failure-related factors year after year, which indicates either a need for recognition or understanding of these factors. Repeating the same errors over and over is unacceptable for professional performance—an improved focus on work methods, problem root cause, and improved performance measurement support that direction.

Evaluating task performance in the manner redescribed requires some new process mechanics, but the output shown is an improvement over the existing method. Even though only a portion of the CC and EV models are used, they capture much of the operational reality needed. The same is true of the new conceptual view of variable duration tasks floating through work chains rather than fixed boxes in time slots as the traditional model represents. It is recognized that the industry has rejected the two techniques described, but they have no relevance when compared against lesser validity techniques. If this industry is going to improve its success rates, it will have to take a hard look at processes that help improve output success.

It is recognized that support is needed for this class of model modification, which includes increased task-level data and the ability to manipulate that data as outlined. Finally, as project management processes evolve, essential knowledge is gained through the status-tracking activity. Sources of both good and bad practices emerge. Mature organizations use this type of information to "learn" and improve. The alternative is not to learn or improve. A mature status quantification process, as described here, helps to uncover the root causes of performance variation and, therefore, often helps to avoid making the same mistakes over and over. Even though manual manipulation and interpretation of the chains seem laborious, they are amenable to machine automation and interpretation. This would be an excellent area for machine learning systems to supply new skills.

As outlined, planning and status-tracking processes will not resolve all the reasons for project failure but will provide a mechanism to evaluate the sources and support improved examination. Collecting status data does not fix the project but is a critical element in the management process needed for proper corrective action. From the examples outlined in this chapter, three new model views of a project in execution add insights into what a more representative status-tracking model brings to the delivery process. Some examples of this are as follows:

- The vision of project work as a set of variable-sized tasks floating through the life cycle and structured around the management work chains is an improved reality view of execution compared to the fixed task model.

- Task overruns in a median-defined model do not necessarily represent poor performance.
- Task execution planned around median estimates will produce stretch goals for the worker and motivation to execute reasonably soon. This is also known as the relay race concept.
- Although buffers may not be viewed positively by external stakeholders, they represent the most appropriate method for handling task variability; it is better to have these as protection elements and move away from a culture of individual task padding.
- Traditional status tracking is based on simple plans versus actual performance measures; it is time to recognize that EV concepts better measure work performance, and their interpretation can be adapted, as shown in this chapter.
- Status tracking is not a monolithic single variable. It has many threads required to represent status. The total status view is the sum of these. The universe of items that could be added to the status view is a collection of input and output delivery goals.

The above summary concludes the theory side of a modified model view. It is essential to recognize that the traditional tools do not align with this view. MSP was used throughout the chapter to demonstrate subsets of the EV and other calculations, but these are only partial tools needed to support this process.

The software in current use is based on a 70-year-old view of the waterfall approach and needs rehab. The status mechanics shown had to be manually produced, but the process outlined could be automated if the plan architectures were standardized to include defined chains and buffers. To convert the existing tools to support this view better, it will be necessary to redesign the plan structure so that chains are defined as a new task type and linked to their task elements. Once that is done, the raw status logic shown in the sample plan can be automatically generated. The resource gap and other components can be automated with proper raw actual performance data. The algorithmic view of this modified process is complex, but advanced machine learning can add significant value to this analysis class.

Sample Problem

Use the sample file named 22 STATUS REVIEW.XLS to evaluate these mechanics. This file is produced using standard fixed task values. Use the existing fixed duration value as the "100" level and modify the duration of all tasks with a specified beta distribution and a profile range. For example, if the defined duration was 30, a "70" equivalent would be 21, while a "150" value would be 45. Use the formula recalculated for the duration of the plan. It is recommended that the duration modification be performed in a spreadsheet and the computed values be imported into the MSP table. Assume the compressed 50/50 value equals "70" as the calculated duration. Execute the new plan with the duration shown and then with compressed values and buffers as described in the chapter. Execute the traditional view, CC view, and simulation model. Use these outputs to develop a presentation for the external stakeholders regarding when this project will be completed (start the project on January 1 to help show calendar duration). Complete this action by using your reality and variability knowledge from the last three chapters.

REFERENCE

Goldratt, E.M. (1997). *Critical Chain*, North River Press, Great Barrington, MA.

CHAPTER 23

Hybrid Conceptual Model

This chapter aims to conceptualize a PM model that deals with the gaps outlined previously in the current models. Today's arguments seem more politically based, with Agile and waterfall being the typical targets. Both have merits but also shortcomings that need to be dealt with by the PM and project team. The term *hybrid* implies some conceptual amalgamation of processes. That statement requires some further explanation.

The output objective here is to present a model structure that fits the needs of all project types. This model should have a flexible structure based on specific project characteristics and goal profiles. At present, there is no known accepted structure for such a model. At this stage in the text, this audience has been exposed to various classic ideas regarding management approaches to plan and execute a project.

23.1 INTRODUCTION

The key question for the future is what a project management model should look like and what it should contain. The material described here is drawn heavily from the book *Optimizing Project Work, Management, and Delivery* by Richardson and Jackson (2024), which represents a year-long examination of management models and a resulting hybrid structure that includes multiple models pieced together.

It seems timely to stretch the current view of the project life cycle and describe a more reality-based approach with needed attributes not

DOI: 10.1201/9781003532989-28

found in the current models. At this stage in the evolution of project management, it is hard to produce a coherent list of all changes that need to be addressed. On the positive side, there are recognized positive points in the waterfall, critical chain, and iterative techniques, so these become targets for the discussion. The agile school has developed a strong following over the past several years. Positive statistical evidence was presented in Chapter 3 on project success factors, indicating delivery performance improvements from Agile, particularly in customer satisfaction. Meanwhile, the Critical Chain theory has never been given a chance for reasons described in the previous chapter. The same can be said for Earned Value (EV) as a performance measurement technique. These ideas have potential value but are not standalone solutions for various reasons. Some industry pundits claim that each school of thought outlined above answers some specific question, and the same can be said about others who reject each of them. The operative definition of hybrid may be an attempt to find common ground.

The first statement about finding a common solution to this is listed below:

> Each project is different, with unique issues that affect how it should be managed. Understanding the target profile and environmental issues needs to be the starting point for designing the proper structure and process.

Let us suppose the environment of the project is not a fixed entity. In such cases, a more flexible way of looking at projects that allows different tools and techniques needs to be employed, rather than a cookbook of fixed methods found in current models. If one were to attempt to summarize the weaknesses in current models, the negative attributes would be:

1. All projects are assumed to fit the single locally defined model.
2. The view of work is either heavily scoped or little scoped (one view of work).
3. Project work is scheduled by fixed work units or groups of less structured work groups (sprints).
4. None of the models recognizes the proper role or decision processes in selecting a project from a portfolio of options.

5. Resource management planning is marginally and poorly dealt with in all models.
6. The organizational infrastructure role is not integrated within either model but is highly relevant to success.
7. The change control process creates management issues that are not addressed adequately.

Many other items could be added to this list, and all model schools of thought would argue with these conclusions. Regardless of one's bias, there are recognized gaps in the popular models.

23.2 WORKFLOW VIEWS

One of the most significant design differences with current models is their internal view of work structure. The waterfall-based approach describes work as a defined unit to be completed. At the same time, the iterative school believes that work cannot be determined that way and is best processed by multiple cycles with an output review after each. Projects are very complex, and no single approach to the work fits all aspects. Some areas would be best served one way, while iteration might be best in others. That capability is a design requirement. In the current model view, as the story goes, if all you have is a single hammer, the world looks like a nail. That is the case with the current mindset. Work design needs flexibility to match the proper view for specific segments.

In the current single model option, the execution process is reasonably defined, and the attempt is to fit the project into that model when it doesn't. A flexible framework has the potential advantage of utilizing multiple work structures that fit the environment. This dictates the need to be able to use known dialect processes where they fit best. For example, the waterfall model has fixed-time tasks linked into an overall defined work unit structure. This has been the prevalent approach for the past 70 or so years until recently, when the flaw in fixed scope definition became recognized. Unfortunately, this has led many to conclude that all aspects of the waterfall approach are wrong, and therefore, all work should be iterative. Projects and their optimum execution method will never be black or white. It is the shades of grey that need to be recognized.

Suppose there is to be an improved approach to project modeling and execution. In that case, a more analytical approach to the solution will be required, which is neither waterfall nor iterative. Also, some concepts related to the Critical Chain model are compelling but require retooling various internal processes. EV also has a relevant view of performance measurement, but it is based on the waterfall mode, so it would need some redesign to be used. This summarizes the design quandary that Richardson and Jackson (2024) described.

The range of potential work methods that one might wish to select ranges across the spectrum from the most conservative waterfall to Critical Chain and some flavor of iterative. Each method has merit based on the output goals, but each has significant process change implications for other management processes that must be considered. Selling any new solution to the static management culture is one of the biggest challenges for new management ideas such as this. Examples of this resistance have been observed earlier in the Critical Chain and EV chapter discussions, where classic status reporting processes would have to be retooled if a new status tracking process were introduced. Both of these methods failed to be accepted at least partially based on that resistance rather than a rejection of the idea's merit.

In addition to the lower-level work execution options, other items need to be included in a new hybrid-based approach. Management of the following factors needs to be considered as key project success drivers (note the brackets at the end of each factor):

1. A visible management gap exists in the decision process that spawns project approval within a competing portfolio. Organizations must have a global perspective and rank projects according to value. [*Portfolio management*]

2. Existing technology offers improved definitions of requirements that would significantly enhance the waterfall approach. [*Requirements definition*]

3. Resources are often not managed well regarding timely availability or skill. Schedules automatically slip when resources are not available on schedule. [*Resource management*]

4. One of the most significant management gaps in the traditional model is the padding of time estimates and a culture of procrastination and multitasking. Note that this is not an implicit attribute of any formal model but is not a common operational practice to cover variability. [*Work estimating practices*]

5. Robust techniques for status and completion prediction of schedule and budget creation. [*Project control processes*]

At the highest level, these are the primary modified process targets for a new model. One must be sensitive to the notion that these are the *what* items to resolve, and more needs to be understood regarding the specific *how* aspect. Fifty years from now, there will be new technology-related answers to this set of questions. By then, projects may be driven by simply telling an automated robot what you want to accomplish. For now, rethinking proper work processes and supporting architecture are the key targets to pursue.

23.3 WORK MANAGEMENT STRATEGY

The future model must recognize that work within a single project structure might be best pursued simultaneously using various techniques—fixed scope, iteration, and critical chain are the targets selected here. An early view of this model research effort was to define a method to support multiple work queues based on predictive and iterative goals.

To recognize the value of iterative practices, the following list contains six work characteristics that are productive and seem to be worthy of general usage consideration:

1. Using higher-level scope definition and allowing the team to flesh out details during the execution (sprint-like work groups).
2. Streamlined documentation techniques (templates, automation, etc.).
3. Brief daily communication sessions.
4. Use dedicated resources with a focus on short-term delivery.
5. Decreased emphasis on task schedule and cost, with more on an overall status view linked to customer satisfaction.

Each idea represents changes in how a project is pursued, but each has shown potential value if used properly.

23.4 WORK TYPES OVERVIEW

One of the more significant management elements embedded in each classic model involves how defined work is processed. When a task is executed in the wrong manner, gaps occur that often bring less desirable results.

The following list represents the primary operational differences compared for each of the four best-known models:

1. Waterfall—tasks are arrayed in a defined linked network. Time estimates often include padding to cover time variances. Scope changes are not specified in the project plan, but additional work will be added upon approval of the new work (scope change). Tracking actual task calendar status and control is a key activity. The plan resembles a bus schedule with fixed dates.
2. Agile—the work produces outcomes based on loosely defined features within fixed-time sprint packages. Improved user satisfaction is a key observed result. Control of the budget and overall schedule is limited to a macro-level.
3. Critical Chain—project work is viewed as a chain of defined tasks with time estimates set for 50/50 probability. The primary goal of this model is to execute the defined work chain as fast as possible. Resource availability and management are key focuses, and buffers are the primary control variable.
4. Modified Scrum—this dialect of Agile is designed for use with a more predictive scope environment. The new model views execution as a sprint-oriented general model. This execution option uses agile principles with a defined deliverable single-sprint completion process. This work option is assumed to satisfy a MoSCoW-type multilevel scope definition.

The challenge with this work option list is finding a way to support all these work types to be utilized concurrently within a single project structure.

23.5 RESOLVING THE PREDICTIVE MODEL ISSUES

In some situations, the predictive waterfall view of the project fits reasonably well, and the base model could be adapted with significant improvement. There are mature processes related to the following:

1. The project scope and work tasks are well-defined.
2. Critical Chain and EV processes can be added to the base structure.

3. Linkage to the project portfolio selection process is established.
4. Evicting automation support is geared to this model.
5. Users, the project team, and management understand the model processes and accept its output even if it is wrong.

The potential general advantages of the iterative approach are as follows:

1. Time is saved in the scope definition.
2. The project team has more flexibility in work management.
3. Sprints have work targets.
4. The concept of control is more user-driven than resource-related.

The Critical Chain model is credited with the following general advantages:

1. The psychology of the worker is a mandatory factor for the future model.
2. The variable time model best fits reality for predictive work.
3. Work chains represent implementable management processes.
4. Use of buffers to cover overruns.
5. Tracking project status via buffers rather than fixed-time schedules.

A challenge to deal with is melding these advantages into a single view. Packaging these items into a single design structure requires the following considerations:

1. The project structure will be customized to match a prioritized list of delivery goals (i.e., schedule, budget, functionality, resources, risk).
2. Project success will be judged by a prioritized list of factors rather than the traditional three-variable approach.
3. Resource management will be elevated to a priority management process to ensure no availability gaps for defined work units.
4. Critical Chain compressed task time theory will be used as a common task-estimating guideline; overruns will not be viewed negatively.

5. Requirements using the MoSCoW multi-level concept will be favored over a singular discrete value.
6. Work units will be grouped into chains as sprint-like management focus groupings.
7. Status reporting will be more macro-level and probability-based.
8. Buffers will be used to protect dates, no individual task padding.

From this modified process view, the general environment would be to have resources in place as needed, track the completion of tasks (or work chains) using agile Kanban-type progress flow-through charts, and challenge the team to move forward quickly. Fixed work unit schedules would be eliminated. The items listed here highlight what a new hybrid model should contain to deal with defined management gaps. At the working level, new options are now determined to execute the defined work, which needs flexibility in its definition. Each existing model defines a work queue performed using its internal process, whether in an agile, iterative, or predictive form.

As we move away from focusing on how the lower-level work will be executed, other global unanswered views need to be described. Richardson and Jackson (2024) identified six decision layers required for a hybrid project model. These are as follows:

1. *Project Approval*—an integrated process that deals with organizational vision and strategy, plus a method for managing the operational selection of project targets.
2. *Plan development*—a mid-level function to evaluate how best to package an approved target (i.e., phases, programs, timing).
3. *Delivery packaging*—developing the high-level Work Breakdown Structure (WBS), which includes elaborating requirements. Here, the project profile is documented along with delivery goal priorities.
4. *Delivery methods planning*—this process further defines the delivery format based on factors derived in the previous step. This new core process identifies the proper way to execute a specific task defined by the planning process. Multiple work types are described here.

5. *Execution*—selected work units are completed according to the delivery plan. This can result in integrating multiple execution workstream types within the same project.

6. *Monitor and control*—execution results are tracked according to the control needs established from the project profile.

Each management layer outlined above is linked as a downward flow stack from top to bottom. For example, suppose a project is not doing well. In that case, that information is known at the strategic level, where a decision might be made to cancel the low-performing project and introduce another higher-priority candidate. Each internal layer links to decisions made at the next highest layer.

The monitoring and control layer is more complex than a single-view model, so presenting a unified status view across multiple work streams is challenging. This is envisioned as a combination of EV, Critical Chain (CC) buffer logic, and Agile/Scrum techniques, describing throughput at high levels for management and work unit level for the project team.

Beyond the model decision structure, there is also a need to recognize best practices and processes found in the current models and incorporate these into the hybrid model. Richardson and Jackson (2024) labeled these *success recipes*, and they deal with survey results related to project failure factors.

23.6 TASK WORK OPTIONS

If a hybrid model is going to be able to execute the breadth of required work types, there has to be an environmental match with the work option. Here are some general guidelines:

1. Waterfall work units—Clear scope requirement, standard deliverables.

2. Iterative Scrum—a broad scope definition, with the primary goal being customer satisfaction; iterative sprints will deliver output.

3. Critical Chain—Time is the critical goal; work chains are the management focus.

4. Modified Scrum combines iteration with fixed goal sprints; the MoSCoW scoping technique drives this process (more explained below).

Option four is the new view of the work required to complete the hybrid structure. Here is a brief description of how a modified Scrum task fits. The first step is to identify a scope item that needs to be produced. It is decided that this can be best performed by setting loose definitions with a MoSCoW three-level output goal for the work. That work task is planned just like a regular task. Let's say the planned duration is ten days. The task would be placed on the project work queue list like any other task, except it will be executed using a single agile sprint cycle. Successful output is judged by how well the MoSCoW three-tier output goal is achieved. If it is not met, the sprint must be extended until the minimal goal level is met. The main difference with this technique is how the initial scope is defined, and how a MoSCoW specification level will determine the success of the sprint. This type of work definition is the frontier of project management today. It fills the gap between Agile and waterfall, potentially providing the best options. Utilizing this multiple-stream type of work technique is one of the key success strategies for the new hybrid model.

23.7 CONSTRUCTING THE HYBRID MODEL

The model described here does not exist in reality. However, one can see the logic of such a model and recognize how the existing models all have a limited view. Good PMs would realize the value of having these work options. Also, the vertical decision structure fits the evolving nature of project decisions. This view highlights how one level of the decision cycle leads to the next lower level. The layered structure also highlights the roles that the organization plays in success and the role that effective resource management plays. None of the existing models shows these characteristics well.

To minimize visible changes in the model view, the project will be represented by a WBS outline to describe the general scope of the effort. Figure 23.1 shows a hypothetical view of the management view. The execution box contains multiple work options to be accomplished by one of the four work types. The vertical stack of boxes implies a layered set of decisions starting at the strategic level and moving downward through the life cycle to completion, which technically could be added as a bottom box. The two horizontal boxes describe processes from the resource and host organization side.

Fixed scope and Critical Chain model logic are used to schedule those respective work types, while iterative and Scrum work units will be placed on a sprint-type backlog to be scheduled. That "community" of tasks will

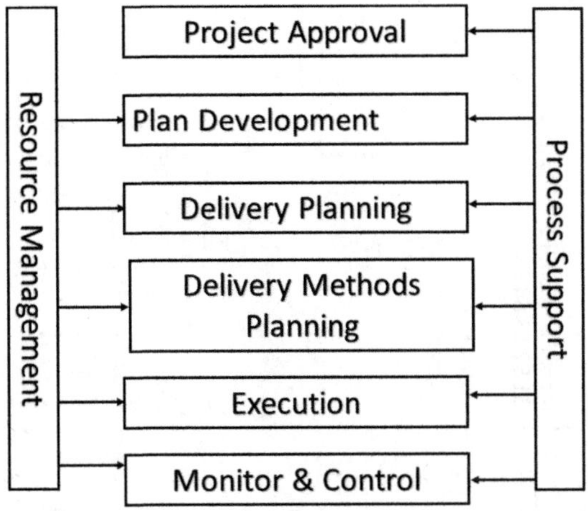

FIGURE 23.1 Integrated model block diagram.

Source: Richardson and Jackson, 2024.

be tracked using their unique tool sets, with Kanban being the most favored tool for iterative status tracking. Although not represented in the diagram, buffers are used between work types to protect the task linkages. The overall project schedule can contain traditional, chain, Scrum, and agile work units, all embedded into a single view. We are no longer arguing about which is better. We are looking at each option where it fits best. Figure 23.2 shows a schematic view of the integrated work model.

The various work streams are shown on either side of the WBS. These are integrated into multiple status streams based on Gantt and Kanban views.

23.8 CONTROL MODEL

The concept of project control has changed in this four-stream workflow. Techniques to manage iterative work streams in a fixed project structure require a modified process. In the agile view, scope completion occurs when the goal is achieved (or, as some say, when the time or money runs out). In this new model case, there will have to be a defined stop point with some specified output goal because other work streams fit into the overall project and have completion requirements. In this sense, these work streams are not pure iterative Agile. The same is true for the Modified Scrum tasks. The MoSCoW view of output scope definition remains a technique for defining stop points. Richardson and Jackson (2024) offer more details regarding the mechanics of the combined work streams.

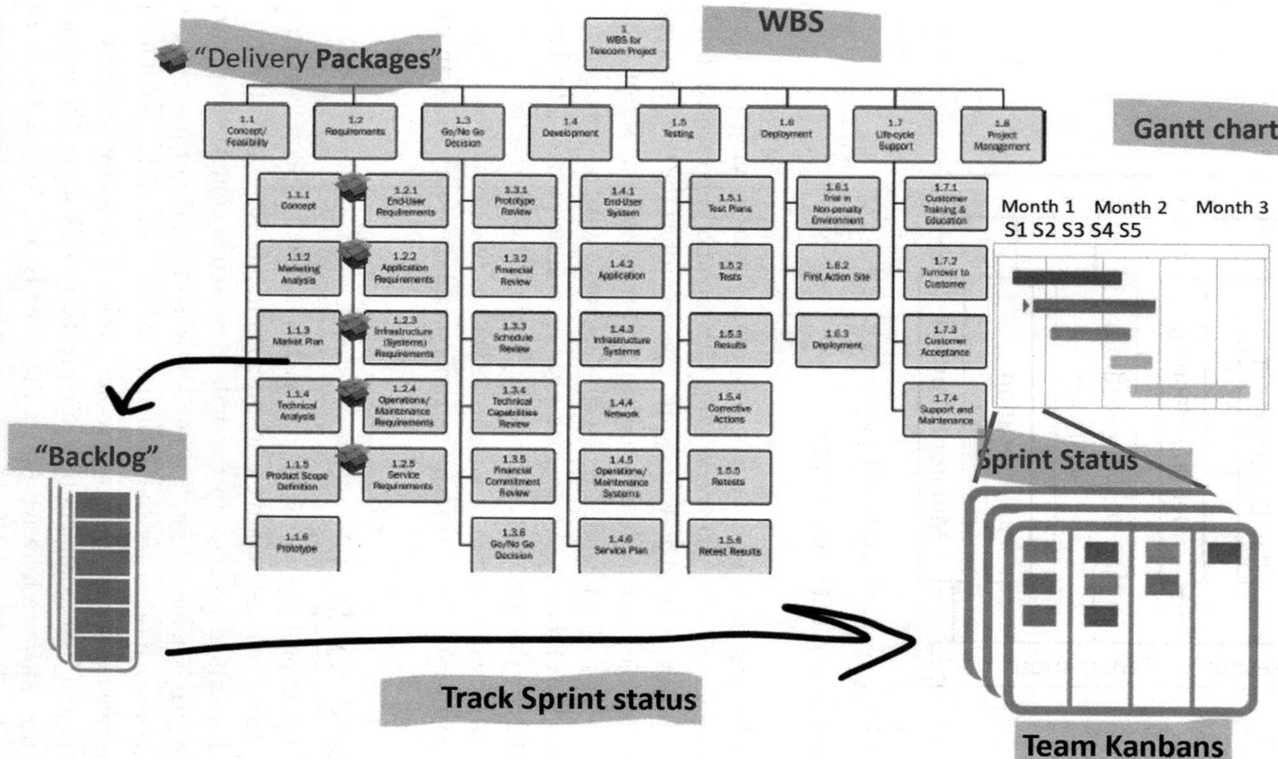

FIGURE 23.2 Integrated model structure.

Source: Richardson and Jackson, 2024.

Given this modified workstream process, it must be recognized that the traditional view of project control must be modified. Also, when tasks are padded, the concept of control should not be used to measure performance against these bloated times. The changes outlined here will require significant retooling techniques to evaluate project performance. The EV variable time model outlined in Chapter 22 is a candidate to use for analysis of the modified work units. Of course, the typical contemporary response is that an Artificial Intelligence (AI)-based capability will come along and solve all of this. There is reason to believe that this evolving technology will offer improved methods to evaluate status, and some of that capability was demonstrated in the modified EV discussion.

More work is required to develop the mechanical processes related to the four work types. Our best hope is to orchestrate models so that an automated examination can produce a usable status evaluation. An EV mindset is needed to measure performance, and work is required to develop the process for the iterative area. There is great promise of improved productivity and user satisfaction if this delivery model can be better defined. The target goals are clear, but the mechanics need further work.

QUESTIONS

1. What are some of the driving forces that suggest a new management model is needed?
2. What are some of the shortcomings found in the waterfall model?
3. What value do you see in the MoSCoW approach to scope definition?
4. How would you manage each concurrently if a future hybrid model had four work types embedded in the project?
5. What is the relationship of strategic project planning to the operational level?
6. If a project were to have multiple work types underway concurrently, how would you present the status to senior management?

7. Name five things wrong with the waterfall and Agile methodologies that must be addressed in a hybrid model.

REFERENCE

Richardson, G.L., & Jackson, B.M. (2024), *Optimizing Project Work, Management, and Delivery*, CRC Press, Boca Raton, FL.

CHAPTER 24

Concluding Remarks

We are now at the end of our journey of *Project Management Tools and Techniques*. Much material has been covered here, and more needs to be developed to support improvement in project success. No author seems to feel that all his story has been told, which is the case here. We have been professionally involved in this topic for many decades and became interested in the inner workings of this topic as we attempted to explain it to various types of students and industry environments, both in the United States and internationally. This experience taught us that learning and using book theory in practice is complex. The models and techniques described here are valid concepts. However, translating them into the operational culture will still be challenging.

Project delivery models help conceptualize the required scope and work and communicate various aspects to a broad internal and external audience. There is one hanging issue that we leave with you. The project delivery methodology question is not about judging whether the waterfall model, Agile, or any other offering is the correct answer. The more appropriate question is what needs to be done to the management process to improve output success. None of the existing models fully supports that goal. Comments made throughout the chapters offered tidbits to describe shortcomings in that view, and each represents a prime candidate for improvement. There is a logical improvement pathway from scope definition with the Work Breakdown Structure (WBS), task estimating, resource allocation, and technical sequencing that could be heavily automated using new technology. Formal team training would be worthwhile for understanding the basic model, using automated tools, and developing a

DOI: 10.1201/9781003532989-29

viable delivery improvement strategy. There is little reason to believe that technical training in one area automatically makes one proficient in a different venue, yet that seems to be a common belief. Despite some apparent similarities, the project life cycle is complex and unique from other technical areas.

One new tool and technique idea that could blossom in this area over the next ten years is the existence of a hybrid model that integrates improved methods from multiple model views. This involves tool mechanics and a more flexible view of scope definition, project planning, and work estimating. If this idea could find fascinating roots, it would require extensive training to implement and change many work processes outlined in the text.

Like so many technology-based scenarios in the past, improvement comes from someone who believes in the topic. These are typically creative thinkers who invest their total focus on a target, whether that be light bulbs, phones, cars, rockets, or the Internet. If individuals of this bent are not in place, the related culture will continue to use kerosene lamps, horses, and Pony Express mail. The problem with today's project environment is the lack of these culturally changing individuals. It is not the lack of basic theory! We need working examples and a mechanism to train the new project teams to use them. Simply collecting skilled knowledge workers into a project structure and expecting improvement is like the Titanic moving its chairs on the deck after it hit the iceberg and expecting it to continue floating. We must change the process to achieve output improvement. This industry is short on creative thinkers and senior management who will work to create sustainable methods. The Japanese quality movement of the mid-1900s illustrated what can happen to a broken process with proper leadership and theory. Yes, it took several years to evolve, and patience is one of the main problems. Current project management practice is at that exact point in its maturity cycle, and the thought leaders are not apparent. The agile school found its leaders in the grassroots layer of the industry, and its growth result is impressive but still incomplete. We hope the material recorded here will stimulate this improvement path through new tools, processes, and delivery results. The introductory material in this textbook said that project management was an art-seeking science. As this subject continues to mature, it becomes increasingly more complex with increased science and technology. That trend shows no sign of changing. We wish each PM success in all future project endeavors.

Appendix A

MICROSOFT PROJECT COMMANDS

This technical appendix contains a condensed summary of MS Project 2021 commands and reference materials. Prior editions of Project and Project 365 versions have a very similar menu structure. Various sections of the text focus on subsets of this utility. Note: A downloadable copy of this can be obtained from the Publisher's web repository (see Chapter 1 for details).

The Basic Startup Command section contains 14 keystrokes that represent the core startup steps for project plan creation. Also, the *Project N Motion* video 6MSP1 mirrors these steps and is a recommended supplement to review. [6MSP1]

A. BASIC PROJECT PLAN CREATION COMMANDS

1. Setup Options—File/Options (look through these pages to set various profile type variables).
2. Hide and Insert data columns—Right click/Hide & Insert (many MSP variables to choose from).
3. Insert WBS column—Right click/Insert column; also, the Format/WBS option.
4. Copy and Paste raw activity data—Standard Windows operation (check import results to ensure proper values since project does logical screening on this operation).
5. Indent structure to obey WBS hierarchy—Task tab/green arrows to define WBS structure.

6. Task scheduling option—Recommend manual setting Task/Automatic Schedule (or globally set through File/Options).

7. *Gener*ate WBS schematic—Save current project; select WBS Schedule Pro add-in utility from menu (assuming it has been installed); select WBS Chart; toggle between format options; try detailed view; select format and review various layout styles; format/edit fields and boxes. Can save this view as a Picture format for external use.

8. Critical Path (red bars) definition—Format/check Critical Task box (small check box).

9. Size Gantt bars area—Right click on timeline/Zoom/Entire project.

10. Gantt Chart Wizard for general formatting (add icon through the File/Options—see below).

11. Bar labels—Format/Format (dates); use Gantt chart wizard first, then manually adjust.

12. Project/Project Information—Set start date, calendar, and status dates.

13. Recurring task (PM)—Task/Task/Recurring (this is for tasks that repeat periodically).

14. Task Mode column—Right click/Task mode (this confirms scheduling logic).

B. QUICK MENU COMMANDS

Add Gantt Chart Wizard to list using keystroke checklist instructions.

C. GLOBAL OPTIONS—FILE/OPTIONS

General—Change date to mm/dd (this saves display space but may need to be changed for some functions that require the year value to show).

Display—Drop $.00 (decimals).

Advanced—Customize main menu ribbon as desired (Automatic Schedule is a reasonable choice).

D. ADDING RESOURCE ESTIMATES TO THE PLAN

1. Cost values—Cost should be generated by allocating resources to the task.
2. Cost through resource allocation—Right click on task/Assign resources, or Resource/Assign Resources (this is the easiest).
3. Define project buffers—Format the buffer in another color bar to distinguish—right click on bar for single bar format.
4. Set Baseline—Sets a fixed mirror copy for status comparison—Project/Set baseline.
5. Capacity Management—Resource leveling—Format/View (select option. See text Chapter 21 for more details on resource analysis).
6. Resource Sheet—Defines project resources (resource pool).
7. Resource Graph—Shows individual resource status in bar graph form.
8. Resource Usage table—Shows detailed work allocations and overallocations by task and individual.
9. Showing Baseline bar in Gantt bar area—Use the Gantt Chart Wizard (Other, Baseline style 3); use Format/Format/Bar Styles for refinement from here. Moving this bar to line 2 helps to separate views.

E. STATUS TRACKING COMMANDS

1. Set status date—Project/Project Information
2. Status gridline—Format/Gridlines/status date (Be sure to select a line type and color)
3. Start and Finish dates—Unhide the columns to show planned and baseline Start and Finish
4. Tracking status—%Complete column, Baseline column, adjust durations as needed, watch Smart Tags
5. Showing Baseline Gantt bar versus plan—Format/Format/Bar styles (use baseline start and baseline finish); row number may be changed for appearance

6. Showing Baseline data in table—All Baseline data uses "baseline" to define (Baseline start)

F. RESCHEDULING COMMANDS

1. Recording actual start and finish times for tasks—Use Actual Start and Actual Finish
2. Set status date—Project/Project information (set status date)
3. Project/Update project—Causes bars to move past status date

G. CUSTOMIZING GANTT BAR FORMATTING NOTES

1. It is easier to work with bar formatting if you first run the Gantt Chart Wizard to establish the base bar formats and then customize them as desired from that point.
2. In the Gantt Chart Wizard, select "Other/Baseline style 3" to get red and grey task bar formats for critical and baseline activities, respectively. Also, select dates as a format option.
3. To further customize the bars, go to Format/Format/Bar styles.
4. Set task and critical task bars to full size with proper colors for better readability.
5. Within the Format/Format Bars option, go to the text tab for bar labeling (dates are one common item to show on the bars, but other variables may be appropriate for specific views. These are status informational items).
6. Note: Date labels for baseline items are all prefixed with "baseline" (this can be confusing. Just remember that Baseline dates are different from current plan dates).

H. RESOURCE ALLOCATION STATUS EVALUATION NOTES

1. Learn to navigate through Gantt, Resource usage, Resource sheet, and Resource Graph views to collect the overall status picture of resources.

2. The Resource Sheet defines the project resource pool that can be allocated to specific project tasks. As a result of that allocation, project costs are created based on the attached rate information.

3. "Red man" symbols in the information column of the Gantt view indicate an overallocation. The same overallocation flag can be seen from an information tag in the Resource Sheet.

4. Resource usage—This table view provides a detailed look at time allocation for each resource. The status analysis problem involves finding resource overallocations.

I. EARNED VALUE DATA ELEMENTS

This is a computationally complex topic area. Follow the mechanics shown in the text reference chapter.

J. MENU RIBBON SUMMARY

Menus	Functions
File	New; Open; Save; Save As; Print; Close; Options
Task	Operations to manipulate data
Resource	Entry to various resource views
Report	Entry to reporting options
Project	Entry to Start, Status, calendar, and Baseline parameters
View	Entry to various view options
Add-Ins	Entry point for Full Monte and Schedule Inspector (third-party utilities)
WBS Schedule Pro	Menu tab for Critical Tools WBS Schedule Pro utility (3rd party)
Format	Various formatting options

Note: Additional MS Project reference materials are available from www.microsoft.com, www.mpuweg.com, and many other Internet technical sources.

Appendix B MS Project Lab Assignment

The purpose of this assignment is to provide enough level of complexity to illustrate a full life cycle view of project planning and tracking. Also, this example shows the data entry sequence required to generate accurate values under controlled conditions. Embedded in the multiple phases are key steps found in the core planning and control actions necessary to develop a project plan and then migrate it through the execution phase. The latter segments focus on tracking performance. The step-by-step instructions carefully sequence the proper steps for each phase, and these must be followed carefully in order to arrive at the provided solution values. Even small keystroke errors can change the results.

Supporting Data Files. The two data files defined below are needed to feed the model at various points. Use this data as instructed. Do not jump the defined sequence. Copying and pasting data from Excel (XLS) to Project (MPP) will save data input time. However, copying and pasting data into the plan must be done carefully and checked again, because Project is picky about this type of process. The following two raw XLS data files are used in the assignment:

A. Project resources—LAB RESOURCES.XLS (the project resource pool)

B. Project activity list and resource allocation data—LAB ACTIVITIES. XLS (contains both plan and later status data).

Various data elements from these two file sources will be used in various places throughout the assignment to supply needed values. Specifics are embedded in the running instruction dialogue for each phase. It is important to only extract the data elements as they are defined. Failure to follow this rule will likely negatively affect outcome values.

It will also be helpful to have the keystroke summary sheet handy through this exercise. This can be retrieved from the publisher's website (see Appendix E for instructions). It will be important to save interim status versions as specified throughout to support backup and recovery from processing errors. The bolded status data values are supplied as important checkpoint values.

Note: All the highlighted/bolded numbers are given to help keep you on track through the steps of the assignment. *Follow the detailed instructions carefully.*

PHASE I. PROJECT PLAN DEVELOPMENT [REVIEW CHAPTER 10]

The goal for this step is to create the initial project schedule from defined scope data.

1. Open MS Project and change various setup options:
    a. Select File/Options/Display. Change "Decimal digits" to "00."
    b. Select File/Options/General; change the date format to "MM/DD." Both changes will save table format space.
    c. Select File/Options/Schedule and change "New tasks created"; change option to "Auto Scheduled."
    d. Go to File/Options/Schedule; scroll down this page for several lines and find the row labeled "Actual costs are always calculated by Project." **Uncheck** this box. This is an important calculation step for later.
2. Hide the Start, Finish, and Resource Names columns.

3. Set up the columns in Project as required for Step #4.

4. Import the following data from LAB ACTIVITIES.XLS: WBS ID, Task Name, Duration, and PRED—note: there may be copy and paste issues with this process, so check the imported data for accuracy. Manually repair any import errors. Use Paste Special/Text Data for pasting.

5. Save this version as "**Case File 1.1**."

6. Verify that the tasks are being scheduled by Automatic Mode—It will help to verify this if the Task Mode column is left visible.

7. After importing the activities list and predecessors, insert a Task at the top row to represent the total project. Label this as "Total Project-<your name>".

8. Indent the activity structure to obey the WBS codes. The top line should have a WBS code of "1." Check your predecessor code values to be sure that you imported this correctly. Also, check your indentions to see that the summary tasks show properly (WBS Schedule Pro can help you evaluate this). The total project duration should now show as **130.9**. If you do not see this value, do not go further until you find the data or indentation error.

9. Show the Critical Path—[Format/Critical Tasks].

10. Insert Start and Finish dates on the bars—The Gantt Chart Wizard would be the easiest way to do this, but the Format/Format/Bar Styles option will also work.

11. Go to Format/Format/Bar Styles and change the label format on summary bars. Use any other format changes you like that make your output cleaner.

12. Insert a new task in front of the existing task #28. Title the new task as Project Buffer with a **25-day** duration. Check the predecessor value for House complete to link to the buffer task.

13. The buffer task should have a WBS value of 1.8, and the House complete task should have a code of 1.9. Change the color of the buffer task (#28) to bright green—[right click on bar/Format Bar/...]. The project duration should now show as **155.9**.

14. Set the project start date to **04/07/2020**—[Project/Project Information]. The project completion date should now be **11/10/2020** (See task 29).
15. Save this version as "**Case File 1.2.**"

PHASE II. RESOURCE ALLOCATION [REVIEW CHAPTER 11]

The goal for this step is to allocate resources to the defined schedule and generate a first-cut project budget.

16. Open the MS Project Resource Sheet view [View/Gantt Chart/Resource sheet], and import the resource plan from the LAB Resources.XLSX. Beware that some of the data may not import properly, so manually correct any data import errors.
17. Insert a column titled "Cost." Values should be zero at this point.
18. Insert a column titled "Resource Names.
19. Import the Resource Allocation values from LAB ACTIVITIES. XLSX column titled "Resource Names" into the MS Project Resource Names field. You can copy and paste this data, but be sure to check the values to ensure that they copy to the correct WBS line. Remember, these resource allocations create the direct project cost. The total project cost should now be **$27,884**. Save the file as "**Case File 2.1.**"
20. Note that WBS items 1.4.6 and 1.4.7 do not have resources assigned. Right-click on these and manually assign the following:

    1.4.6 Two field workers and five rolls of carpet

    1.4.7 Four painters

21. The project cost is now increased to **$32,304**, and Duration is **155.9 Days**.
22. Save this version as "**Case File 2.2.**" This is an important step because the baseline will be set next, and it will be difficult to back up after that point.
23. Set project baseline—[Project/Set Baseline].

24. Use the Gantt Chart Wizard and choose "**Other**," then "**Baseline Style 3**" and "dates." Then select Format Bar Styles and move the baseline to "Row 2," and expand the size of the various Gantt bars to improve readability.
25. With the step above, you should now see an indication that the baseline is in place. Black bars appear in parallel to the regular red and blue status bars for each task. Save this version as "**Case File 2.3**."

Discussion Questions (Part II)

a. Do you see any resource allocation issues? If so, where?
b. How many field workers do you need at the max point?
c. Why are the baseline and cost column values the same?

PHASE III. TRACKING ACTUAL STATUS [REVIEW CHAPTER 18]

The goal for this section is to track actual task completion.

26. Table formatting: Hide the data columns for Predecessors, Task Mode, and Resource Names. Add new table columns for Actual Start, Actual Finish, Baseline Cost, Baseline Duration, and %Complete. Clean up all columns by centering data and shrinking column width as appropriate.
27. Import ***%Complete*** data from LAB ACTIVITIES.XLSX. As in other cases, check the pasted values to ensure that they are imported to the correct WBS code.
28. Import Actual Duration data from LAB ACTIVITIES.XLSX into the Duration column.
29. Open data columns for Actual Start and Actual Finish. Manually enter the data from the table below. This is a somewhat tedious exercise, but this data can aid in improved estimating for future projects.

Actual Completion Data

WBS	Task Name	Actual Start	Actual Finish
1	**Total Project**		
1.1	Site Work		
1.1.1	Clearing	4/7	5/9
1.1.2	Excavation	5/9	5/18
1.2	**Foundation**		
1.2.1	Frame Floors	6/2	6/12
1.2.2	Pour Foundation	6/12	7/7
1.3	**Rough-Ins**		
1.3.1	Electrical	7/7	7/18
1.3.2	Plumbing	7/18	7/29
1.4	**Interior Finishes**		
1.4.1	Frame, Ceiling, and Roof	7/29	

30. Format the Baseline "Project Completion" Gantt (Diamond) icon to show its date to the left of the icon using "Baseline Finish" for the label. Change the format of the icon to make it more visible. Save this version as **Case File 2.4.**

31. The project plan now shows **43%** complete, with a Duration of **169.9 Days**, and Cost of **$34,128**. Recognize that ID #30 is a buffer (not a real task), so the calculated duration or cost is not technically correct at this point.

PHASE IV. RESCHEDULING TASKS BASED ON A STATUS DATE

The goal for this section is to reschedule the project based on the current task status and a defined status date.

a. Format the Task Progress bars to black—this shows the portion of those bars %Complete graphically (note an example of this at 1.4.1).

b. Set project status date to **8/5/2020** ***(be careful with the year as you enter this date. Use the red gridline to confirm accuracy).***

c. Set a red status gridline [Format/gridlines/status date]. If a red line does not show on the screen, you have made a data error in one of the two previous steps. Check the calendar year to confirm that you have set the status date properly.

d. Save this as **Case File 3.1.**

32. This step will reschedule the project (Gantt bars and data) by moving the unfinished work downstream past **8/5/2020** (the defined status date)

    Go to Project/Update Project and select the "Reschedule uncompleted work" check box. Click OK. Note that all Gantt bars with unfinished work have now been moved past the status date (even partial bars are moved).

33. The next step deals with the duration calculation issue. Recognize that 19 days of the original buffer are now "eaten" as a result of the task overruns. To correct the plan view, go to WBS 1.8 and change the buffer duration to 6 days. To this point the buffer has protected the plan from overruns. The plan now shows a completion date of **11/10/2020**, with six days of buffer remaining. This view yields a traditional project status with comparisons of baseline to current values. Save this version as **Case File 3.2**.

Discussion Questions (Part IV)

a. Write a brief executive summary outlining the project status. This story should be based on the final plan (File 3.2).

b. Explain the logic of plan rescheduling to a non-technical (user) manager.

PHASE V. CALCULATING EARNED VALUE PARAMETERS [REVIEW CHAPTER 14]

MS Project handles Earned Value (EV) parameters in a somewhat obtuse and complex way. The example here is designed mostly to show that it can perform the calculations. The real user issue here is to explain the role of IV in project status tracking and forecasting. For this example, Actual Costs will be manually fed to the program to simulate the accounting system role. ***Set up the following EV parameter columns to contain your status.***

34. Open up columns for BCWS (PV), BCWP (EV), SPI, CPI, EAC, VAC, and TCPI. Organize the table layout with any other non-EV status data that you desire. Move the Gantt section out of the way.

This segment will essentially be all table data and little Gantt graphics.

Important Checks: Go to File/Options/Schedule; scroll down this page for several lines and find the row labeled "Actual costs are always calculated by Project." Make sure this box is unchecked.

35. Import the Actual Cost data from the LAB ACTIVITIES.XLSX spreadsheet.
36. Remember to adjust the buffer to handle the project overrun. Your planned completion date is **11/10/2020**.
37. The Baseline Cost is now **$32,304**, and the Actual Cost is **$37,134**.
38. Save this version as **Case File 3.3**.
39. Interpret the resulting parameters from Step 3.3
    a. What is the projected project cost? Does this value agree with what is shown in the Cost column of the plan?
    b. What is the EV projected cost overrun?
    c. According to the EV parameter calculations, what is the projected project duration? Show calculation.
    d. What is the worst-performing task thus far? Why?
    e. What is the best-performing task thus far? Why?
    f. How much of the project has been completed at this point? Compare the EV values with %Complete.
    g. The cost performance figures for each task are not impressive; however, the overall difference in cost versus baseline values does not seem so bad. What do you think might be hiding this?
    h. Note that the actual cost shown is less than the baseline cost, which implies that the project is doing well. However, most of the tasks have overrun, and the SPI is less than 1.0. How can this be?

FINAL REVIEW

Go back through the major sections and review the results. Note the output formats and goals for that section. From a life cycle viewpoint, recognize the following milestone processes:

a. Section I—basic first cut plan with no constraints

b. Section II—resource allocation to the schedule

c. Section III—tracking plan versus actual performance

d. Section IV—rescheduling process

e. Section V—mechanics for generating basic EV parameters

Appendix C Templates

Here's a list of resources offering free project management templates that you can download and use:

1. **Project Manager**: https://www.projectmanager.com/pm-templates
    - Offers free templates like:
        - Construction Schedule
        - Agile Sprint Planner
        - Product Development Plan
        - IT Project Plan
2. **Smartsheet**: https://www.smartsheet.com/content-center/best-practices/project-management-guide/templates-project-management
    - Templates include:
        - Project Task List
        - Event Plan & Budget
        - Gantt Chart Templates
        - Work Schedule

3. **ClickUp**: https://clickup.com/blog/project-management-templates/
    - Free templates available:
        - Project Timeline
        - High-Level Project Plan
        - Agile Project Tracker
        - Daily Planner
4. **HubSpot**: https://www.hubspot.com/business-templates/project-management
    - Includes:
        - Project Budget Tracker
        - Timeline and Milestone Planner
        - Risk Management Templates
        - Proposal Templates
5. **Atlassian**: https://www.atlassian.com/project-management/templates
    - Examples:
        - Project Plan Template
        - Status Report
        - Retrospective Template
        - Task Management Board
6. **Monday**: https://monday.com/
    - Free templates for:
        - Team Project Tracker
        - Work Breakdown Structure (WBS)
        - Marketing Campaign Planner
        - Software Development Plan

7. **Project Management**: https://www.projectmanagement.com/templates/#_=_
    - Download templates like:
        - Risk Register
        - Stakeholder Analysis
        - Communication Plan
        - Work Breakdown Structure
8. **Vertex42**: https://www.vertex42.com/ExcelTemplates/excel-project-management.html
    - Excel-based templates such as:
        - Gantt Chart
        - Project Timeline
        - Resource Management
        - Milestone Tracker
9. **Project Management Institute (PMI)**: https://www.pmi.org/insights
    - Templates include:
        - AI Governance Plan
        - Strategic Planning Templates
        - Budget Forecast Sheets

Index

Pages in *italics* refer to figures and pages in **bold** refer to tables.

Q

R

V

W

X

9781032874791